PRAISE FOR IAN BARBOUR AND
Ethics in an Age of Technology

"Barbour complements theory with practice, continuing the genius for comprehensive and sure-footed synthesis and criticism he has shown for the last half-century. Facing the new millennium, few questions or none are more important than whether and how technology can be put to increasingly moral uses, protecting human and environmental values. Few analyses are more important than Barbour's."

   —HOLMES ROLSTON, III, author of *Science & Religion*
      and *Environmental Ethics*

"Brings to the most important topics in Barbour's earlier work, *Technology, Environment, and Human Values,* important new foci on computers, genetic engineering, and nuclear weapons, as well as an updated discussion of energy, environmental degradation, population, and the debate over values. [Barbour] brings to this intensely researched and eminently fair analysis his own paradigm of human and environmental liberation rooted in a Biblical faith perspective. Together with volume 1 of his Gifford Lectures, *Ethics* is surely the culmination of one of the most extraordinary careers of the twentieth century. As physicist, philosopher, theologian, ethicist, environmentalist, technologist, worshiper in the Biblical tradition, Ian G. Barbour embodies the best of the two—and the twenty—worlds he spans. If there is a paradigm for the relationship between science and religion, its exemplar is Barbour."

   —ROBERT JOHN RUSSELL, founder and director,
      The Center for Theology and the Natural Sciences, Berkeley

ETHICS
IN AN AGE OF
TECHNOLOGY

THE GIFFORD LECTURES
ABERDEEN, SCOTLAND,

First Series, 1989–90
RELIGION IN AN AGE OF SCIENCE

Second Series, 1990–91
ETHICS IN AN AGE OF TECHNOLOGY

ETHICS
IN AN AGE OF
TECHNOLOGY

The Gifford Lectures
1989–1991
Volume 2

Ian G. Barbour

HarperSanFrancisco
An Imprint of HarperCollins*Publishers*

HarperCollins Web Site: http://www.harpercollins.com
HarperCollins®, ▦®, and HarperSanFrancisco™ are trademarks of
HarperCollins Publishers Inc.

FIRST EDITION

Library of Congress Cataloging-in-Publication Data
Barbour, Ian G.
    Ethics in an age of technology / Ian G. Barbour.—1st HarperCollins ed.
      (The Gifford lectures, 1989–91, v. 2)
    Includes bibliographical references and index.
    ISBN 0-06-060934-6 (cloth). — ISBN 0-06-060935-4 (pbk.)
      1. Technology—Moral and ethical aspects.   2. Technology—Religious aspects—
Christianity.   3. Appropriate technology.   I. Title.   II. Series: Gifford lectures ;
1990–1991.
BJ59.B37   1993
170—dc20   91-59045

99 ❖ RRD(H) 10 9 8 7 6

To Deane

Contents

PART ONE

CONFLICTING VALUES

PART TWO

CRITICAL TECHNOLOGIES

PART THREE

TECHNOLOGY AND THE FUTURE

Acknowledgments

It was a great pleasure to return to the University of Aberdeen to give this second series of Gifford Lectures. I am grateful indeed for the hospitality of members of the Philosophy Department and the Faculty of Divinity, especially Robin Cameron, Nigel Dower, David Fergusson, and Michael Partridge.

I am deeply indebted to the people who subsequently read a draft of the manuscript and offered suggestions for revising it: Deane Barbour, Nigel Dower, Frederick Ferré, Edward Langerak, Carl Mitcham, Robert Russell, and Roger Shinn. Helpful comments on individual chapters came from Gene Bakko, John Barbour, Peter Hartel, Chuck Huff, Roger Kirchner, Ernest Simmons, and Norman Vig.

Many of the topics considered here were discussed in courses and seminars in the program in Science, Technology, and Public Policy at Carleton College. The insights of students and faculty colleagues from a wide range of disciplines contributed to the formulation of several of these chapters. Earlier versions of portions of chapters 1, 2, 3, and 9 were presented in my *Technology, Environment, and Human Values* (New York: Praeger Publishers, 1980).

In response to the joy of months shared in Aberdeen, and a life shared for forty-five years, this volume is dedicated to Deane.

CARLETON COLLEGE
NORTHFIELD, MINNESOTA

Abbreviations

ABM	Antiballistic Missile
AEC	Atomic Energy Commission
AI	Artificial Intelligence
AT	Appropriate Technology
CBA	Cost-Benefit Analysis
CFCs	Chlorofluorocarbons
CO_2	Carbon Dioxide
DOE	Department of Energy
DNA	Deoxyribonucleic Acid
EC	European Community
EPA	Environmental Protection Agency
GNP	Gross National Product
ICBM	Intercontinental Ballistic Missile
IPM	Integrated Pest Management
IT	Intermediate Technology
LDC	Less Developed Country
LGU	Land-Grant University
LWR	Light Water Reactor
NAS	National Academy of Sciences
NASA	National Aeronautics and Space Administration
NATO	North Atlantic Treaty Organization
NIH	National Institutes of Health
NPT	Nonproliferation Treaty
OECD	Organisation for Economic Co-operation and Development
OPEC	Organization of Petroleum Exporting Countries
OTA	Office of Technology Assessment
PC	Personal Computer
PCBs	Polychlorinated Biphenyls
PV	Photovoltaic
R & D	Research and Development
SALT	Strategic Arms Limitation Talks
SDI	Strategic Defense Initiative
START	Strategic Arms Reduction Talks
TA	Technology Assessment

TNC	Transnational Corporation
UN	United Nations
UNEP	United Nations Environmental Program
UNESCO	United Nations Educational, Scientific, and Cultural Organization
UNICEF	United Nations Children's Fund
VDT	Visual Display Terminal
WCC	World Council of Churches

Preface

The first volume in this series examined the challenges to religion presented by the methods and theories of science. This second volume deals with the challenges to ethics arising from technology and applied science. Since the Industrial Revolution in the eighteenth century, it has been widely assumed that science-based technology would automatically lead to progress and improvement in human life. Modern technology has indeed brought increased food production, improved health, higher living standards, and better communications. But its environmental and human costs have been increasingly evident.

During the 1980s, the Challenger space shuttle explosion, the Chernobyl nuclear accident, the Bhopal chemical disaster, and the Exxon Valdez oil spill reminded us that large-scale systems are vulnerable to human errors and technical malfunctions with far-reaching consequences. But most of the harmful impacts of technology came from the normal operations of agricultural and industrial systems that deplete our resources and pollute our air, water, and land. Chemical-intensive agriculture has raised productivity, but it has also led to extensive water pollution. Fossil fuels put huge quantities of carbon dioxide into the atomsphere, threatening worldwide disruption from global warming. Toxic chemical wastes are starting to contaminate water supplies in many countries. Continued population growth is accelerating environmental damage and the depletion of resources.

The human impacts of technology are no less dramatic. Family farms are replaced by large, mechanized corporate farms. Machine-tool operators are replaced by automated machines. Efficiency and productivity are commendable goals, but they tend to crowd out consideration of meaningful work and participation in decisions on the job. Moreover, the benefits of technology fall very unevenly. Disparities between rich and poor around the world are growing. Some nations have crop surpluses, while a billion people are malnourished, more than ever before in history. According to the UNICEF report, *State of the World's Children 1992*, 250,000 children die every week from hunger and easily preventable diseases. Despite the end of the Cold War, a large fraction of the world's research is still devoted to military technology. The threat of nuclear war continues to hang over humanity as additional nations seek nuclear weapons.

Six themes recur throughout the volume:

1. Impacts Distant in Space and Time. Acid rain from German factories harms Scandinavian forests; coal-burning plants in the United States damage Canadian lakes and trees. Brazil clears rain forests in order to export timber and beef to industrial countries, leading to the extinction of thousands of rare species that are an irreplaceable genetic heritage and a potential source of new medications. Chlorofluorocarbon (CFC) refrigerants released in any nation deplete the ozone layer, subjecting people half a world away to more solar radiation that causes skin cancer. Radioactive wastes from today's nuclear power plants will endanger anyone exposed to them ten thousand years from now. Genetic engineering of germ-line cells in plants, animals, and human beings will affect generations far in the future. We are acquiring powers of life and death far beyond those of any previous generation. We face unprecedented choices for which traditional ethics give us little guidance. The evaluation of technology today must be global, anticipatory, and interdisciplinary.

2. *Inequitable Distribution of Costs and Benefits.* Frequently one group benefits from a technology while other groups bear the brunt of the risks and indirect costs. A chemical plant may benefit consumers and stockholders, while its effluents, emissions, and toxic wastes put local citizens at risk. Giant tomato harvesters bring profits to food processing companies and large landowners, but smallholders lose their land and farm workers lose their jobs. Biotechnology research is directed mainly to the diseases of affluent societies, while tropical diseases affecting far larger populations are neglected. Computers and information are sources of social power, and access to them varies greatly within nations and between nations. Technology has contributed to the enormous disparities between rich and poor countries. One U.S. citizen consumes as much of the world's limited resources as fifty citizens of India.

3. *The Concentration of Economic and Political Power.* Technology is both a product and an instrument of social power. It tends to reinforce exisiting social structures. In the Third World, the Green Revolution favored large landowners who could afford tractors and fertilizer, and this led to the further concentration of land ownership. In Western nations, absentee or corporate farm ownership is common, and food processing companies sometimes control the whole food cycle: farm inputs, crop or feedlot contracts, processing, marketing, and restaurant chains. Economic power translates into political power through lobbying and election campaign contributions. Strong lobbies have promoted policies and subsidies favorable to oil, coal, and nuclear power, while solar energy and conservation measures have received little support from industry. Large-scale capital-intensive technologies require large investments and the centralization of management, making participation by workers and regulation by legislatures more difficult.

4. *Manipulative Attitudes.* A more subtle danger is the extension of technological attitudes to all of life until human beings and other creatures are treated as objects to be exploited and controlled. It is easier to seek technical fixes for social problems than to ask whether major changes in social institutions are needed. When a legitimate interest in material progress becomes an

addictive consumerism, human experience is impoverished and community life and personal relationships are eroded. Humility and respect for nature are left behind in the quest for unlimited power. Genetic modification of animals can benefit humanity, but it can also lead us to view them as mere commodities for our use. Some people are already talking about going beyond the cure of genetic diseases to redesign human beings. As more intelligent computers are built, we are beginning to think of ourselves as nothing but information processors. Feminists have pointed out that technology has been a predominantly male enterprise and that in our culture men are identified with reason and mastery—to the exclusion of intuition and the nurture of relationships, which are assumed to be characteristic of women.

5. *The Redirection of Technology.* Some social analysts accept technological determinism, the thesis that technology determines social change and society simply responds to its demands. I reject this view and argue that technology is always a social construction. Technological designs are not inevitable; they are the product of choices by individuals and organizations. Even though industrial corporations and government bureaucracies have great political power, citizens in a democracy can influence public policies through elected representatives, court challenges, advisory panels, and environmental and public interest groups.

Half the world's scientists and engineers today are employed in defense-related research, and many of the remainder are working on projects that will provide luxuries for the privileged. I will urge the redirection of technology to basic human needs, especially food, health, and shelter. If a quarter of the trillion dollars spent annually on arms around the world were put into sustainable agriculture, renewable energy, reforestation, and family planning, the prospects for human welfare and the global environment would be dramatically transformed. In many situations, decentralized intermediate-scale technologies are a promising alternative to both traditional methods and centralized large-scale technologies.

6. *New Value Priorities.* What kinds of technology do we really want? The answer will depend on our value priorities and our vision of the good life. The environmental and resource crises of today are a reflection of fundamental attitudes toward nature and society. The biblical tradition could make a strong contribution to an ethic of respect for all creatures and concern for future generations, since it envisages a created order that spans time and space. This tradition also points to forms of personal fulfillment that are more frugal and less resource-consumptive than prevailing patterns in industrial societies. Above all, it advocates justice and more equitable access to the world's resources. Provided that population growth is curbed, global resources are sufficient for every need, but not for every greed.

The book is divided into three parts. Part 1 explores *conflicting values* that are relevant to the appraisal of technology. Chapter 1 presents widely divergent contemporary views of technology and asks about their underlying assumptions. Chapter 2 analyzes two groups of human values affected by technology: values most significant in individual life (food and health, meaningful work, and personal fulfillment), and those characterizing society (social justice, participatory freedom, and economic development). Chapter 3 deals with three environmental values: resource sustainability, environmental protection, and respect for all forms of life. The scientific, philosophical, and religious grounds for defending each of these values are examined. Technological policies can be discussed in the public arena in terms of such values, which people with various philosophical and religious commitments can understand and support, though I will try to show that the Christian tradition offers a distinctive perspective on each of them. I will suggest that justice, participation, and sustainability are particularly important today—and often neglected—in policy choices concerning technology. These opening chapters provide the ethical framework for the discussion of specific policy decisions in subsequent chapters.

Part 2 offers case studies of *three critical technologies:* agriculture, energy, and computers. In agriculture (chapter 4), environmental constraints such as soil erosion and water pollution from fertilizers and pesticides must be addressed, but the human consequences of mechanization and agribusiness as they affect rural life and family farms must also be considered. In energy (chapter 5), reliance on fossil fuels poses severe problems of global justice, environmental quality, and sustainability; nuclear energy is sustainable but expensive and vulnerable to large-scale accidents and the diversion of plutonium to nuclear weapons. Computers (chapter 6) have low environmental impacts but raise major questions about meaningful work, inequitable access to information, centralization versus decentralization in decision making, and the prospect for artificial intelligence. Most of my examples are taken from the United States, with which I am most familiar, but I have also included comparisons with Europe and Japan. Each case study includes a section on the Third World. I do not take up medical technologies such as life-prolonging equipment, organ transplants, and new reproductive techniques, because a much more extensive literature is already available on medical ethics than on ethical issues in other kinds of technology.

Part 3 looks to *the future of technology.* Chapter 7 deals with the unprecedented powers of recent technology: damage to global environments (including global warming and the rapid extinction of endangered species); the genetic engineering of plants, animals, and humans; and the creation of more destructive conventional weapons and the proliferation of nuclear weapons. The chapter calls for international action in each of these areas. Chapter 8 maintains that citizens can participate in the democratic control of technology, despite the technical character of these policy choices and the difficulties in assessing and regulating particular technologies. Chapter 9 explores the strengths and weaknesses of the appropriate technology movement and the possibilities for more efficient technologies and more frugal life-styles in industrial nations. It closes by suggesting four sources of change: education, political action, crises that evoke new perceptions, and alternative visions of the good life.

The challenge for our generation is to redirect technology toward realizing human and environmental values on planet earth.

Part One

CONFLICTING VALUES

CHAPTER 1

Views of Technology

Technology, the source of the problem, will once again prove to contain within itself the germs of a solution compatible with the betterment of man's lot and dignity.

CHARLES SUSSKIND[1]

Our enslavement to the machine has never been more complete.

JOHN ZERMAN AND ALICE CARNES[2]

What we call Man's power over Nature turns out to be a power exercised by some men over other men with Nature as its instrument.

C. S. LEWIS[3]

Appraisals of modern technology diverge widely. Some see it as the beneficent source of higher living standards, improved health, and better communications. They claim that any problems created by technology are themselves amenable to technological solutions. Others are critical of technology, holding that it leads to alienation from nature, environmental destruction, the mechanization of human life, and the loss of human freedom. A third group asserts that technology is ambiguous, its impacts varying according to the social context in which it is designed and used, because it is both a product and a source of economic and political power.[4]

In this chapter, views of technology are grouped under three headings: Technology as Liberator, Technology as Threat, and Technology as Instrument of Power. In each case the underlying assumptions and value judgments are examined. I will indicate why I agree with the third of these positions, which emphasizes the social construction and use of particular technologies. The issues cut across disciplines; I draw from the writings of engineers, historians, sociologists, political scientists, philosophers, and theologians. The human and environmental values relevant to the appraisal of technology are further analyzed in chapters 2 and 3. These three chapters provide the ethical categories and principles for examining policy decisions about particular technologies in later chapters.

Technology may be defined as *the application of organized knowledge to practical tasks by ordered systems of people and machines.*[5] There are several advantages to such a broad definition. "Organized knowledge" allows us to include technologies based on practical experience and invention as well as those based on scientific theories. The "practical tasks" can include both the production of material goods (in industry and agriculture, for instance) and the provision of services (by computers, communications media, and biotechnologies, among others). Reference to "ordered systems of people and machines" directs attention to social institutions as well as to the hardware of technology. The breadth of the definition also reminds us that there are major differences among technologies.

I. TECHNOLOGY AS LIBERATOR

Throughout modern history, technological developments have been enthusiastically welcomed because of their potential for liberating us from hunger, disease, and poverty. Technology has been celebrated as the source of material progress and human fulfillment.

1. THE BENEFITS OF TECHNOLOGY

Defenders of technology point out that four kinds of benefits can be distinguished if one looks at its recent history and considers its future:

1. Higher Living Standards. New drugs, better medical attention, and improved sanitation and nutrition have more than doubled the average life span in industrial nations within the past century. Machines have released us from much of the backbreaking labor that in previous ages absorbed most of people's time and energy. Material progress represents liberation from the tyranny of nature. The ancient dream of a life free from famine and disease is beginning to be realized through technology. The standard of living of low-income families in industrial societies has doubled in a generation, even though relative incomes have changed little. Many people in developing nations now look on technology as their principal source of hope. Productivity and economic growth, it is said, benefit everyone in the long run.

2. Opportunity for Choice. Individual choice has a wider scope today than ever before because technology has produced new options not previously available and a greater range of products and services. Social and geographical mobility allow a greater choice of jobs and locations. In an urban industrial society, a person's options are not as limited by parental or community expectations as they were in a small-town agrarian society. The dynamism of technology can liberate people from static and confining traditions to assume responsibility for their own lives. Birth control techniques, for example, allow a couple to choose the size and timing of their family. Power over nature gives greater opportunity for the exercise of human freedom.[6]

3. More Leisure. Increases in productivity have led to shorter working hours. Computers and automation hold the promise of eliminating much of the monotonous work typical of earlier industrialism. Through most of history, leisure and cultural pursuits have been the privilege of the few, while the mass of humanity was preoccupied with survival. In an affluent society there is time for continuing education, the arts, social service, sports, and participation in community life. Technology can contribute to the enrichment of human life and the flowering of creativity. Laborsaving devices free us to do what machines cannot do. Proponents of this viewpoint say that people can move beyond materialism when their material needs are met.

4. Improved Communications. With new forms of transportation, one can in a few hours travel to distant cities that once took months to reach. With electronic technologies (radio, television, computer networks, and so on), the speed, range, and scope of communication have vastly increased. The combination of visual image and auditory message have an immediacy not found in the linear sequence of the printed word. These new media offer the possibility of instant worldwide communication, greater interaction, understanding, and mutual appreciation in the "global village." It has been suggested that by dialing coded numbers on telephones hooked into computer networks, citizens could participate in an instant referendum on political issues. According to its defenders, technology brings psychological and social benefits as well as material progress.

In part 2 we will encounter *optimistic forecasts* of each of the particular technologies examined. In agriculture, some experts anticipate that the continuing Green Revolution and the genetic engineering of new crops will provide adequate food for a growing world population. In the case of energy, it is claimed that breeder reactors and fusion will provide environmentally benign power to replace fossil fuels. Computer enthusiasts anticipate the Information Age in which industry is automated and communications networks enhance commercial, professional, and personal life. Biotechnology promises the eradication of genetic diseases, the improvement of health, and the deliberate design of new species—even the modification of humanity itself. In subsequent chapters we will examine each of these specific claims as well as the general attitudes they reveal.

2. OPTIMISTIC VIEWS OF TECHNOLOGY

Let us look at some authors who have expressed optimism regarding technology. Melvin Kranzberg, a prominent historian of technology, has presented a very positive picture of the technological past and future. He argues that urban industrial societies offer *more freedom* than rural ones and provide greater choice of occupations, friends, activities, and life-styles. The work week has been cut in half, and human wants have been dramatically fulfilled.[7] Emanuel Mesthene, former director of the Harvard Program in Technology and Society, grants that every technology brings risks as well as benefits, but he says that our task is the rational management of risk. Some technologies poison the environment, but others reduce pollution. A new technology may displace some workers but it also creates new jobs. Nineteenth-century factories and twentieth-century assembly lines did involve dirty and monotonous work, but the newer technologies allow greater creativity and individuality.[8]

A *postindustrial society,* it is said, is already beginning to emerge. In this new society, according to the sociologist Daniel Bell, power will be based on knowledge rather than property. The dominant class will be scientists, engineers, and technical experts; the dominant institutions will be intellectual ones (universities, industrial laboratories, and research institutes). The economy will be devoted mainly to services rather than material goods. Decisions will be made on rational-technical grounds, marking "the end of ideology." There will be a general consensus on social values; experts will coordinate social planning, using rational techniques such as decision theory and systems analysis. This will be a future-oriented society, the age of the professional managers, the technocrats.[9] A bright picture of the coming technological society has been given by many "futurists," including Buckminster Fuller, Herman Kahn, and Alvin Toffler.[10]

Samuel Florman is an articulate engineer and author who has written extensively *defending technology* against its detractors. He insists that the critics have romanticized the life of earlier centuries and rural societies. Living standards were actually very low, work was brutal, and roles were rigidly defined. People have much greater freedom in technological societies. The automobile, for example, enables people to do what they want and enhances geographical and class mobility. People move to cities because they prefer life there to "the tedium and squalor of the countryside." Florman says that worker alienation in industry is rare, and many people prefer the comfortable monotony of routine tasks to the pressures of decision and accountability. Technology is not an independent force out of control; it is the product of human choice, a response to public demand expressed through the marketplace.[11]

Florman grants that technology often has undesirable side effects, but he says that these are amenable to *technological solutions.* One of his heroes is Benjamin Franklin, who "proposed technological ways of coping with the unpleasant consequences of technology."[12] Florman holds that environmental and health risks are inherent in every technical advance. Any product or process can be made safer, but always at an economic cost. Economic growth and lower prices for consumers are often more important than additional safety, and absolute safety is an illusory goal. Large-scale systems are usually more efficient than small-scale ones. It is often easier to find a "technical fix" for a social problem than to try to change human behavior or get agreement on political policies.[13]

Florman urges us to rely on *the judgment of experts* in decisions about technology. He says that no citizen can be adequately informed about complex technical questions such as acid rain or radioactive waste disposal. Public discussion of these issues only leads to anxiety and erratic political actions. We should rely on the recommendations of experts on such matters.[14] Florman extols the "unquenchable spirit" and "irrepressible human will" evident in technology:

For all our apprehensions, we have no choice but to press ahead. We must do so, first, in the name of compassion. By turning our backs on technological change, we would be expressing our satisfaction with current world levels of hunger, disease, and privation. Further, we must press ahead in the name of the human adventure. Without experimentation and change our existence would be a dull business. We simply cannot stop while there are masses to feed and diseases to conquer, seas to explore and heavens to survey.[15]

Some theologians have also given very positive appraisals of technology. They see it as a source not only of higher living standards but also of *greater freedom and creative expression.* In his earlier writings, Harvey Cox held that freedom to master and shape the world through technology liberates us from the confines of tradition. Christianity brought about the desacralization of nature and allowed it to be controlled and used for human welfare.[16] Norris Clarke sees technology as an instrument of human fulfillment and self-expression in the use of our God-given intelligence to transform the world. Liberation from bondage to nature, he says, is the victory of spirit over matter. As cocreators with God we can celebrate the contribution of reason to the enrichment of human life.[17] Other theologians have affirmed technology as an instrument of love and compassion in relieving human suffering—a modern response to the biblical command to feed the hungry and help the neighbor in need.

The Jesuit paleontologist Pierre Teilhard de Chardin, writing in the early years of nuclear power, computers, and molecular biology, expressed *a hopeful vision of the technological future.* He envisioned computers and electronic communication in a network of interconnected consciousness, a global layer of thought that he called "the noosphere." He defended eugenics, "artificial neo-life," and the remodeling of the human organism by manipulation of the genes. With this new power over heredity, he said, we can replace the crude forces of natural selection and "seize the tiller" to control the direction of future evolution. We will have total power over matter, "reconstructing the very stuff of the universe." He looked to a day of interplanetary travel and the unification of our own planet, based on intellectual and cultural interaction.[18]

Here was an inspiring vision of a planetary future in which *technology and spiritual development* would be linked together. Teilhard affirmed the value of secular life in the world and the importance of human efforts in "building the earth" as we cooperate in the creative work of God. Technology is participation in divine creativity. He rejected any note of despair, which would cut the nerve of constructive action. At times he seemed to have unlimited confidence in humanity's capacity to shape its own destiny. But his confidence really lay in the unity, convergence, and ascent of the cosmic process of which humanity and technology are manifestations. The ultimate source of that unity and ascent is God as known in the Christ whose role is cosmic. For Teilhard, eschatological hope looks not to an intervention discontinuous from history, but to the fulfillment of a continuing process to which our own actions contribute.

Teilhard's writings present us with a magnificent sweep of time from past to future. But they do not consider the institutional structures of economic power and self-interest that now control the directions of technological development. Teilhard seldom acknowledged the tragic hold of social injustice on human life. He was writing before the destructive environmental impacts of technology were evident. When Teilhard looked to the past, he portrayed humanity as an integral part of the natural world, interdependent with other creatures. But when he looked to the future, he expected that because of our technology and our spirituality we will be increasingly separated from other creatures. Humanity will move beyond dependence on the organic world. Though he was ultimately theocentric (centered on God), and he talked about the redemption of the whole cosmos, many of his images are anthropocentric (centered on humanity) and imply that other forms of life are left behind in the spiritualization of humankind that technology will help to bring about.

3. A REPLY TO THE OPTIMISTS

Subsequent chapters will point to inadequacies of these views, but some major criticisms can be summarized here.

First, the *environmental costs and human risks* of technology are dismissed too rapidly. The optimists are confident that technical solutions can be found for environmental problems. Of course, pollution abatement technologies can treat many of the effluents of industry, but often unexpected, indirect, or delayed consequences occur. The effects of carcinogens may not show up for twenty-five years or more. The increased death rates among shipyard workers exposed to asbestos in the early 1940s were not evident until the late 1960s. Toxic wastes may contaminate groundwater decades after they have been buried. The hole in the ozone layer caused by the release of chlorofluorocarbons had not been anticipated by any scientists. Above all, soil erosion and massive deforestation threaten the biological resources essential for human life, and global warming from our use of fossil fuels threatens devastating changes in world climates.

Second, environmental destruction is symptomatic of a deeper problem: *alienation from nature.* The idea of human domination of nature has many roots. Western religious traditions have often drawn a sharp line between humanity and other creatures (see chapter 3). Economic institutions treat nature as a resource for human exploitation. But technological enthusiasts contribute to this devaluation of the natural world if they view it as an object to be controlled and manipulated. Many engineers are trained in the physical sciences and interpret living things in mechanistic rather than ecological terms. Others spend their entire professional lives in the technosphere of artifacts, machines, electronics, and computers, cut off from the world of nature. To be sure, sensitivity to nature is sometimes found among technological optimists, but it is more frequently found among the critics of technology.

Third, technology has contributed to *the concentration of economic and political power.* Only relatively affluent groups or nations can afford the latest technology; the gaps between rich and poor have been perpetuated and in many cases increased by technological developments. In a world of limited resources, it also appears impossible for all nations to sustain the standards of living of industrial nations today, much less the higher standards that industrial nations expect in the future. Affluent nations use a grossly disproportionate share of the world's energy and resources. Commitment to justice within nations also requires a more serious analysis of the distribution of the costs and benefits of technology. We will find many technologies in which one group enjoys the benefits while another group is exposed to the risks and social costs.

Fourth, *large-scale technologies* typical of industrial nations today are particularly problematic. They are capital-intensive rather than labor-intensive, and they add to unemployment in many parts of the world. Large-scale systems tend to be vulnerable to error, accident, or sabotage. The near catastrophe at the Three Mile Island nuclear plant in 1979 and the Chernobyl disaster in 1986 were the products of human errors, faulty equipment, poor design, and unreliable safety procedures. Nuclear energy is a prime example of a vulnerable, centralized, capital-intensive technology. Systems in which human or mechanical failures can be disastrous are risky even in a stable society, quite apart from additional risks under conditions of social unrest. The large scale of many current systems is as much the product of government subsidies, tax and credit policies, and particular corporate interests as of any inherent economies of scale.

Fifth, greater *dependence on experts* for policy decisions would not be desirable. The technocrats claim that their judgments are value free; the technical elite is supposedly nonpolitical. But those with power seldom use it rationally and objectively when their own interests are at stake. When social planners think they are deciding for the good of all—whether in the French or Russian revolutions or in the proposed technocracy of the future—the assumed innocence of moral intentions is likely to be corrupted in practice. Social controls over the controllers are always essential. I will suggest that the most important form of freedom is participation in the decisions affecting our lives. We will return in chapter 8 to this crucial question: How can both experts and citizens contribute to technological policy decisions in a democracy?

Lastly, we must question the linear view of the *science-technology-society relationship,* which is assumed by many proponents of optimistic views. Technology is taken to be applied science, and it is thought to have an essentially one-way impact on society. The official slogan of the Century of Progress exposition in Chicago in 1933 was: "Science Finds—Industry Applies—Man Conforms." This has been called "the assembly-line view" because it pictures science at the start of the line and a stream of technological products pouring off the end of the line.[19] If technology is fundamentally benign, there is no need for government interference except to regulate the most serious risks. Whatever guidance is needed for technological development is supplied by the expression of consumer preferences through the marketplace. In this view, technologies develop from the "push" of science and the "pull" of economic profits.

I accept the basic framework of private ownership in *a free market economy*, but I believe it has severe limitations that require correction through political processes. When wealth is distributed unevenly, the luxuries of a few people carry much more weight in the marketplace than the basic needs of many others. Many of the social and environmental costs of industrial processes are not included in market prices. Because long-term consequences are discounted at the current interest rate, they are virtually ignored in economic decisions. Our evaluation of technology, in short, must encompass questions of justice, participation, environmental protection, and long-term sustainability, as well as short-term economic efficiency.

II. TECHNOLOGY AS THREAT

At the opposite extreme are the critics of modern technology who see it as a threat to authentic human life. We will confine ourselves here to criticisms of the human rather than environmental consequences of technology.

1. THE HUMAN COSTS OF TECHNOLOGY

Five characteristics of industrial technology seem to its critics particularly inimical to human fulfillment.[20]

1. Uniformity in a Mass Society. Mass production yields standardized products, and mass media tend to produce a uniform national culture. Individuality is lost and local or regional differences are obliterated in the homogeneity of industrialization. Nonconformity hinders efficiency, so cooperative and docile workers are rewarded. Even the interactions among people are mechanized and objectified. Human identity is defined by roles in organizations. Conformity to a mass society jeopardizes spontaneity and freedom. According to the critics, there is little evidence that an electronic, computerized, automated society will produce more diversity than earlier industrialism did.

2. Narrow Criteria of Efficiency. Technology leads to rational and efficient organization, which requires fragmentation, specialization, speed, the maximization of output. The criterion is efficiency in achieving a single goal or a narrow range of objectives; side effects and human costs are ignored. Quantitative criteria tend to crowd out qualitative ones. The worker becomes the servant of the machine, adjusting to its schedule and tempo, adapting to its requirements. Meaningful work roles exist for only a small number of people in industrial societies today. Advertising creates demand for new products, whether or not they fill real needs, in order to stimulate a larger volume of production and a consumer society.

3. Impersonality and Manipulation. Relationships in a technological society are specialized and functional. Genuine community and interpersonal interaction are threatened when people feel like cogs in a well-oiled machine. In a bureaucracy, the goals of the organization are paramount and responsibility is diffused, so that no one feels personally responsible. Moreover, technology has created subtle ways of manipulating people and new techniques of electronic surveillance and psychological conditioning. When the technological mentality is dominant, people are viewed and treated like objects.

4. Uncontrollability. Separate technologies form an interlocking system, a total, mutually reinforcing network that seems to lead a life of its own. "Runaway technology" is said to be like a vehicle out of control, with a momentum that cannot be stopped. Some critics assert that technology is not just a set of adaptable tools for human use but an all-encompassing form of life, a pervasive structure with its own logic and dynamic. Its consequences are unintended and unforeseeable. Like the sorcerer's apprentice who found the magic formula to make his broom carry water but did not know how to make it stop, we have set in motion forces that we cannot control. The individual feels powerless facing a monolithic system.

5. Alienation of the Worker. The worker's alienation was a central theme in the writing of Karl Marx. Under capitalism, he said, workers do not own their own tools or machines, and they are powerless in their work life. They can sell their labor as a commodity, but their work is not a meaningful form of self-expression. Marx held that such alienation is a product of capitalist ownership and would disappear under state ownership. He was optimistic about the use of technology in a communist economic order, and thus he belongs with the third group below, the contextualists, but his idea of alienation has influenced the pessimists.

More recent writers point out that *alienation* has been common in state-managed industrial economies too and seems to be a product of the division of labor, rationalization of production, and hierarchical management in large organizations, regardless of the economic system. Studs Terkel and others have found in interviews that resentment, frustration, and a sense of powerlessness are widespread among American industrial workers. This contrasts strongly with the greater work autonomy, job satisfaction, and commitment to work found in the professions, skilled trades, and family-owned farms.[21]

Other features of technological development since World War II have evoked widespread concern. The allocation of more than two-thirds of the U.S. federal research and development budget to military purposes has diverted expertise from environmental problems and urgent human needs. Technology also seems to have contributed to the impoverishment of human relationships and a loss of community. The youth counterculture of the 1970s was critical of technology and sought harmony with nature, intensity of personal experience, supportive communities, and alternative life-styles apart from the prevailing industrial order. While many of its expressions were short-lived, many of its characteristic attitudes, including disillusionment with technology, have persisted among some of the younger generation.[22]

2. RECENT CRITICS OF TECHNOLOGY

To the French philosopher and social critic Jacques Ellul, technology is *an autonomous and uncontrollable force* that dehumanizes all that it touches. The enemy is "technique"—a broad term Ellul uses to refer to the technological mentality and structure that he sees pervading not only industrial processes, but also all social, political, and economic life affected by them. Efficiency and organization, he says, are sought in all activities. The machine enslaves people when they adapt to its demands. Technology has its own inherent logic and inner necessity. Rational order is everywhere imposed at the expense of spontaneity and freedom.

Ellul ends with a *technological determinism,* since technique is self-perpetuating, all-pervasive, and inescapable. Any opposition is simply absorbed as we become addicted to the products of technology. Public opinion and the state become the servants of technique rather than its masters. Technique is global, monolithic, and unvarying among diverse regions and nations. Ellul offers us no way out, since all our institutions, the media, and our personal lives are totally in its grip. He holds that biblical ethics can provide a viewpoint transcending society from which to judge the sinfulness of the technological order and can give us the motivation to revolt against it, but he holds out little hope of controlling it.[23] Some interpreters see in Ellul's recent writings a very guarded hope that a radical Christian freedom that rejects cultural illusions of technological progress might in the long run lead to the transformation rather than the rejection of technology. But Ellul does not spell out such a transformation because he holds that the outcome is in God's hands, not ours, and most of his writings are extremely pessimistic about social change.[24]

The political scientist Langdon Winner has given a sophisticated version of the argument that technology is *an autonomous system* that shapes all human activities to its own requirements. It makes little difference who is nominally in control—elected politicians, technical experts, capitalist executives, or socialist managers—if decisions are determined by the demands of the technical system. Human ends are then adapted to suit the techniques available rather than the reverse. Winner says that large-scale systems are self-perpetuating, extending their control over resources and markets and molding human life to fit their own smooth functioning. Technology is not a neutral means to human ends but an all-encompassing system that imposes its patterns on every aspect of life and thought.[25]

The philosopher Hans Jonas is impressed by *the new scale of technological power* and its influence on events distant in time and place. Traditional Western ethics have been anthropocentric and have considered only short-range consequences. Technological change has its own momentum, and its pace is too rapid for trial-and-error readjustments. Now genetics gives us power over humanity itself. Jonas calls for a new ethic of responsibility for the human future and for nonhuman nature. We should err on the side of caution, adopting policies designed to avert catastrophe rather than to maximize short-run benefits. "The magnitude of these stakes, taken with the insufficiency of our predictive knowledge, leads to the pragmatic rule to give the prophecy of doom priority over the prophecy of bliss."[26] We should seek "the least harm," not "the greatest good." We have no right to tamper genetically with human nature or to accept policies that entail even the remote possibility of the extinction of humanity in a nuclear holocaust.

Another philosopher, Albert Borgmann, does not want to return to a pretechnological past, but he urges the selection of technologies that encourage *genuine human fulfillment.* Building on the ideas of Heidegger, he holds that authentic human existence requires the engagement and depth that occur when simple things and practices focus our attention and center our lives. We have let technology define the good life in terms of production and consumption, and we have ended with mindless labor and mindless leisure. A fast-food restaurant replaces the family meal, which was an occasion of communication and celebration. The simple pleasures of making music, hiking and running, gathering with friends around the hearth, or engaging in creative and self-reliant work should be our goals. Borgmann thinks that some large-scale capital-intensive industry is needed (especially in transportation and communication), but he urges the development of small-scale labor-intensive, locally owned enterprises (in arts and crafts, health care, and education, for example). We should challenge the rule of technology and restrict it to the limited role of supporting the humanly meaningful activities associated with a simpler life.[27]

In *Technology and Power,* the psychologist David Kipnis maintains that those who control a technology have power over other people and this affects personal attitudes as well as social structures. Power holders interpret technological superiority as moral superiority and tend to look down on weaker parties. Kipnis shows that military and transportation technologies fed the conviction of colonists that they were superior to colonized peoples. Similarly, medical knowledge and specialization have led doctors to treat patients as impersonal cases and to keep patients at arms length with a minimum of personal communication. Automation gave engineers and managers increased power over workers, who no longer needed special skills. In general, "power corrupts" and leads people to rationalize their use of power for their own ends. Kipnis claims that the person with technological knowledge often has not only a potent instrument of control but also a self-image that assumes superiority over people who lack that knowledge and the concomitant opportunities to make decisions affecting their lives.[28]

Some Christian groups are critical of *the impact of technology on human life.* The Amish, for example, have resolutely turned their backs on radios, television, and even automobiles. By hard work, community cooperation, and frugal ways, they have prospered in agriculture and have continued their distinctive life-styles and educational patterns. Many theologians who do not totally reject technology criticize its tendency to generate a Promethean pride and a quest for unlimited power. The search for omnipotence is a denial of creaturehood. Unqualified devotion to technology as a total way of life, they say, is a form of idolatry. Technology is finally thought of as the source of salvation, the agent of secularized redemption.[29] In an affluent society, a legitimate concern for material progress readily becomes a frantic pursuit of comfort, a total dedication to self-gratification. Such an obsession with things distorts our basic values as well as our relationships with other persons. Exclusive dependence on technological rationality also leads to a truncation of experience, a loss of imaginative and emotional life, and an impoverishment of personal existence.

Technology is *imperialistic and addictive,* according to these critics. The optimists may think that, by fulfilling our material needs, technology liberates us from materialism and allows us to turn to intellectual, artistic, and spiritual pursuits. But it does not seem to be working out that way. Our material wants have escalated and appear insatiable. Yesterday's luxuries are today's necessities. The rich are usually more anxious about their future than the poor. Once we allow technology to define the good life, we have excluded many important human values from consideration.

Several theologians have expressed particular concern for the impact of technology on *religious life.* Paul Tillich claims that the rationality and impersonality of technological systems undermine the personal presuppositions of religious commitment.[30] Gabriel Marcel believes that the technological outlook pervades our lives and excludes a sense of the sacred. The technician treats everything as a problem that can be solved by manipulative techniques without personal involvement. But this misses the mystery of human existence, which is known only through involvement as a total person. The technician treats other people as objects to be understood and controlled.[31] Martin Buber contrasts the I–It relation of objective detachment with the I–Thou relation of mutuality, responsiveness, and personal involvement. If the calculating attitude of control and mastery dominates a person's life, it excludes the openness and receptivity that are prerequisites of a relationship to God or to other persons.[32] P. H. Sun holds that a high-tech environment inhibits the life of prayer. Attitudes of power and domination are incompatible with the humility and reverence that prayer requires.[33]

3. A REPLY TO THE PESSIMISTS

In replying to these authors, we may note first that there are *great variations among technologies,* which are ignored when they are lumped together and condemned wholesale. Computerized offices differ greatly from steel mills and auto assembly lines, even if they share some features in common. One survey of journal articles finds that philosophers and those historians who trace broad trends (in economic and urban history, for example) often claim that technology determines history, whereas the historians or sociologists who make detailed studies of particular technologies are usually aware of the diversity of social, political, and economic interests that affect the design of a machine and its uses.[34] I will maintain that the uses of any technology vary greatly depending on its social contexts. To be sure, technological systems are interlocked, but they do not form a monolithic system impervious to political influence or totally dominating all other social forces. In particular, technology assessment and legislation offer opportunities for controlling technology, as we shall see.

Second, technological pessimists neglect possible avenues for *the redirection of technology*. The "inevitability" or "inherent logic" of technological developments is not supported by historical studies. We will note below some cases in which there were competing technical designs and the choice among them was affected by various political and social factors. Technological determinism underestimates the diversity of forces that contribute to technological change. Unrelieved pessimism undercuts human action and becomes a self-fulfilling prophecy. If we are convinced that nothing can be done to improve the system, we will indeed do nothing to try to improve it. This would give to the commercial sponsors of technology the choices that are ours as responsible citizens.

Third, technology can be *the servant of human values*. Life is indeed impoverished if the technological attitudes of mastery and power dominate one's outlook. Calculation and control do exclude mutuality and receptivity in human relationships and prevent the humility and reverence that religious awareness requires. But I would submit that the threat to these areas of human existence comes not from technology itself but from preoccupation with material progress and unqualified reliance on technology. We can make decisions about technology within a wider context of human and environmental values.

III. TECHNOLOGY AS INSTRUMENT OF POWER

A third basic position holds that technology is neither inherently good nor inherently evil but is an ambiguous instrument of power whose consequences depend on its social context. Some technologies seem to be neutral if they can be used for good or evil according to the goals of the users. A knife can be used for surgery or for murder. An isotope separator can enrich uranium for peaceful nuclear reactors or for aggression with nuclear weapons. But historical analysis suggests that most technologies are already molded by particular interests and institutional goals. Technologies are social constructions, and they are seldom neutral because particular purposes are already built into their design. Alternative purposes would lead to alternative designs. Yet most designs still allow some choice as to how they are deployed.

1. TECHNOLOGY AND POLITICAL POWER

Like the authors in the previous group, those in this group are critical of many features of current technology. But they offer hope that technology can be used for more humane ends, either by political measures for more effective guidance within existing institutions or by changes in the economic and political systems themselves.

The people who make most of the decisions about technology today are not a technical elite or technocrats trying to run society rationally or disinterested experts whose activity was supposed to mark "the end of ideology." The decisions are made by managers dedicated to *the interests of institutions,* especially industrial corporations and government bureaucracies. The goals of research are determined largely by the goals of institutions: corporate profits, institutional growth, bureaucratic power, and so forth. Expertise serves the interests of organizations and only secondarily the welfare of people or the environment.

The interlocking structure of *technologically based government agencies and corporations,* sometimes called the "technocomplex," is wider than the "military-industrial complex." Many companies are virtually dependent on government contracts. The staff members of regulatory agencies, in turn, are mainly recruited from the industries they are supposed to regulate. We will see later that particular legislative committees, government agencies, and industries have formed three-way alliances to promote such technologies as nuclear energy or pesticides. Networks of industries with common interests form lobbies of immense political power. For example, U.S. legislation supporting railroads and public mass transit systems was blocked by a coalition of auto manufacturers, insurance companies, oil companies, labor unions, and the highway construction industry. But citizens can also influence the direction of technological development. Public opposition to nuclear power plants was as important as rising costs in stopping plans to construct new plants in almost all Western nations.

The historian Arnold Pacey gives many examples of *the management of technology for power and profit.* This is most clearly evident in the defense industries with their close ties to government agencies. But often the institutional biases associated with expertise are more subtle. Pacey gives as one example the Western experts in India and Bangladesh who in the 1960s advised the use of large drilling rigs and diesel pumps for wells, imported from the West. By 1975, two thirds of the pumps had broken down because the users lacked the skills and maintenance networks to operate them. Pacey calls for greater public participation and a more democratic distribution of power in the decisions affecting technology. He also urges the upgrading of indigenous technologies, the exploration of intermediate-scale processes, and greater dialogue between experts and users. Need-oriented values and local human benefits would then play a larger part in technological change.[35]

2. THE REDIRECTION OF TECHNOLOGY

The political scientist Victor Ferkiss expresses hope about the redirection of technology. He thinks that both the optimists and the pessimists have neglected the diversity among different technologies and *the potential role of political structures* in reformulating policies. In the past, technology has been an instrument of profit, and decisions have been motivated by short-run private interests. Freedom understood individualistically became license for the economically powerful. Individual rights were given precedence over the common good, despite our increasing interdependence. Choices that could only be made and enforced collectively—such as laws concerning air and water pollution—were resisted as infringements on free enterprise. But Ferkiss thinks that economic criteria can be subordinated to such social criteria as ecological balance and human need. He believes it is possible to combine centralized, systemwide planning in basic decisions with decentralized implementation, cultural diversity, and citizen participation.[36]

There is a considerable range of views among *contemporary Marxists.* Most share Marx's conviction that technology is necessary for solving social problems but that under capitalism it has been an instrument of exploitation, repression, and dehumanization. In modern capitalism, according to Marxists, corporations dominate the government and political processes serve the interests of the ruling class. The technical elite likewise serves the profits of the owners. Marxists grant that absolute standards of living have risen for everyone under capitalist technology. But relative inequalities have increased, so that class distinctions and poverty amidst luxury remain. Marxists assign justice a higher priority than freedom. Clearly they blame capitalism rather than technology for these evils of modern industrialism. They believe that alienation and inequality will disappear and technology will be wholly benign when the working class owns the means of production. The workers, not the technologists, are the agents of liberation. Marxists are thus as critical as the pessimists concerning the consequences of technology within capitalism but as enthusiastic as the optimists concerning its potentialities—within a proletarian economic order.

How, then, do Western Marxists view the human effects of *technology in Soviet history?* Reactions vary, but many would agree with Bernard Gendron that in the Soviet Union workers were as alienated, factories as hierarchically organized, experts as bureaucratic, and pollution and militarism as rampant as in the United States. But Gendron insists that the Soviet Union did not follow Marx's vision. The means of production were controlled by a small group within the Communist party, not by the workers. Gendron maintains that in a truly democratic socialism, technology would be humane and work would not be alienating.[37] Most commentators hold that the demise of communism in Eastern Europe and the Soviet Union was a product of both its economic inefficiency and its political repression. It remains to be seen whether any distinctive legacy from Marxism will remain there after the economic and political turmoil of the early nineties.

We have seen that a few theologians are technological optimists, while others have adopted pessimistic positions. A larger number, however, see technology as *an ambiguous instrument of social power.* As an example consider Norman Faramelli, an engineer with theological training, who writes in a framework of Christian ideas: stewardship of creation, concern for the dispossessed, and awareness of the corrupting influence of power. He distrusts technology as an instrument of corporate profit, but he believes it can be reoriented toward human liberation and ecological balance. Technology assessment and the legislative processes of democratic politics, he holds, can be effective in controlling technology. But Faramelli also advocates restructuring the economic order to achieve greater equality in the distribution of the fruits of technology.[38] Similar calls for the responsible use of technology in the service of basic human needs have been issued by task forces and conferences of the National Council of Churches and by the World Council of Churches (WCC).[39] According to one summary of WCC documents, "technological society is to be blessed for its capacity to meet basic wants, chastised for its encouragement of inordinate wants, transformed until it serves communal wants."[40]

Egbert Schuurman, a Calvinist engineer from Holland, rejects many features of current technology but holds that it can be *transformed and redeemed* to be an instrument of God's love serving all creatures. Western thought since the Renaissance has increasingly encouraged "man the master of nature"; secular and reductionistic assumptions have prevailed. Schuurman says that technology was given a messianic role as the source of salvation, and under the rule of human sin it has ended by enslaving us so we are "exiles in Babylon." But we can be converted to seek God's Kingdom, which comes as a gift, not by human effort. Receiving it in joy and love, and responding in obedience, we can cooperate in meaningful service of God and neighbor. Schuurman holds that technology can be redirected to advance both material and spiritual well-being. It has "a magnificent future" if it is incorporated into God's work of creation and redemption. A liberated technology could do much to heal the brokenness of nature and society. Unfortunately, he gives us few examples of what such a technology would be like or how we can work to promote it.[41]

The American theologian Roger Shinn has written extensively on Christian ethics and gives attention to *the structures of political and economic power* within which technological decisions are made. He agrees with the pessimists that various technologies reinforce each other in interlocking systems, and he acknowledges that large-scale technologies lead to the concentration of economic and political power. But he argues that when enough citizens are concerned, political processes can be effective in guiding technology toward human welfare. Policy changes require a combination of protest, political pressure, and the kind of new vision that the biblical concern for social justice can provide.[42]

This third position seems to me more consistent with *the biblical outlook* than either of the alternatives. Preoccupation with technology does become a form of idolatry, a denial of the sovereignty of God, and a threat to distinctively human existence. But technology directed to genuine human needs is a legitimate expression of humankind's creative capacities and an essential contribution to its welfare. In a world of disease and hunger, technology rightly used can be a far-reaching expression of concern for persons. The biblical understanding of human nature is realistic about the abuses of power and the institutionalization of self-interest. But it also is idealistic in its demands for social justice in the distribution of the fruits of technology. It brings together celebration of human creativity and suspicion of human power.

The attitudes toward technology outlined in this chapter can be correlated with the typology of historic Christian attitudes toward society set forth by H. Richard Niebuhr.[43] At the one extreme is *accommodation to society*. Here society is considered basically good and its positive potentialities are affirmed. Niebuhr cites the example of liberal theologians of the nineteenth century who had little to say concerning sin, revelation, or grace. They were confident about human reason, scientific and technological knowledge, and social progress. They would side with our first group, those who are optimistic about technology.

At the opposite extreme, Niebuhr describes Christian groups advocating *withdrawal from society*. They believe that society is basically sinful. The Christian perfectionists, seeking to maintain their purity and to practice radical obedience, have withdrawn into monasteries or into separate communities, as the Mennonites and Amish have done. They would tend to side with our second group, the critics of technology.

Niebuhr holds that the majority of Christians are in three movements that fall between the extremes of accommodation and withdrawal. A *synthesis of Christianity and society* has been advocated historically by the Roman Catholic church. Aquinas held that there is both a revealed law, known through scripture and the church, and a natural law, built into the created order and accessible to human reason. Church and state have different roles but can cooperate for human welfare in society. This view encourages a qualified optimism about social change (and, I suggest, about technology).

Another option is the view of Christian life and society as *two separate realms*, as held in the Lutheran tradition. Here there is a compartmentalization of spiritual and temporal spheres and different standards for personal and public life. Sin is prevalent in all life, but in personal life it is overcome by grace; gospel comes before law as the Christian responds in faith and in love of neighbor. In the public sphere, however, sin must be restrained by the secular structures of authority and order. This view tends to be more pessimistic about social change, but it does not advocate withdrawal from society.

The final option described by Niebuhr is a *transformation of society* by Christian values. This position has much in common with the Catholic view and shares its understanding that God is at work in history, society, and nature as well as in personal life and the church. But it is more skeptical about the exercise of power by the institutional church, and it looks instead to the activity of the layperson in society. Calvin, the Reformed and Puritan traditions, the Anglicans, and the Methodists all sought a greater expression of Christian values in public life. They had great respect for the created world ordered by God, and they called for social justice and the redirection of cultural life. This position holds that social change (including the redirection of technology) is possible, but it is difficult because of the structures of group self-interest and institutional power. I favor this last option and will develop it further in subsequent chapters.

3. THE SOCIAL CONSTRUCTION OF TECHNOLOGY

How are science, technology, and society related? Three views have been proposed (see Fig. 1).

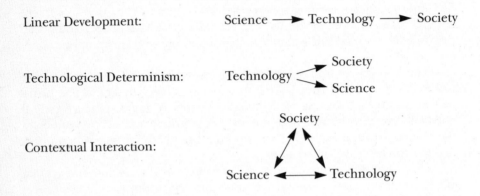

Linear Development: Science ⟶ Technology ⟶ Society

Technological Determinism: Technology ⟨ Society / Science

Contextual Interaction: Society / Science ⟷ Technology

Fig. 1. Views of the Interactions of Science, Technology, and Society.

1. Linear Development. In linear development it is assumed that science leads to technology, which in turn has an essentially one-way impact on society. The deployment of technology is primarily a function of the marketplace. This view is common among the optimists. They consider technology to be predominantly beneficial, and therefore little government regulation or public policy choice is needed; consumers can influence technological development by expressing their preferences through the marketplace.

2. Technological Determinism. Several degrees and types of determinism can be distinguished. Strict determinism asserts that only one outcome is possible. A more qualified claim is that there are very strong tendencies present in technological systems, but these could be at least partly counteracted if enough people were committed to resisting them. Again, technology may be considered an autonomous interlocking system, which develops by its own inherent logic, extended to the control of social institutions. Or the more limited claim is made that the development and deployment of technology in capitalist societies follows only one path, but the outcomes might be different in other economic systems. In all these versions, science is itself driven primarily by technological needs. Technology is either the "independent variable" on which other variables are dependent, or it is the overwhelmingly predominant force in historical change.

Technological determinists will be pessimists if they hold that the consequences of technology are on balance socially and environmentally harmful. Moreover, any form of determinism implies a limitation of human freedom and technological choice. However, some determinists retain great optimism about the consequences of technology. On the other hand, pessimists do not necessarily accept determinism, even in its weaker form. They may acknowledge the presence of technological choices but expect such choices to be misused because they are pessimistic about human nature and institutionalized greed. They may be pessimistic about our ability to respond to a world of global inequities and scarce resources. Nevertheless, determinism and pessimism are often found together among the critics of technology.

3. Contextual Interaction. Here there are six arrows instead of two, representing the complex interactions between science, technology, and society. Social and political forces affect the design as well as the uses of particular technologies. Technologies are not neutral because social goals and institutional interests are built into the technical designs that are chosen. Because there are choices, public policy decisions about technology play a larger role here than in either of the other views. Contextualism is most common among our third group, those who see technology as an ambiguous instrument of social power.

Contextualists also point to *the diversity of science-technology interactions.* Sometimes a technology was indeed based on recent scientific discoveries. Biotechnology, for example, depends directly on recent research in molecular biology. In other cases, such as the steam engine or the electric power system, innovations occurred with very little input from new scientific discoveries. A machine or process may have been the result of creative practical innovation or the modification of an existing technology. As Frederick Ferré puts it, science and technology in the modern world are both products of the combination of theoretical and practical intelligence, and "neither gave birth to the other."[44] Technology has its own distinctive problems and builds up its own knowledge base and professional community, though it often uses science as a resource to draw on. The reverse contribution of technology to science is also often evident. The work of astronomers, for instance, has been dependent on a succession of new technologies, from optical telescopes to microwave antennae and rockets. George Wise writes, "Historical studies have shown that the relations between science and technology need not be those of domination and subordination. Each has maintained its distinctive knowledge base and methods while contributing to the other and to its patrons as well."[45]

In the previous volume, I discussed the "*social construction of science*" thesis, in which it is argued that not only the direction of scientific development but also the concepts and theories of science are determined by cultural assumptions and interests. I concluded that the "strong program" among sociologists and philosophers of science carries this historical and cultural relativism too far, and I defended a reformulated understanding of objectivity, which gives a major role to empirical data while acknowledging the influence of society on interpretive paradigms.

The case for "*the social construction of technology*" seems to me much stronger. Values are built into particular technological designs. There is no one "best way" to design a technology. Different individuals and groups may define a problem differently and may have diverse criteria of success. Bijker and Pinch show that in the late nineteenth century inventors constructed many different types of bicycles. Controversies developed about the relative size of front and rear wheels, seat location, air tires, brakes, and so forth. Diverse users were envisioned (workers, vacationers, racers, men and women) and diverse criteria (safety, comfort, speed, and so forth). In addition, the bicycle carried cultural meanings, affecting a person's self-image and social status. There was nothing logically or technically necessary about the model that finally won out and is now found around the world.[46]

The historian John Staudenmaier writes that

contextualism is rooted in the proposition that technical designs cannot be meaningfully interpreted in abstraction from their human context. The human fabric is not an envelope around a culturally neutral artifact. The values and world views, the intelligence and stupidity, the biases and vested interests of those who design, accept and maintain a technology are embedded in the technology itself.[47]

Both the linear and the determinist view imply that technology determines *work organization*. It is said that the technologies of the Industrial Revolution imposed their own requirements and made repetitive tasks inevitable. The contextualists reply that the design of a technology is itself affected by social relations. The replacement of workers by machines was intended not only to reduce labor costs but also to assert greater control by management over labor. For instance, the spinning mule helped to break the power of labor unions among skilled textile workers in nineteenth-century England. Some examples in the choice of designs for agricultural harvesters, nuclear reactors, and computer-controlled manufacturing are discussed in later chapters.

Other contextualists have pointed to the role of technology in *the subordination of women*. Engineering was once considered heavy and dirty work unsuitable for women, but long after it became a clean and intellectual profession, there are still few women in it. Technology has been an almost exclusively male preserve, reflected in toys for boys, the expectations of parents and teachers, and the vocational choices and job opportunities open to men and women. Most technologies are designed by men and add to the power of men.

Strong *gender divisions* are present among employees of technology-related companies. When telephones were introduced, women were the switchboard operators and record keepers, while men designed and repaired the equipment and managed the whole system. Typesetting in large printing frames once required physical strength and mechanical skills and was a male occupation. But men continued to exclude women from compositors' unions when linotype, and more recently computer formatting, required only typing and formatting skills.[48] Today most computer designers and programmers are men, while in offices most of the data are entered at computer keyboards by women. With many middle-level jobs eliminated, these lower-level jobs often become dead ends for women.[49] A study of three computerized industries in Britain found that women were the low-paid operators, while only men understood and controlled the equipment, and men almost never worked under the supervision of women.[50]

Note that contextualism allows for a *two-way interaction* between technology and society. When technology is treated as merely one form of cultural expression among others, its distinctive characteristics may be ignored. In some renditions, the ways in which technology shapes culture are forgotten while the cultural forces on technology are scrutinized. The impact of technology on society is particularly important in the transfer of a technology to a new cultural setting in a developing country. Some Third World authors have been keenly aware of technology as an instrument of power, and they portray a two-way interaction between technology and society across national boundaries.

IV. CONCLUSIONS

Let me try to summarize these three views of technology in relation to the conflicting values (identified in italics) that are discussed in the next two chapters. There are many variations within each of the three broad positions outlined above, but each represents a distinctive emphasis among these values.

The optimists stress the contribution of technology to *economic development*. They hold that greater productivity improves standards of living and makes *food and health* more widely available. For most of them, the most important form of *participatory freedom* is the economic freedom of the marketplace, though in general they are also committed to political democracy. These authors say that social justice and environmental protection should not be ignored, but they must not be allowed to jeopardize economic goals. The optimists usually evaluate technology in a utilitarian framework, seeking to maximize the balance of costs over benefits.

The pessimists typically make *personal fulfillment* their highest priority, and they interpret fulfillment in terms of human relationships and community life rather than material possessions. They are concerned about individual rights and the dignity of persons. They hold that *meaningful work* is as important as economic productivity in policies for technology. The pessimists are dedicated to *resource sustainability* and criticize the high levels of consumption in industrial societies today. They often advocate *respect for all creatures* and question the current technological goal of mastery of nature.

The contextualists are more likely to give prominence to *social justice* because they interpret technology as both a product and an instrument of social power. For them the most important forms of *participatory freedom* are opportunities for participation in political processes and in work-related decisions. They are less concerned about economic growth than about how that growth is distributed and who receives the costs and the benefits. Contextualists often seek *environmental protection* because they are aware of the natural as well as the social contexts in which technologies operate.

I am most sympathetic with the contextualists, though I am indebted to many of the insights of the pessimists. Four issues seem to me particularly important in analyzing the differences among the positions outlined above.

1. Defense of the Personal. The pessimists have defended human values in a materialistic and impersonal society. The place to begin, they say, is one's own life. Each of us can adopt individual life-styles more consistent with human and environmental values. Moreover, strong protest and vivid examples are needed to challenge the historical dominance of technological optimism and the disproportionate resource consumption of affluent societies. I admire these critics for defending individuality and choice in the face of standardization and bureaucracy. I join them in upholding the significance of personal relationships and a vision of personal fulfillment that goes beyond material affluence. I affirm the importance of the spiritual life, but I do not believe that it requires a rejection of technology. The answer to the destructive features of technology is not less technology, but technology of the right kind.

2. The Role of Politics. Differing models of social change are implied in the three positions. The first group usually assumes a free market model. Technology is predominantly beneficial, and the reduction of any undesirable side effects is itself a technical problem for the experts. Government intervention is needed only to regulate the most harmful impacts. Writers mentioned in the second section, by contrast, typically adopt some variant of technological determinism. Technology is dehumanizing and uncontrollable. They see runaway technology as an autonomous and all-embracing system that molds all of life, including the political sphere, to its requirements. The individual is helpless within the system. The views expressed in the third section presuppose a "social conflict" model. Technology influences human life but is itself part of a cultural system; it is an instrument of social power serving the purposes of those who control it. It does systematically impose distinctive forms on all areas of life, but these can be modified through political processes. Whereas the first two groups give little emphasis to politics, the third, with which I agree, holds that conflicts concerning technology must be resolved primarily in the political arena.

3. The Redirection of Technology. I believe that we should neither accept uncritically the past directions of technological development nor reject technology *in toto* but redirect it toward the realization of human and environmental values. In the past, technological decisions have usually been governed by narrowly economic criteria, to the neglect of environmental and human costs. In a later chapter we will look at technology assessment, a procedure designed to use a broad range of criteria to evaluate the diverse consequences of an emerging technology—*before* it has been deployed and has developed the vested interests and institutional momentum that make it seem uncontrollable. I will argue that new policy priorities concerning agriculture, energy, resource allocation, and the redirection of technology toward basic human needs can be achieved within democratic political institutions. The key question will be: What decision-making processes and what technological policies can contribute to human and environmental values?

4. *The Scale of Technology.* Appropriate technology can be thought of as an attempt to achieve some of the material benefits of technology outlined in the first section without the destructive human costs discussed in the second section, most of which result from large-scale centralized technologies. Intermediate-scale technology allows decentralization and greater local participation in decisions. The decentralization of production also allows greater use of local materials and often a reduction of impact on the environment. Appropriate technology does not imply a return to primitive and prescientific methods; rather, it seeks to use the best science available toward goals different from those that have governed industrial production in the past.

Industrial technology was developed when capital and resources were abundant, and we continue to assume these conditions. Automation, for example, is capital-intensive and labor saving. Yet in *developing nations* capital is scarce and labor is abundant. The technologies needed there must be relatively inexpensive and labor-intensive. They must be of intermediate scale so that jobs can be created in rural areas and small towns, to slow down mass migration to the cities. They must fulfill basic human needs, especially for food, housing, and health. Alternative patterns of modernization are less environmentally and socially destructive than the path that we have followed. It is increasingly evident that many of these goals are desirable also in industrial nations. I will suggest that we should develop a mixture of large- and intermediate-scale technologies, which will require deliberate encouragement of the latter.

The redirection of technology will be no easy task. Contemporary technology is so tightly tied to industry, government, and the structures of economic power that changes in direction will be difficult to achieve. As the critics of technology recognize, the person who tries to work for change within the existing order may be absorbed by the establishment. But the welfare of humankind requires a creative technology that is economically productive, ecologically sound, socially just, and personally fulfilling.

CHAPTER 2

Human Values

Empirical knowledge of our biological nature will allow us to make optimum choices among the competing criteria of progress.

EDWARD O. WILSON[1]

The principles of justice are . . . the principles that free and rational persons concerned to further their own interests would accept in an initial position of equality.

JOHN RAWLS[2]

What does the Lord require of you but to do justice, and to love kindness, and to walk humbly with your God?

MICAH 6:8

By what criteria can we judge the consequences of technology? How can we balance the competing values that enter into technological policy decisions? The authors quoted above look respectively to science, philosophy, and religion for the justification of their value commitments. We will consider each of these approaches in turn, examining its strengths and weaknesses as a basis for ethics in a technological age.

I will suggest that the most significant values in individual life affected by technology are food and health, meaningful work, and personal fulfillment. The values in social life that are most relevant to the appraisal of technology are social justice, participatory freedom, and economic development. Environmental values are taken up in the next chapter. Later in the volume I will ask how the values discussed here may be expressed in particular technological decisions in public policy, in engineering design, and in the life of consumers and citizens. This chapter and the next one provide an analytical and methodological framework for the practical choices that are discussed in subsequent chapters.

A *value* may be defined as a general characteristic of an object or state of affairs that a person views with favor, believes is beneficial, and is disposed to act to promote.[3] To hold a value is to have a favorable attitude toward its realization; in this respect, values resemble preferences or desires. However, subscription to a value also includes beliefs about benefits or moral obligations that can be used to justify or defend it or recommend it to others; such beliefs are open to rational reflection and discussion, as individual preferences are not. When a choice is defended in terms of values, reasons are given and general principles are invoked. Principles of right and wrong in human actions, and good and evil in the consequences of actions, constitute the domain of ethics.

I. SCIENCE AND HUMAN VALUES

A widely held view posits the *total independence* of science and ethics. Many scientists affirm the neutrality and objectivity of science. They believe that scientific theories are totally value free. They defend an absolute distinction between facts and values. In the positivist version of this view, science is rational and objective, while value judgments are emotional and subjective. Proponents of the independence thesis often draw a sharp line between science and technology and assert that scientists have no responsibility for the applications of their discoveries. They point out that the uses of science are unpredictable and that scientists are not qualified to make judgments outside their limited area of technical expertise.

The historian of science Loren Graham calls this view *"restrictionism"* because it restricts science to a clearly demarcated sphere and rejects any links between science and values. It insulates science from social criticism. Such compartmentalism prevents the misuse of science by "expansionists" who extend scientific ideas to justify political convictions (such as the Social Darwinists' arguments for economic competition or the Nazi arguments for eugenics or the political uses of Lysenko's genetics in the Soviet Union). Moreover, says Graham, the notion of value-free research provides support for the autonomy of science. Freedom of inquiry is protected by the internal regulation and self-government of the scientific community, which supposedly bases decisions only on the professional judgment of peers. But Graham argues that there is no value-free science. He portrays a spectrum from the relatively minor role of values in physics to the "inescapably value-laden" concepts and implications of the biological and social sciences.[4]

The philosopher Stephen Toulmin claims that the inescapable *connections between science and ethics* make the independence thesis untenable. First, many concepts in science—especially in biology and the social sciences—reflect value judgments. Toulmin cites as examples concepts of "function" and "adaptation" in biology, and "normal" and "deviant" behavior in psychology. Second, the line between science and technology is not at all sharp. The possible uses of many scientific discoveries are sufficiently foreseeable that scientists have a responsibility to inform the public about them. Third, the process of research often raises ethical issues, especially in the case of experiments on animals or human subjects, or possible threats to public safety. Lastly, the ideal of scientific objectivity is itself dubious, according to Toulmin, for we are participants and not onlookers in the world, as evidenced in disciplines as diverse as quantum physics, ecology, and cultural anthropology.[5]

Claims concerning *the autonomy of science* have also been strongly criticized. While science does have an impressive system of self-regulation (through grant allocation, peer review, and publication and reward structures), it is by no means free from external pressures. The "scientific establishment" has strong connections to government agencies as well as to funding sources in industry. According to the critics, the claim of scientific neutrality is an evasion of responsibility that allows other groups to determine the directions of scientific advance as well as technological application.[6]

These interpretations suggest that science and ethical values interact and cannot be totally isolated from each other. But some authors want to carry the integration of science and ethics much further by *deriving ethical principles from science.* We look first at attempts to base ethics on the values intrinsic to scientific inquiry. Then we consider arguments that ethical principles can be derived from evolutionary biology. If these arguments are valid, they would give us some guidance in decisions about technology.

1. VALUES INTRINSIC TO SCIENCE

It has been proposed that values *inherent in scientific inquiry* can be extended as basic ideals for society. Jacob Bronowski gives an eloquent statement of this position. Scientific activity, he says, itself requires tolerance, disinterestedness, rationality, freedom of thought, and the right of dissent. Scientific knowledge is universal, and its pursuit demands cooperation, honesty, and loyalty to truth. Bronowski urges us to extend these "scientific values" to the social order in such forms as political freedom, respect for other persons, and international cooperation.[7] In a similar vein, the sociologist Robert Merton maintains that the highest social values can be found in the institutionalized ethics of science, which he describes under the headings of universalism, organized skepticism, disinterestedness, and communality.[8]

Such portrayals of values intrinsic to science have come under attack from several quarters. For one thing, *the actual motives of scientists* are more mixed than this idealized picture acknowledges. Scientists are in fact seldom disinterested or devoted to the truth alone; like other people, they often seek professional recognition, personal success, and higher pay. Incentive and reward structures in science favor conformity to prevailing cognitive frameworks. Disputes over priority show the diversity of motives operating in science. Scientists talk about universality and cooperation, but they often practice secrecy—out of loyalty to their employer or nation, for instance, or in order to ensure the priority of a discovery.[9] With the emergence of "big science," the goals of research are increasingly set by industry or government whose objective is not truth for its own sake but knowledge as an instrument of power.[10]

Even if the "scientific values" were accepted as representing the ideal if not the practice of scientific activity, it seems dubious that an adequate *social ethic* could be derived from them. The values found in science are insufficient as criteria for a humane political order. Historically, scientific work has been compatible with a wide range of political and philosophical viewpoints. Bronowski himself, in a revised preface to his book, writes, "I would, were I beginning again, give some space also to a discussion of those values which are not generated by the practice of science—the values of tenderness, of kindliness, of human intimacy and love." To these personal virtues I would add justice and other social values. The pursuit of scientific knowledge is only one among many human endeavors, and it cannot supply the norms for activities with very different goals.

The theologian H. Richard Niebuhr holds that there is an important *ethical dimension* in scientific inquiry, but he says that it represents only a limited range of values. In the scientific community there is an implicit covenant of truth telling as well as a commitment to truth seeking. Without trust and common loyalties, says Niebuhr, there can be no community. Niebuhr describes the scientist's commitment to communication and universalism, but he also notes the conflicting loyalties to nation or group that lead to secrecy.[11] The philosopher Michael Polanyi finds ethical elements in the cooperation that occurs within the scientific community, the universal intent of scientists, and their personal responsibility for their claims.[12]

Honesty is still characteristic of scientists, even though several highly publicized cases of fraud have occurred recently and the pressures to publish have increased the temptation to claim credit for what is not one's own work.[13] I would maintain, however, that such honesty is not primarily the result of the personal virtue of the individual scientist but rather of the institutional structure of science and the requirement that results be reproducible by other scientists. In short, important ethical values are intrinsic to science, but we should not expect them to give us an adequate social ethic or to provide motivation for ethical choices outside the institutions of science.

2. EVOLUTIONARY ETHICS

The second attempt to derive ethics from science is based on *evolutionary biology*. Julian Huxley, C. H. Waddington, and other biologists have proposed that human values can be inferred from the character and the direction of evolutionary trends. The past history of nature can be extrapolated to determine the direction in which humanity should proceed in the future. Evolutionary history shows a trend toward intelligence, self-awareness, cooperation, and group loyalty. These characteristics have been selected by the evolutionary process because they contributed to survival, so we can be confident that in seeking them we will be acting in harmony with the order of nature.[14]

But one difficulty with this proposal is that evolution displays *contrasting characteristics*. Cooperation and mutual aid are present, but so is intense competitive struggle. It is perhaps not surprising that a wide variety of ethical conclusions have been drawn from evolution. Herbert Spencer and the Social

Darwinists took the struggle for survival to be a justification for competitive capitalism. Thomas Huxley believed that nature's cruel struggle is precisely the opposite of what humanity should seek, while Kropotkin found in the co-operation among differing species a model for cooperative anarchism. Nietz-sche thought evolution supports the idea of the "superman," and Hitler tried to justify genocide and racial purity by appeal to the idea of the survival of the fittest. The phenomena of evolution are so diverse that one can find in them support for almost any ethical position.

A number of biologists have said that attempts to derive ethical principles from evolutionary history are invalid because they ignore *the distinctiveness of human nature and culture.* G. G. Simpson and T. H. Dobzhansky hold that self-awareness and the capacity for reflective choice distinguish human beings from all other creatures. Even if clear trends were apparent in the past, the op-tion is open for us to try to initiate new trends consistent with the goals we consciously select. Human culture, including technology, presents choices for which prehuman history can give us little guidance.[15] George Williams says that in seeking models for humanity we must reject many forms of behavior found in other life forms, including cruelty, wasteful predation, infanticide, and cannibalism. He wants us to disavow the genetic selfishness that natural selection encouraged.[16]

Many philosophers hold that it is impossible to derive *prescriptive conclusions* from purely *descriptive premises.* Hume argued that "ought" statements cannot be deduced from "is" statements because they are logically different kinds of assertions. G. E. Moore said that anyone who tries to derive values from facts commits "the naturalistic fallacy." In their detailed critiques of evolutionary ethics, Anthony Flew and Anthony Quinton claim that the "direction" of evo-lution is very ambiguous, but that even if there were a clear direction, we could not say in advance that we will approve of it whatever it turns out to be. We still have to evaluate it by independent criteria. Flew points out that Julian Huxley said that if nature *were* as ruthlessly competitive as his grandfather, T. H. Huxley, had thought, he too would advocate opposing the values seen in nature instead of adopting them—which shows that even he judged nature by independent criteria. Whatever the trends we find in prehuman and human evolution, we are free to choose whether to continue them or to try to modify them.[17]

Thomas Nagel holds that evolution can explain *the origins* of our reflective capacities but not *the conclusions* we reach with these capacities. We can, for ex-ample, point to the evolutionary origins and physiological foundations of the intellectual abilities that we use in mathematics and scientific inquiry, but we cannot expect a biological justification for a particular mathematical theorem or scientific theory. Similarly, says Nagel, if ethics is critical rational reflection on our prereflective ethical intuitions, evolution might explain the starting point and the capacities we use in ethical reflection but not the conclusions we reach. The defense and criticism of both scientific theories and ethical principles are quite independent of inquiry into the historical evolution of human capacities.[18]

Sociobiologists have made strong claims concerning *the evolutionary origins of ethics,* as we saw in the previous volume. E. O. Wilson and others have shown that altruistic behavior, in which insects and animals sacrifice themselves for the group, favors the survival of relatives that share a common set of genes. Such altruism confers an adaptive advantage and has therefore been encouraged by the selective process. In some of his writings, Wilson proposes the derivation of human ethical norms from evolutionary science. "Empirical knowledge of our biological nature will allow us to make optimum choices among the competing criteria of progress."[19] He says that a complete neural understanding of the brain will provide "a firm foundation for ethics."

Wilson is less interested in deriving ethical principles from evolution than in explaining the evolutionary origins and biological basis of human behavior. He holds that ethical principles are the intellectual rationalization of emotions that arise from the upwelling of deep impulses and instincts encoded genetically in the brain, especially in the hypothalamus and limbic systems. Ethical norms that encouraged social cooperation clearly had great survival value because they enhanced the unity of the group. Cultural systems allow such cooperation to be extended beyond genetic relatives.[20]

The philosopher Michael Ruse accepts a sociobiological explanation for the *existence* of ethical norms, but he says that their *specific content* cannot be inferred from evolutionary history. He defends the is/ought distinction and insists that we cannot assume that the natural is inherently good. Shared moral rules supporting social cooperation are adaptive, but they are products of a contingent evolutionary history, not eternal principles. We might have evolved without a need for love and sympathy. "What I would argue is that morality is a kind of game that conscious reflective beings play (that they are locked into playing by their genes) in order to achieve such relatively smooth sociality."[21] Because we believe in right and wrong we are biologically fitter.

Ruse concludes that all values are *subjective,* but the fact that we think they are *objective* can be explained by sociobiology. Values are actually human constructions that we project on the world, he says, but in order to take them seriously we have to believe that they are objective. There has been evolutionary selection favoring the "collective illusion" of objectivity. "Darwinian theory shows that in fact morality is a function of (subjective) feelings; but it shows also that we have (and must have) the illusion of objectivity. . . . In a sense, therefore, morality is a collective illusion foisted upon us by our genes."[22] Ruse says that the belief that God is the source of moral rules makes such rules more socially effective and thus serves a useful biological function. It would seem, however, that Ruse's position is self-defeating, for once the secret is out that ethical norms are a collective illusion, we can hardly expect their social effectiveness to continue. I suggest that evolutionary biology does not either establish or discredit ethical principles.

3. THE CONTRIBUTION OF SCIENCE TO ETHICS

The *social sciences* can help us to understand the patterns of personality and human behavior within which people seek to realize the values they hold. As one example, the psychologist Abraham Maslow has suggested a basis for ordering values in a hierarchy of human needs. Maslow finds *five levels of need:*

1. Survival (physiological needs): food, shelter, health.
2. Security (safety needs): protection from danger and threat.
3. Belonging (social needs): friendship, acceptance, love.
4. Self-esteem (ego needs): self-respect, recognition, status.
5. Self-actualization (fulfillment needs): creativity, realization of individual potentialities.

Maslow maintains that these levels form a hierarchy; lower levels must be satisfied before the individual can give attention to higher levels. People are preoccupied with survival and security if these are threatened. A starving person has little interest in artistic creativity or political liberty. The lower levels are more fundamental; at least a minimal fulfillment of them is a precondition for interest in higher levels. Once a need is satisfied, its role in motivation dwindles; the lowest level of ungratified need is the primary influence on behavior.[23]

Maslow's *hierarchy of needs* is very helpful in any consideration of values. In his writing he shows great sensitivity to the multiple dimensions of human experience. There is indeed considerable evidence that under conditions of extreme scarcity people do give priority to survival needs. I would submit, however, that higher-level needs are always present; they do not suddenly emerge when lower needs are satisfied. Hungry people miss those they love as much as the well fed do. Many creative artists have been very poor or have lived in impoverished societies. I will suggest in later chapters that policies designed to meet basic material needs should have top priority but should be accompanied by measures that promote community life and the fulfillment of human potentialities. Moreover, we always have choices concerning the goals we will pursue even though we are constrained by the structures of human nature.

Neither the natural nor the social sciences can provide a self-sufficient basis for deriving ethical principles, but they are relevant to ethics in three ways.

First, the biological and social sciences can show us *the constraints on human behavior* arising from the structures that are given to us. Human choices cannot ignore the limits set by the evolutionary past or by the genetic programs, physiological processes, and psychological and social forces at work in our lives. We are dependent on many conditions beyond our control, even though many choices are still open to us. Our actions must be in harmony with the structures within which we live, even as we assume responsibility for our own decisions.

Second, science can provide increasingly reliable estimates of *the consequences of our decisions.* I will suggest below that analysis of consequences is an important component of ethical reflection, though not the only consideration.

Cost-benefit and risk-benefit analysis, environmental impact assessment, technology assessment, and regulatory standards all depend heavily on scientific judgments concerning the probable effects of alternative policy decisions. While the choice of goals is not a scientific question, the choice of means to achieve goals requires knowledge of technical possibilities and estimation of their direct and indirect consequences. Scientific expertise is an essential component of technological policy decisions, though we can expect from the discussion above that it will seldom be totally free of value judgments or personal and institutional interests.

Third, science contributes to *the worldview within which our decisions are made.* Science is one of the main sources of the changing views of the world and our place in it and of the metaphysical categories by which we interpret the diverse dimensions of our experience, as we saw throughout the first volume in this series. Our unifying vision of reality must be consistent with the findings of science and our understanding of human potentialities. Though we cannot directly derive a set of ethical principles from evolution, we can obtain from it a new perspective on temporality and change and the interdependence of all forms of life. In the past, ethics dealt almost exclusively with duties to human beings and virtually ignored other forms of life. In the next chapter the ethical implications of the ecological understanding of interdependence are explored. In sum, ethics cannot be derived from science alone, but ethics is influenced in significant ways by scientific knowledge.

II. PHILOSOPHY AND HUMAN VALUES

The philosopher is interested in the conceptual clarity, the consistency, and the universality of ethical principles. We start by looking at the strengths and weaknesses of utilitarianism as a principle for policy choice. Then the concepts of justice and freedom are analyzed. A few examples of the application of these concepts and principles in technological policy decisions are offered here, but applications are mainly explored in later chapters.

1. UTILITARIANISM AND ITS CRITICS

Utilitarianism has been not only an important school of thought among philosophers, but also a major influence among social scientists. Cost-benefit analysis and other formal methods used in environmental and technological decisions share the assumptions of utilitarianism. We can present here only the broad outlines of utilitarian philosophy and some of the issues that it raises for technological policy.

The central principle of utilitarianism is *the greatest good for the greatest number.* That action should be chosen which produces the greatest net balance of good over evil consequences. For Jeremy Bentham, the good was identified with pleasure; one should select the alternative that maximizes the balance of pleasure over pain. John Stuart Mill maintained that happiness is a more inclusive and long-lasting good than pleasure.[24] The utilitarian economists in turn sought to maximize total social welfare, aggregated either from individual welfare or from subjective preferences and perceived satisfactions. There are significant differences among these versions, but some observations can be made about their common assumptions.

Most forms of utilitarianism are *anthropocentric*. "The greatest good for the greatest number" has usually been taken to refer exclusively to human beings. Any harm to other creatures is to be considered only insofar as it affects humanity. We will examine in the next chapter a broader rendition of the principle that includes the good of all sentient beings. Although the principle usually is taken to apply only to presently existing persons, it does not actually distinguish present from future generations. But there are difficulties when future persons are included, since at least in principle the largest total good might be achieved by having an enormous population at a low level of well-being. The question of how much weight to attach to future costs and benefits is also problematic, as we shall see.

Utilitarianism faces serious difficulties in attempting to *quantify* "the greatest good." If the good is identified with happiness, can it be measured on a single numerical scale? Utilitarian economists speak of maximizing satisfactions or preferences. But do people really look on diverse kinds of satisfactions as equivalent and substitutable? Can preferences among persons be compared and then aggregated in order to determine whether the total for society has been maximized? Many economists have concluded that the only practical way to measure people's preferences is by their willingness to pay. But the distribution of purchasing power is very uneven, and it is often misleading to assume that everything that is prized can be priced. The concern for quantification has tended to restrict attention to measurable costs and benefits.

Another criticism is that in utilitarianism only the total good, and not its *distribution* among people, is relevant to moral choice. Suppose the extermination of a small minority would make the majority so happy that the total happiness is increased. Suppose total national income can be increased if we accept great poverty for one segment of society. The utilitarian can object to these actions only if it can be demonstrated that there are indirect repercussions that will harm the total welfare, for utilitarianism finds nothing inherently wrong with injustice or inequality as such. In many cases the long-term social costs of setting a precedent by unjust actions might be so serious that they would outweigh any short-term benefits. But such considerations would not always prevent the sacrifice of some individuals for the social good.

Many contemporary philosophers hold that utilitarian principles must be supplemented by a *principle of justice*. If the total good were the only criterion, we could justify a small social gain even if it entailed a gross injustice. But if justice were the only norm, we would have to correct a small injustice even if it resulted in widespread suffering or social harm. It appears, then, that we need to consider both justice and the total good.[25]

I will suggest later that *cost-benefit* and *risk-benefit* analyses, when supplemented by a principle of justice, are often useful techniques if one is comparing a small number of options and there is a narrow range of very specific objectives. But most policy decisions today involve a large number of options and a broad range of impacts, many of which are difficult or impossible to quantify. The trade-offs are multidimensional and cannot be measured in a single unit or aggregated as a numerical total. They involve highly diverse types of value. I will maintain that environmental impact assessment and technology assessment methods allow a broader range of value considerations and thereby escape some of the limitations of utilitarian calculations. I will also argue that policy choices usually entail value judgments among incommensurables, and therefore the basic decisions must be made through political processes, not by technical experts using formal analytic techniques.

A final objection is a broader one. Utilitarianism judges entirely by consequences. But there are some acts, such as murder or experimentation on human subjects without their voluntary consent, which we do not condone even if they have good consequences. An alternative approach to ethics stresses *duty* and *obligation,* the choice of acts that are right in themselves, apart from the calculation of consequences. Theories based on obligations are called *deontological* (from the Greek *deon,* "that which is binding").

Historically there have been many variants of the idea that particular acts can be judged *right* or *wrong* according to universal principles or laws, without attempting to calculate their consequences. The Stoics said that people have a duty to act in accordance with the natural law, the rational and moral order expressed in the structure of the world. Judaism and Christianity stressed obedience to the divine law revealed in scripture. Immanuel Kant held that the right is determined by the unconditional obligation of rational moral law, apart from any consideration of consequences. He maintained that an action is right if the principle it expresses could be universally applied. For Kant, the demand for freedom and justice is based on the equality of persons as autonomous and rational moral agents; individual persons should never be treated merely as means to social ends.[26]

Whereas utilitarianism emphasizes the social good, deontological ethics typically defends *individual rights.* Fundamental rights must not be violated even in the interest of beneficial social consequences. Rights are in general correlated with *duties.* My right to life implies your duty not to violate my life. The language of rights appears to be absolutist and often does lead to inflexible positions. If rights are "inalienable" and "inviolable" and duties are "categorical," there seems to be no room for compromise. However, it is possible to employ a deontological approach with considerable flexibility. For example, one can formulate a universal rule with built-in qualifications that allow for special cases. Moreover, one duty may be outweighed by other duties. When two rights conflict, one of them may be assigned priority. So rights and duties should not be regarded as absolute.[27]

The defense of *individual rights* is indeed important in a technological age in which governments wield vast powers (through electronic surveillance and the control of information, for example), and they frequently defend their actions by pointing to benefits for society. Only a basic respect for persons can lead us to protect a minority from exploitation for the benefit of the majority. But in an ecologically interdependent world, the direct and indirect consequences of our actions are often far-reaching and should not be neglected. Both the protection of the individual and the good of society must be considered in the complex decisions we face today, and there is no simple formula for combining them.

In some cases that we will examine, the social consequences are paramount (for example, nuclear weapons). In other cases individual rights are the main issue (for instance, the confidentiality of computerized personnel records). Sometimes individual rights and the future welfare of society are very difficult to reconcile, as in the debate over population growth. But in general I will use both a broad evaluation of consequences (going beyond utilitarianism and cost-benefit analysis by including nonquantifiable values) and a defense of rights and duties that avoids absolutism.

2. THE CONCEPT OF JUSTICE

Most ideas of justice start from an assumption of *the fundamental equality of persons*. For some people this may be based on a religious conviction of the equal worth of every individual in God's sight. For others it may derive from a doctrine of equal intrinsic human rights ("natural rights") or the requirements of a harmonious social order. Some philosophers have argued from the common nature of persons as rational beings or the universality of basic human capacities. Others take respect for human beings and belief in their equal dignity to be unanalyzable ultimate attitudes. Distributive justice, then, starts with the idea that people should be treated equally because they are fundamentally equal.[28]

Unequal treatment can be justified on a variety of grounds. Special provisions for people with special needs and disabilities is in itself unequal, but the goal is an equal opportunity for a good life. Individuals are selected for positions of leadership, but such positions should be open to anyone with appropriate qualifications. Some differences in income may be justified as an incentive to productivity, from which everyone supposedly benefits, but inequalities of the magnitude that exists in industrial societies today are hardly justifiable for this purpose.[29] The radical inequalities between nations could never be justified by the need for work incentives.

Unequal treatment is justified, in short, only if it helps to *correct some other form of inequality* or if it is *essential for the good of all*. Inequalities of authority are necessary for maintaining the social order, but there can be equal access to the positions and offices that carry such authority. But are there limits to the degree of inequality that we will tolerate for the sake of other social benefits? Are inequalities in some goods and services more significant than in others? Questions of inequality assume added urgency if technology tends to increase the gap between rich and poor and if global scarcities limit the resources available for distribution.

I would maintain that equality is a more compelling value in the distribution of resources to meet *basic human needs* (such as food, health, and shelter—the lowest levels in Maslow's hierarchy) than in the distribution of other goods and services. Food to meet minimum protein and calorie requirements is necessary for life itself; justice in the production and distribution of food to meet these requirements should have the highest priority. Access to health care also is crucial since it so strongly affects life prospects. But some margin beyond bare survival is a prerequisite for a minimally decent human life. Estimates of the minimal material levels for human dignity and self-respect are of course historically and culturally relative; there is no sharp line between physical needs and psychological desires influenced by changing expectations. In the United States, a poverty line has been established for entitlement to food stamps and health care benefits; unemployment insurance and social security also were instituted in the name of justice rather than charity. A project sponsored by the United Nations has tried to establish quantitative measures for basic needs and standards on which there is an emerging world consensus.[30]

The most influential recent treatment of the relation between justice and equality is John Rawls's *A Theory of Justice*. Rawls asks us to imagine a hypothetical "original position" in which a group of people are formulating the basic principles for a social order. No one knows what his or her status will be in the society that is to be established. In agreeing on a "social contract," each person acts from rational self-interest, but impartiality in formulating the rules is guaranteed because these contracting individuals do not know what their own social positions will be. Such a hypothetical situation can help us establish principles for the fair distribution of scarce resources. It is similar to a situation in which the child who cuts the cake does not know which piece he or she will get.

Rawls maintains that persons in such an "original position" would accept two basic principles for the social order:

1. Each person is to have an equal right to the most extensive total system of equal basic liberties compatible with a similar system of liberty for all.

2. Social and economic inequalities are to be arranged so that they are both: (a) to the greatest benefit of the least advantaged, and (b) attached to offices and positions open to all under conditions of fair equality of opportunity.[31]

As Rawls develops it, the second principle requires equality in the distribution of all the primary social goods (income, wealth, power, and self-respect), with the exception noted. Inequalities are allowed only if they maximize benefits to the least advantaged and are attached to offices open to all. Attention to the impact on *the least advantaged* is a product not of altruism but of the rational self-interest of people in the "original position," any of whom might end up in that worst-off status. Rawls suggests that if the least advantaged benefit, it is likely that most other social groups will benefit also. But he rejects the utilitarian view that a loss to some people can be justified by greater gains to others.

Rawls holds that rational contractors would insist that, once a minimal level of material well-being had been reached, *liberty* should have priority over *equality*. The first principle is thereafter to be fulfilled before and independently of the second. Liberty—especially liberty of conscience and political liberty (equal participation in government)—is not to be exchanged for any other benefits, including greater equality. Neither freedom nor justice is subject to trade-offs with other benefits. Political rights should not be sacrificed for the sake of economic gains, except under conditions of extreme scarcity.

Not surprisingly, Rawls has been attacked from the right for being *too egalitarian*. Defenders of free enterprise capitalism say that the degree of equality that Rawls seeks would not provide adequate incentives for the most able persons, and it would protect the indolence of the least able. The enforcement of equality, it is claimed, would require coercive measures and would violate property rights and the acquisition of wealth by legitimate means. If one is really dedicated to freedom, one must set strict limits on the powers of the state, including its power to redistribute legitimately acquired property.[32] But Rawls is attacked from the left for *not being egalitarian enough*. Marxist and socialist critics insist that political equality, which the first principle endorses, is jeopardized by the degree of economic inequality that the second principle allows. For economic power becomes political power in capitalist societies, and inequalities perpetuate themselves.[33]

Rawls's hypothetical *"original position"* has also been criticized. If one starts from separate, autonomous individuals, can an adequate concept of community ever emerge? If one starts by abstracting the individual from all political and historical contexts, can one obtain principles relevant to actual choices in the real world? Despite such limitations, I see the "original position" as a useful analytic device for asking what would be a fair distribution of resources. It is one of the few ways of dealing with justice between generations—simply by asking you to imagine what policies you would recommend if you did not know to which generation you would belong. In later chapters I will use Rawls's second principle (maximizing benefits to the least advantaged) in discussing policies for food production, energy conservation, industrial growth, and so forth.

3. FREEDOM AS PARTICIPATION

One consideration in the evaluation of any technological policy is the extent to which it restricts or extends individual freedom. But freedom has many forms, which may be affected in diverse ways by a policy decision. According to the philosopher Joel Feinberg, freedom can be expressed as a relation between *an agent, a constraint, and an activity*. Explicitly or implicitly, it has a general structure: x is free from y to do z. People have particular kinds of constraints and activities in mind when they defend freedom. Sometimes they emphasize the absence of a constraint, and sometimes they emphasize the opportunity for and choice of an activity they deem important.[34]

The negative side of freedom is *the absence of external constraints:* freedom from coercion or direct interference imposed by other persons or institutions. Locke and the early British tradition of libertarian political philosophy interpreted freedom primarily as the absence of interference by other individuals or by the state. They wanted to protect the individual against abuses of the power of government; they sought the maximum scope for individual initiative in economic affairs and in the use of private property. This view was influential among the authors of the U. S. Constitution and was reinforced by the American experience of the frontier, abundant resources, and the vision of a land of unlimited opportunities for everyone. It seemed that a person free of human constraints could pursue the mastery of nature without interfering with other persons.

The positive side of freedom is *the presence of opportunities for choice.* Freedom to choose among genuine alternatives requires a range of real options and the power to act to further the alternative chosen. Even in the absence of external constraints, unequal power results in unequal opportunity for choice. Some degree of personal autonomy is an essential component of freedom. Many of the conditions for the exercise of choice are internal. People vary widely in their awareness of alternatives, ability to make deliberate choices, and personal initiative and self-direction. But in dealing with public policy, we are concerned mainly about the external conditions, the social structures within which people can have some control over their own futures.

The *negative* and *positive* sides of freedom are inescapably related in any social order. If we try to minimize external constraints while there are great inequalities of economic power, the weak will have little protection from domination by the strong. In a complex society, the actions of one person can greatly affect the choices open to other people. Limitations on the actions of some persons are necessary if other persons are to be able to exercise choice. Positive freedom to achieve desired outcomes exists only within an orderly society. The state is an instrument of order and law, but it is also an instrument of freedom when it restricts some actions to make other actions possible. Emphasis on the positive side of freedom is also consistent with the social character of selfhood defended in the previous volume (chapter 7), in contrast to the more individualistic view of freedom as absence of interference.

Political freedom too has both negative and positive aspects. On the negative side are limits to the powers of government, such as censorship and arbitrary arrest. I will suggest, for example, that citizens must be protected from invasion of privacy through electronic surveillance and the misuse of personal information in computerized databanks. On the positive side are institutions of political self-determination and democratic forms of government whereby each citizen can have a voice in decision-making processes. Civil liberties, such as freedom of speech, assembly, and the press, can be defended both as basic human rights and as preconditions of democracy. The moves toward democracy around the world in the early 1990s involved the right of dissent and freedom to organize opposition parties.

In technological societies, *the right of governments to intervene* to protect health, safety, and welfare has been expanding to include ever-wider areas, as the uses of private property have had more far-reaching public consequences. Such common resources as air and water can only be protected by collective action through regulations or economic incentives. In other cases, governmental powers were expanded to protect citizens from the growing power of private institutions such as industrial corporations and labor unions.

The forms of freedom that are most relevant to technological policy can thus be understood positively as opportunities to participate in the decisions that affect our lives.

1. Participation in the Marketplace. In a free market economy, decisions are decentralized among many producers and consumers. The recent overthrow of communist governments around the world reflects a wide recognition of the economic inefficiencies and bureaucracy of state ownership and central control, as well as a desire for democracy in place of political repression. In a market economy, however, the goals of economic efficiency and social justice are not easily reconciled. Some loss of efficiency accompanies the use of taxes to mitigate extremes of wealth and poverty and to support health and welfare measures designed to ensure that no one lacks the basic necessities of life.

Some types of *government action* entail much greater intervention in the marketplace than others. For example, the individual farmer and the agribusiness corporation make decisions in response to market forces, but some government subsidies and regulations are acceptable in agriculture because it affects so many other areas of public policy: food prices, farm income, foreign trade, soil erosion, water pollution, land use, the quality of rural life, and so forth. Again, a heavy tax on the discharge of industrial pollutants relies on economic incentives and allows a greater variety of responses than strategies that mandate specific abatement technologies or set absolute standards for emissions. Energy conservation proposals range from voluntary restraint, through economic incentives, to mandatory fuel efficiency standards and fuel rationing. In each of these cases the consequences of insufficient regulation must be weighed against the dangers of excessive bureaucracy and the losses in efficiency and private initiative.

2. Participation in Political Processes. Democracy requires a free press, the right to dissent, and provisions for the election of representatives at local and national levels. Officials can be held accountable, and citizen input can take place, through legislative and regulatory hearings and court challenges. But citizens often feel incompetent to deal with complex technological decisions. The risks to human health and safety from nuclear reactors and toxic substances, for instance, are very difficult to evaluate. Yet such decisions should not be left to technical experts alone, since they require the comparison of diverse risks and benefits and the assessment of alternative policies, which are not purely scientific questions. An industry or a government agency that has an interest in promoting a technology usually has far more extensive legal and scientific resources than those opposing it. In chapter 8 some procedures that can facilitate the democratic control of technology will be examined.

3. Participation in Work-Related Decisions. The institutions within which work is carried out vary widely, but they should include some provision for the voices of workers to be heard, such as labor unions, labor-management committees, producer cooperatives, small businesses, or owner-operated farms. We will return to these issues in later chapters in discussing particular agricultural and industrial technologies.

All three types of participation are more difficult in *large-scale technologies* than in those of *intermediate scale.* For example, nuclear energy is complex and centralized; it demands huge capital investments and entails unusual risks that require an exceptional degree of government regulation and strict security measures. By contrast, many forms of solar energy are decentralized; equipment can be locally installed and managed. *Decentralization* counteracts the concentration of economic and political power, and it contributes to diversity and local control. Yet in many cases the *centralization* of authority is necessary. Air and water pollution crosses jurisdictional boundaries. Local governments have been ineffective in controlling pollution because they are dependent on industrial growth for new tax revenues. An increasing national role in environmental regulation and resource conservation is unavoidable, but citizen participation is more difficult at the national level.

To sum up: philosophy can help us clarify ethical principles for evaluating technological choices. It can remind us of the importance of taking into account both the good of society and the rights of individuals. It can give more precise meaning to concepts of justice and freedom (and of course many other ethical concepts). I have defended the idea of justice as the greatest equality compatible with the welfare of the least advantaged. I have suggested that in a technological society the most important form of freedom is participation in the decisions that affect our lives.

III. RELIGION AND HUMAN VALUES

If philosophy makes an important contribution to ethics in clarifying concepts and critically analyzing universal principles, religion can also make distinctive contributions to ethics in both theory and practice. The most fundamental ethical convictions, such as the value of the individual and respect for nonhuman nature, are dependent on one's understanding of ultimate reality. Ethics cannot be divorced from metaphysics. Some naturalistic philosophies have provided a wider framework within which the significance of human life can be defended. But it is primarily in the religious traditions that these wider questions are raised.

Moreover, abstract ethical principles do not in themselves provide motivation for ethical action, which involves the will and the affections as well as the intellect. Whereas Socrates held that to know the good is to do it, the apostle

Paul asserted, "For I do not do the good I want, but the evil I do not want is what I do" (Rom. 7:19). Religion also reminds us that we are not isolated individuals but members of communities with common memories and a shared life. Religious stories affect motivation more strongly than abstract principles do, but the values expressed in them must be subjected to critical reflection and application in changing historical circumstances. In this chapter I will consider only the Christian tradition; in the next chapter some distinctive insights in Asian religions are presented.

1. CHRISTIAN ETHICS

The record of Christian institutions in practicing the ideals they profess has been very mixed. They have often been on the side of the status quo, resisting changes in the social order. To victims of injustice they have sometimes offered only resigned acceptance and the consolation of a future life. Their ethical teachings frequently have been confined to the sphere of family life and personal relationships. Charity has been a substitute for justice. Yet they also have nurtured prophetic leaders who have been in the forefront of social reforms—in hospitals and prisons and the abolition of slavery in the last century, for instance, or in the civil rights and antiwar movements of the 1960s—and they could play a creative role in the future.

The history of Christianity includes three types of ethics, each of which has parallels in other religious traditions. They focus respectively on choice of the good, obedience to the right, and search for the fitting response.[35]

The first type understands ethics as *the choice of the good.* This approach is goal oriented and concentrates on either the virtues required to reach the goal or the values that identify the goal. The focus on *virtues* goes back to Aristotle, for whom the goal was happiness attained through excellent activity and the realization of one's potentialities. Attaining the good, he said, is achieved through cultivating virtue and character. Aquinas gave this a Christian reformulation: the human end is happiness, which consists in the vision of God given as a reward for virtue. Aquinas combined the classical virtues (courage, temperance, wisdom, and justice) with the biblical virtues (faith, love, and hope). Wesley, in turn, saw Christian ethics as the cultivation of distinctive personal virtues. Alasdair MacIntyre, Stanley Hauerwas, and others have recently stressed the importance of virtues and character development.[36]

Several problems arise, however, if *virtues* are the center of attention. If one seeks virtue, the neighbor can become a means for one's own self-realization or future reward. Again, the courageous person does not seem to be aiming at courage or the loving person at love. Moreover, ethical appraisal has to look at the consequences of our actions as well as at the goals that elicit them and the character traits they express.

In dealing with public policy it seems to me more fruitful to consider *values,* which are characteristics of individual and social life, rather than virtues, which are traits of individual character. We should ask about justice in social institutions, for example, and not simply about the just person. (The two are not unrelated, of course, since personal commitment to a value can be viewed as a virtue.) Among philosophers the most common form of ethics as choice of the good is utilitarianism, in which the greatest good of the greatest number is sought. Utilitarians, as we have seen, identify the good with the aggregate of individual satisfactions or welfare, whereas I have advocated identifying it with a wider set of values. I believe that the Christian can take part in the discussion of policy issues in the public arena most effectively by using an ethics of the good expressed in terms of values, an approach that is found in the writings of George Thomas, Harold DeWolf, Roger Shinn, and others.[37]

The second basic type of Christian ethics is *obedience to duty*. We have seen that the central concept in deontological ethics is "the right" rather than "the good." In the Christian tradition, duty is usually identified with obedience to the divine law revealed in scripture. Specific rules and injunctions in scripture are taken to be God's eternal command. Obedience to law has been stressed by Calvin, by the Puritans, and by conservative Protestants today. In the Catholic heritage, the teachings of scripture are supplemented by the ongoing teaching authority of the church, as well as by the "natural law" accessible to human reason (the moral order built into the structures of creation).

Such an ethics of *law, duty, and obedience* has an important place in the Christian community, but it has several limitations. It often becomes a rigid legalism in which the letter of the law is stressed rather than its purpose. Moreover, we tend to react to law as something externally imposed on us. Law provides little motivation for action unless it is supported by the sanctions of eternal rewards and punishments or by the sanctions of church authorities—neither of which is intrinsic to ethical action itself. Finally, specific rules are static and not readily adaptable to changing circumstances or problems. Technology gives us unprecedented powers and presents us with ethical questions about which scripture is silent, such as the potential and dangers of computers, nuclear energy, or genetic engineering.

The third basic type is *an ethics of response*. Here God's nature and purposes are the ultimate norm rather than a specific set of teachings. A common theme has been response to what God as Redeemer has done in the past, especially in the Covenant and in the person of Christ. Human faith and concern for the neighbor are said to be responses in gratitude for what God has done. Paul and Luther held that faith comes before works and that gospel (good news) comes before law. Works should not be neglected, but they are to be viewed as "fruits of the Spirit." Augustine held that the Christian virtues are by-products of love toward God.

An ethics of response may also be called an ethics of *relationship* or an ethics of *context*. Paul Lehmann, for example, says that relationships and not principles are primary in the Christian life. Actions arise from a person's relationship to God within the covenant community, the community of faith; decisions are made within concrete social contexts in which there is inescapable ambiguity.[38]

Historically, an ethics of response has been expressed primarily in person-to-person relationships, but it can readily be extended to social issues. It has usually had more to say about God as Redeemer than about God as Creator, but here too its scope can be broadened.

H. Richard Niebuhr brings together *response to God as Creator, Sustainer, and Redeemer.* In response to God as Creator we can affirm the goodness of the created order and seek to participate in God's ongoing creative action. In response to God as Sustainer we can acknowledge both the possibilities and the constraints within which we live. This is an interdependent world in which given structures limit our actions and cause us pain and suffering. Sometimes we can see God's judgment and governance in such suffering, though we cannot say that everything that happens is predestined as God's will. Niebuhr says that ethics is the search for the fitting action, the appropriate response to all that is going on within a wide community of agents. We have to look at the biological, social, and theological context of our lives. Moreover, God's redemptive activity is not confined to Christ. God is Redeemer in all history, though we can see redemption most clearly in the person of Christ.[39]

If Christian ethics includes response to *what God is doing in the present,* as well as what God has done in the past, the discussion of continuing creation and redemption in the previous volume may help us in reflection on ethics in a technological society. Western ethics has been predominantly anthropocentric. Now we must consider ourselves participants in an interdependent community of life. The next chapter explores the ways in which such a theology of nature can lead to a more adequate environmental ethics.

In comparing the three types of ethics, then, an *ethics of the good* is most helpful in the public discussion of policy issues because it involves the defense of values concerning which people holding diverse religious and philosophical positions can find common ground. In this volume, therefore, I will analyze ethical issues primarily in terms of values. An *ethics of duty* has a role within the church, reminding us of the radical character of the gospel and the human tendency to rationalize self-interest, but it has difficulty in providing guidance for unprecedented situations and in offering motivation to action. An *ethics of response* is more flexible, provides stronger motivation, and places more responsibility on the individual. It is appropriate for reflection within the Christian community, where it can be combined with a consideration of values.

In quoting from the Bible below, I am assuming *a historical understanding of scripture* rather than the literalistic view sometimes taken by proponents of an ethics of law and duty. In the previous volume, I maintained that theology is critical reflection on story, ritual, and religious experience, which are the primary religious phenomena. Scripture and tradition have been continuously reinterpreted in the light of new knowledge, categories of thought, and cultural situations. Because our outlook is limited and our conclusions tend to be distorted by self-interest, ethical reflection must always include self-criticism.

We cannot expect to find in the Bible any easy answers to complex issues today, but we can try to identify in it some of the values that can guide us in our choices.

2. INDIVIDUAL VALUES

Let us consider first the values of food and health, meaningful work, and personal fulfillment. We look at these values from the standpoint of individuals, though of course individuals always live within a social order. Great diversity can be found within biblical and Christian thought, and an adequate treatment of historical developments would require a more extended treatment than is possible here. I can only give a few examples of each of these values and indicate their relevance for technological decisions.

1. Food and Health

The laws of Deuteronomy and the message of the Hebrew prophets refer frequently to basic physiological needs. The biblical God is committed to persons and is concerned about all aspects of their lives. Whereas Greek and medieval thought presented a body-soul dualism in which the body is inferior to the soul, the Bible assumes a unity of the whole person. In the New Testament, love is not a sentiment or emotion but active caring for persons and response to the needs of the neighbor for food, clothing, and health. "For I was hungry and you gave me food . . . I was naked and you clothed me, I was sick and you visited me" (Matt. 25:35). In one of Jesus' parables, the Good Samaritan takes care of the physical needs of a man beaten by robbers (Luke 10:25). The Lord's Prayer recognizes our nutritional as well as our spiritual needs: "Give us this day our daily bread."

Such teachings have inspired Christian groups in modern times to found hospitals and send medical and agricultural missionaries abroad. They have been active in efforts for famine relief and have supported expanded foreign aid for agricultural development. The most effective way to alleviate human suffering today is to broaden access to agricultural and medical technologies. Human health is also the main goal in reducing technological risks (from pollution, toxic substances, industrial accidents, and so forth). The health and safety of workers, consumers, and citizens are at issue in many decisions today. These basic human needs are so obvious that no one would challenge their importance, but disputes do occur about their priority relative to other values. I will suggest that justice requires that technologies for food and basic health services should be given high priority in the allocation of research funds and in public policy decisions.

2. Meaningful Work

Adam is instructed to "till and keep" the garden, though he is told that under the conditions of human sin, toil will not be easy; "in the sweat of your face you shall eat bread" (Gen. 3:19). In the Bible, work is accepted as part of life but not its goal. "Six days you shall labor and do all your work, but the seventh day is a sabbath to the Lord your God" (Exod. 20:9). Proverbs contains many exhortations to diligent work and warnings of the dangers of idleness. Paul enjoined the early church to "work with your hands, as we charged you" (1 Thess. 4:11).[40]

This positive attitude toward daily work was not shared by Greek society, which exalted intellectual and artistic activity and looked down on manual labor as the province of slaves and the lower classes. Some depreciation of labor was evident in the Middle Ages, when the contemplative life of the monk was ranked above the active life of the laity. Luther rejected this hierarchy of "religious" and "secular" vocations and insisted that any useful work is a service to both God and the neighbor. "It looks like a small thing when a maid cooks and cleans and does housework. But because God's command is there, even such a small work must be praised as a service to God far surpassing the holiness and asceticism of all monks and nuns."[41] The Reformation understanding of vocation restored the dignity of all forms of useful work, though it was sometimes accompanied by the more dubious message that one should be content with one's "appointed station" because it had been assigned by God. The later Calvinists supported the Protestant "work ethic" of diligent effort and frugality.

In modern times, technology has had a profound impact on the character and organization of work, the relative power of workers and owners, and working conditions. During the Industrial Revolution, most of the churches accepted the new forms of work, but they tried to correct its worst abuses, such as child labor and long hours in factories and mines. In the twentieth century, major Catholic and Protestant statements have defended labor unions, collective bargaining, and the right to strike. John Paul II, for instance, asserts the dignity of work and "the principle of the priority of labor over capital."[42] Local church councils have tried to prevent plant closures, recognizing that unemployment is a threat to self-respect and to family and community life. Ecumenical conferences have been concerned about the quality of work life, the repetitive character of assembly-line jobs, and the displacement of skilled jobs by automation.[43]

A theology of work today, I suggest, must uphold the dignity of work without romanticizing it. One motive for work is that it is necessary to earn a living and to support a family. Job security, a fair wage, and safe working conditions are thus primary objectives. But work should also be a useful contribution to the community, and this goal should be kept in mind in both individual vocational choice and in social policy decisions. In addition, we should seek a social order in which a larger proportion of jobs allow individual self-realization, creative activity, and the expression of individual capacities. Recognizing that many jobs that need to be done are bound to be rather dull, we can seek a greater voice by workers in decisions affecting their lives at work. We can try to promote more meaningful work, while acknowledging that with a shorter

work week and more leisure time many people will find their greatest fulfillment outside the job in family and community life.

3. Personal Fulfillment

Beyond the requirements of food, health, and work, the Bible seems to legitimate the pursuit of a higher level of material well-being. Preparing to enter
the promised land, the Israelites are told that "the Lord will make you abound
in prosperity" (Deut. 28:11). Possessions are not innately evil, and asceticism is
not praised in the Hebrew scriptures. But the prophets also strongly criticize
the luxuries of the rich. Proverbs recommends a middle path:

> Give me neither poverty nor riches;
> feed me with the food that is needful for me,
> lest I be full, and deny thee,
> and say, "Who is the Lord?"
> or lest I be poor, and steal,
> and profane the name of my God. (30:8–9)

The New Testament includes stronger warnings about the dangers of material possessions. "How hard it will be for those with riches to enter the Kingdom of God!" (Mark 10:23). Jesus warns about anxiety concerning possessions,
for "where your treasure is, there will be your heart also" (Luke 12:34). Cov-
etousness is listed nineteen times in the New Testament among the tempta-
tions to be avoided. "Love of money is the root of all evils" (1 Tim. 6:10). To be
sure, attachment to riches and neglect of the needs of the poor are condemned, rather than riches in themselves. Yet throughout Christian history
admonitions about the dangers of affluence have been repeated, though the
message has frequently been compromised by the church's own wealth or dependence on people with wealth.

Judaism has continued to hold a broad view of personhood and the many
dimensions of human fulfillment. It has affirmed the goodness of life; monas-
ticism and world-renouncing asceticism have been rare in its history. A person's duty is to sanctify daily life. Judaism encourages a proper self-respect; the
individual is free to choose the good, to turn anew, and to work for God's
kingdom on earth. Study and learning—both of the Torah and of worldly
knowledge—are held in high regard. Personal fulfillment can occur only in
the context of the community and through the realization of harmonious relationships. The God of Israel is interested in the fabric of the community's
life; the rabbis, like the prophets, attacked the barriers that separate people.
Judaism holds up the vision of social harmony as well as justice in a more humane society.

In Christianity, too, the good life is identified with personal existence in
community. Fulfillment consists of right relationship to God and neighbor.
Bread is essential, but "man does not live by bread alone." The New Testament
speaks of reconciliation, fellowship, and mutual support among persons in

the early church. Christian writers have often defended the integrity of family and communal life and the goal of cooperation rather than competition. Interpersonal community assumes new significance today as an alternative to both the isolation of individuals and the impersonal collectivism to which technological societies are prone. In technological decisions, one consideration should be the consequences for community life (for example, when an urban neighborhood is disrupted by a highway or when rural life is affected by large corporate farms).

Technology has expanded enormously the array of consumer products available to us. It has given to millions a level of comfort and convenience once available only to a few. At the same time it has increased our preoccupation with consumer goods and has created new wants that often seem insatiable. To sell its products, industry spends billions on advertising, appealing to material self-interest and the status conferred by possessions. "Keeping up with the Joneses" becomes an obsession that governs our lives. Part of the Christian message for today, I believe, is a critique of the tendency of material possessions to monopolize our attention. Such a critique of consumerism and materialism will be effective only if it is accompanied by an alternative vision of human fulfillment and a dedication to social justice in a world of great inequality.

The dominant images of our society identify happiness with material possessions. But in the Christian perspective, human beings are more than producers and consumers, and fulfillment has many facets. Churches have strongly supported education at all levels; many colleges and universities were founded by church groups. Most branches of Christianity have had a strong interest in music, art, and literature and have encouraged artistic imagination and creativity. Technology has brought increased leisure but has done little to encourage the constructive use of leisure. Television is a passive form of entertainment; it seldom enhances distinctive individual capacities or human interaction. The church itself is a community in which people interact at many levels: in celebration and in worship, in social action and in meditation, in mutual support and in spiritual growth. Religious experience, the experience of the holy, is itself the highest form of personal fulfillment, but it need not be divorced from other aspects of individual and communal life. Views of personal fulfillment are, of course, highly dependent on interpretations of human nature, which I will consider at the end of this chapter.

In a technological society we need this witness to the priority of the personal and the importance of community. In a time of scarce resources, affluent societies can seek less resource-intensive sources of satisfaction, such as personal relationships, community participation, education, and the enjoyment of nature. Here religious faith speaks to the crisis of meaning that underlies the compulsive pursuit of affluence. The biblical view of human fulfillment can strengthen the search for more creative life-styles today and for the kinds of technology that might accompany them.

3. SOCIAL VALUES

The previous set of values was presented from the standpoint of the individual. The Christian tradition also supports three values that refer primarily to characteristics of the social order: social justice, participatory freedom, and economic development.[44]

1. Social Justice

Both the laws of the Torah and the message of the prophets present the demands of justice that have been central in Judaism throughout its history. Micah gives this summary: "What does the Lord require of you but to do justice, and to love kindness, and to walk humbly with your God?" (Mic. 6:8). Speaking in the name of a God of justice, Amos denounces the inequalities of his day:

For three transgressions of Israel, and for four, I will not revoke the punishment; because they sell the righteous for silver, and the needy for a pair of shoes—they that trample the head of the poor into the dust of the earth, and turn aside the way of the afflicted. . . . But let justice roll down like waters, and righteousness like an ever-flowing stream. (Amos 2:6–7; 5:24)

Justice is a demand of the biblical law, but it is also a community's response to a God of justice and righteousness. The Hebrew prophets condemned the oppression of the poor as both a violation of the Covenant with God and a violation of human relationships. As in the philosophical discussion cited earlier, the central biblical meaning of justice is equal treatment. Distributive inequalities are harshly judged in the light of the fundamental equality of all persons before God. As with Rawls, the treatment of the dispossessed serves as a test case for justice in society, though the prophets go further in their impassioned concern for the specific victims of injustice.

The biblical God usually seems to be on the side of the poor. The Exodus brought the liberation of slaves from Egypt, and God is portrayed as continuing to seek the liberation of the oppressed. Jesus opened his ministry with a quotation from Isaiah: "The spirit of the Lord is upon me, because he has anointed me to preach good news to the poor. He has sent me to proclaim release to the captives and recovery of sight to the blind, to set at liberty those who are oppressed" (Luke 4:18). Jesus spoke more often of love than of justice, but of course he lived in an occupied land that had little control of its national destiny, and he called primarily for the personal response of individuals to the coming Kingdom. Subsequent interpreters have understood justice as love expressed in the social order, extending but not replacing the radical demands of love between individuals.[45] Social justice is often mentioned in medieval and Reformation thought and is prominent in recent church pronouncements.

Technology raises many problems of justice. Inequalities between rich and poor nations in the distribution of food, health care, energy use, and resource consumption are huge and increasing. The developing nations have two-thirds of the world's population but only 6 percent of the world's scientists and engineers and 3 percent of all expenditures for scientific research and development.[46] In industrial nations one social group usually receives most of the benefits of a technology, while the risks fall mainly on another group. The owners of a factory seldom live downwind of its smokestacks or downstream of its effluents. We will encounter in subsequent chapters many issues of justice between groups, classes, genders, races, nations, and generations.

2. Participatory Freedom

We have said that freedom always has a negative and a positive side: the absence of external constraints and the presence of opportunities for choice. Since there are many kinds of constraint and choice, there are many forms of freedom. The Exodus story of liberation from slavery has always symbolized the imperative of political and religious freedom to the Jewish community. In its subsequent history as an oppressed minority it frequently has been on the side of freedom against institutions of political and religious power. There is considerable latitude for variations in individual belief within Judaism. But its outlook is not individualistic; persons are portrayed as social beings, interrelated selves in a community. Jewish life has always tried to balance the rights of the individual against the good of society. Throughout its history, it has given expression to a concept of freedom as self-determination and individual responsibility.

In his first sermon, Jesus said he had come "to set at liberty those who are oppressed." The context suggests that he, like Isaiah, was referring to the constraints imposed by economic and social conditions. Paul said "Christ set us free" (Gal. 5:1), which he elaborated as freedom from observance of ritual laws and freedom from bondage to sin, making possible a free response to God's love made known in Christ. But the record of the institutional church in relation to religious and political freedom has been very mixed. At times it has worked for the liberation of the socially oppressed; at other times it has itself become an instrument of oppression, acquiring a vested interest in the structures of political and economic power. In its own internal life it has often been highly authoritarian; through much of its history, religious intolerance has prevailed. Yet the Protestant reformers defended individual conscience, and advocates of religious freedom and the separation of church and state have been prominent in all Christian traditions since the eighteenth century.

Modern ideas of *political freedom* have many historical roots. Democratic governments in ancient Greece accorded political rights to citizens, though not to slaves. Ideas of due process are indebted to the Roman and medieval legal systems. The idea of inalienable rights can be found in Stoic as well as medieval conceptions of natural law. But modern views of civil liberties and political democracy also reflect the indirect influence of biblical assumptions concerning the dignity and fundamental equality of all persons. In American history, one can point to the experience of the free churches and the Puritans in participatory self-government at the local level. In Eastern Europe in 1989, meetings of the nascent movements for democracy were usually held in churches because they were the only institution not under government control.

The realistic biblical estimate of *human nature* fits well with democratic ideals. In order to entrust citizens with political power, we have to respect the capacities of ordinary people; but if we acknowledge human sinfulness we will avoid entrusting anyone with too much power, and checks and balances will always be needed. As Reinhold Niebuhr put it: "Man's capacity for justice makes democracy possible, but man's inclination to injustice makes democracy necessary."[47]

We have seen that technology increases the interdependence of people over wide spans of space and time. Each person's life is affected by the actions of many others. I have suggested that we must therefore think of freedom today not primarily as the absence of constraints but as the presence of opportunities to participate in the decisions that affect our lives—primarily through political processes. In a technological society, freedom has to be balanced against the common good and the requirements of order.

3. Economic Development

The Bible contains relatively few teachings on the economic order as such. In the Hebrew scriptures the right of *private property* was accepted but not absolutized. Every seventh year (the sabbatical year), slaves were to be freed, debtors forgiven, and the land was to lie fallow. (These injunctions were apparently followed at some periods of Israel's history but not at others.) The demands of justice and stewardship set limits on how wealth was to be used. Jesus' disciples and the early Christian communities shared their resources, as did the subsequent monastic communities. But Aquinas defended private property as an encouragement to diligent use, though he also asserted the principle of stewardship and use for the common good. The church in the Middle Ages prohibited interest on loans (usury) and advocated a "just price" in financial transactions.

Calvin and his followers did not object to interest on loans, and they actively encouraged the growth of trade and commerce, though they advocated restraint on profits and responsibility for the common good. The contribution of Protestantism to *the rise of capitalism* has been extensively debated. In its early stages the Protestant ethic called for thrift as well as hard work, and the combination often led to financial success. When the restraints were forgotten and prosperity was seen as a predestined sign of God's favor, the ethic could be distorted to the point that some people sought wealth partly to prove their virtue.[48] Such economic individualism was reinforced by John Locke's idea that there is a natural right to property and that property rights provide an essential protection for individual freedom and self-determination. Adam

Smith provided the final justification for free-enterprise capitalism; he said that the pursuit of private profit will automatically result in the common good because there is "a harmony of interests." Left to itself, the "invisible hand" of the marketplace will guide the economy so that everyone will benefit.

The combination of capitalism, new industrial technologies, and relatively cheap raw materials and energy in the Industrial Revolution resulted in higher productivity and *rapid economic growth*. The twentieth century saw the rise of giant corporations, first nationally and then multinationally. Even within free enterprise economies, governments assumed a larger role in providing public services and social security benefits and in regulating monopolies, health risks, and environmental impacts. Recent Catholic and Protestant statements have insisted that Christianity cannot be identified with either capitalism or socialism, and that each has characteristic dangers. In practice capitalism has provided more rapid economic growth and greater freedom in economic choices, while socialism has done more for distributional equality and justice.

Until recently it has been assumed that *economic growth* is a desirable goal, even if its benefits are inequitably distributed. Though the rich benefit most from economic growth, many gains supposedly "trickle down" to those at the bottom. The economy was compared to a pie that is growing in size; even those with the smallest slices were better off, without any change in their relative shares. By the early 1970s, however, a number of writers were claiming that such economic growth cannot continue because of resource and environmental constraints. The writers of *The Limits to Growth* argued that if industrial production continues to grow exponentially, global limits would be exceeded within a few decades.[49] Technological optimists replied that technical advances will extend environmental and resource limits. New technologies will turn previously useless raw materials into useful resources and will reduce pollution per unit of production.

In chapter 7 I will examine this "limits to growth" controversy. I will suggest that only *selective economic growth* is sustainable. The mix of goods and services must shift toward human services and toward less polluting and resource-intensive products. In industrial nations, sustainability requires greater support of recycling and conservation technologies and some willingness to adopt less resource-consumptive life-styles. In developing nations, economic growth is essential to human well-being, but it must be achieved through environmentally informed plans for sustainable development. In any case, economic growth is not valuable in itself, but only for the sake of other values to which it can contribute, including food and health, meaningful work, and personal fulfillment; it must therefore always be considered in relation to these other values. It is for these reasons that I advocate *economic development* rather than economic growth as the third social value to be sought in technological policy.

The *gross national product* (GNP) measures economic growth, but it is a misleading indicator of a nation's well-being. It says nothing about income distribution; GNP may rise because an elite is more wealthy, even though the majority of the population is sinking into deeper poverty. It says nothing about how the money is being spent; huge defense budgets and large hospital bills for treating lung cancer are counted in the GNP, while home child care is not. The GNP says nothing about the costs in depleted natural resources and environmental degradation.

The United Nations Development Program has used a broader measure, the *Human Development Index,* which combines three indicators: real income (adjusted for purchasing power), life expectancy, and literacy. Compared to other nations, the United States ranks nineteenth on this index, though it is first in GNP.[50] The importance of wide access to health care and education is evident in the state of Kerala in India: life expectancy is eleven years longer than the national average and adult literacy twice that for the nation, even though income per capita is only two-thirds the national average.[51] Herman Daly and John Cobb have proposed an even broader index that includes measures of environmental degradation and resource depletion.[52] Pursuit of economic growth alone leaves unanswered the crucial questions: Whose growth? What kind of growth? I take *economic development* to include these wider considerations.

4. HUMAN NATURE

In addition to offering a distinctive perspective on each of these individual and social values, the biblical tradition holds a view of human nature that has particular relevance for a technological age.

First, the biblical view of human nature combines realism with idealism.[53] *Human sinfulness* refers in part to the attitudes of individuals. Sin is estrangement from God, from one's true self, from the neighbor, and from nature. But sinfulness also refers to social institutions. Every group tends to rationalize its own self-interest. If individuals and groups tend to abuse power, economic and political power must always be subject to checks and balances. We should be hesitant to turn over policy decisions to technical experts, however well-intentioned. We should also be cautious about large-scale systems in which human fallibility and institutional self-interest can have catastrophic consequences. When technology gives us the power to destroy ourselves and to alter human genes, humility and caution may be essential to protect the human future.

But the biblical tradition is also idealistic in its affirmation of *creative human potentialities*. Through technology we can use our God-given intellectual capacities to promote humane goals. I will argue that, if used intelligently and carefully, the resources of the created order are indeed adequate for human and nonhuman needs. Moreover, the biblical tradition affirms that a more just social order is possible. Political processes are a vehicle for the restraint of self-interested power, but they also offer positive opportunities for achieving greater social justice and human welfare. These assumptions would not lead us to reject technology but rather to redirect it toward basic human needs, meaningful work, economic development, and other goals outlined above.

Second, the Judeo-Christian understanding of *the person in community* provides correctives for some of the harmful tendencies of modern technology. The defense of personal and interpersonal existence can help us to resist the depersonalization that has accompanied industrialism. Synagogues and churches can evoke awareness of dimensions of human experience not accessible to technical reason, and they can nourish human compassion and tenderness. They can cultivate the arts and music in their liturgical celebration. Awareness of the sacred and recognition of mystery and human limits provide antidotes to the arrogant assumption of technological omnipotence. Receptivity and acknowledgment of grace are attitudes that contrast with control and manipulation.

Our religious traditions invite reflection on the ends of life and the nature of genuine fulfillment. Their images of human fulfillment go beyond consumption; a person's life cannot be adequately measured by material possessions, nor a nation's by its gross national product. Respect for human dignity today includes sensitivity to the effects of technology on people, resistance to attempts at manipulation and control, and concern for the quality of human relationships. In these traditions the person is never viewed in isolation, but always as part of a wider community. The character of the community's life must be an important consideration as we move toward a postindustrial society.

Third, religion can be a source of *individual and social transformation.* At the personal level, the biblical message holds out the possibility of release from guilt and anxiety, liberation from self-centeredness, and a life of genuine relatedness and openness when the power of love breaks into our lives. Healing, wholeness, and reconciliation can take place between persons, in communities of mutual acceptance, and in the relations between groups. Such experiences and their expression in story and ritual can be a more powerful influence on cultural change than abstract values or philosophical principles.

Today, visions of *alternative futures* can be a source of hope and renewal. Such a future would involve new definitions of what is necessary to sustain a good life, social patterns in which cooperation replaces competition, and life-styles that avoid the compulsion to consume. The ideal of simplicity can be recovered in affluent societies, not in a spirit of ascetic self-denial, but because global resources are limited—and because there are positive values in a simpler life. Once again, this does not imply rejecting technology but rather designing it according to criteria differing from those prevalent today.

Finally, the Hebrew prophets brought a double message of *judgment and hope* during times of national crisis and international conflict not unlike our own. On the one hand they spoke of God's judgment on human greed. They even saw military defeat and national catastrophe as forms of divine judgment on the materialism, idolatry, and injustice of national life. Their first word was a call to repentance and humility. But the other side of the prophetic message is hope and openness to new possibilities. Beyond judgment and repentance there is reconciliation and redemption, creative renewal in response to God's activity. The ultimate symbol of this hope is the vision of a Kingdom of peace and unity on earth. The prophetic imagination pictured a future harmony that would include all humankind and all nature. All people would be at peace with their neighbors and with the created order. The image of the Kingdom gathers up the themes of human fulfillment and the fulfillment of nature.

The idea of *the Kingdom* took many forms in subsequent history. Some members of the early church expected it to come very soon on earth. Others visualized the Kingdom as another world, a heavenly realm unrelated to this world. More commonly it has been understood both as the goal of history and as beyond history. The Kingdom (or Divine Reign, if we want to avoid a gendered term) is indeed an imaginative vision, but it is not just an idle dream. As do all visions of the future, it influences the way we interpret the present. It leads us to see the world in a new way. We can see both judgment and renewal in history, and we can act in response to them. The recovery of these biblical themes, I suggest, could contribute significantly to a more just, participatory and sustainable society, within which technology may be truly liberating.

IV. CONCLUSIONS

While I believe that a revitalized biblical faith could contribute significantly to new social goals and patterns of behavior, I am aware that in many parts of the institutional church there is great resistance to change. In addition, Christian writers have historically had little to say about nature and the environment. Moreover, public policies in a pluralistic culture cannot be built on the assumptions of any one religious tradition. Christians must make common cause with persons holding other philosophical and religious positions in working together for particular social policies.

In later chapters, then, I will frequently be drawing on *science*, especially for its estimates of the consequences of alternative technological policies and for its understanding of the interdependence of humanity and the nonhuman world. I will be drawing on *philosophy* in referring to both the good of society and individual human rights. Utilitarian consequences will be considered, but within a broader framework than cost-benefit analysis allows. I will refer back to the philosophical analysis of justice and freedom. From *the Christian tradition* I will draw on distinctive insights concerning each of the six values discussed here, though these values are often shared by others and can be discussed in the arena of public debate. The final chapter will take up again the biblical understanding of human nature, hope for the future, and the motivation for action arising from an ethics of response.

I submit that in a world of technological power, scarce resources, and increasing gaps between rich and poor, *social justice* is the crucial value in most policy decisions today. I will employ Rawls's criterion of benefit to the least advantaged, which is consistent with the biblical demand for social justice. The basic needs of millions of people are today unmet. Justice in access to *food and health* should therefore be our highest priority. But I also will refer at several points to *meaningful work* and a new vision of *personal fulfillment,* which I consider essential to human dignity in a technological society. *Participatory freedom* will come up repeatedly in examining ways of controlling technology. *Economic development* is affected by almost all technologies; our task will be to see it in relation to other values. Most of the difficult technological decisions we face involve conflicts among these six values and with the environmental values discussed in the next chapter.

CHAPTER 3

Environmental Values

A thing is right when it tends to preserve the integrity, stability, and beauty of the biotic community.

ALDO LEOPOLD[1]

What it needs for the most part is not so much a "new ethic" as a more general adherence to a perfectly familiar ethic. For the major sources of our ecological disasters—apart from ignorance—are greed and shortsightedness.

JOHN PASSMORE[2]

More science and more technology are not going to get us out of the present ecologic crisis until we find a new religion, or rethink an old one.

LYNN WHITE[3]

Much of the contemporary concern about technology is directed toward its environmental impacts and the attitudes that have led to such impacts. Where can we turn for an environmental ethic that will encourage the protection of the environment from these destructive features of technology? The authors quoted above look respectively to science, philosophy, and religion. In successive sections I examine the potential contribution to environmental ethics from each of these disciplines. I ask about the grounds for three environmental values: resource sustainability, environmental protection, and respect for all forms of life. The conclusions reached on these issues will affect the analysis of particular technological policy decisions throughout the remainder of the volume.

I. SCIENCE AND ENVIRONMENTAL VALUES

We consider first the changing scientific views of nature in recent centuries and the development of ecology. Some attempts to use ecology as the foundation of a new biocentric ethics are also presented. In keeping with my observations in the previous chapter, I will conclude that environmental ethics must be informed by contemporary science but cannot be based on science alone.

1. NEW VIEWS OF NATURE

The *mechanistic view of nature* in the seventeenth and eighteenth centuries was one factor in the growth of exploitative attitudes. Newton and his followers said that nature is constituted by impersonal masses and forces operating according to deterministic laws. Newton himself thought of nature as a complex machine designed by God, but his more secular successors had no scruples about exploiting it. If nature is a machine, it has no inherent rights or interests, and we need not hesitate to manipulate and use it.[4]

Moreover, the Newtonian world view perpetuated a sharp separation of *humanity* from *the nonhuman world*. It was claimed that apart from the human mind, the world consists of particles in motion. Newton accepted the Cartesian dualism of mind and matter. Descartes asserted that animals are machines without minds or feelings; he extolled practical knowledge that would make us "the lords and masters of nature," and he thought that our unique rationality justified such sovereignty. Whereas Newton held that understanding is the goal of science, others gave greater prominence to the control of nature. Francis Bacon said that the goal of science is the conquest of nature, for "knowledge is power." "Let the human race recover the right over nature which belongs to it by divine bequest." Bacon's *New Atlantis* called for a state-funded research establishment and a scientific elite through which humanity's rightful supremacy would be systematically extended.[5]

In the emerging *industrial technology* of the eighteenth century, domination over nature was increasingly achieved in practice as well as in theory. To the leaders of the Industrial Revolution, the environment was primarily a source of raw materials. In the new capitalism, private ownership of resources fostered the treatment of the natural world as a source of commercial profit. Along with rising standards of living came increasing burdens on the environment. Ever since antiquity, deforestation, overgrazing, and soil erosion have been occurring; but the technologies that developed in the last two centuries produced pollution and consumed natural resources at unprecedented rates. Mechanistic science, dualistic philosophy, industrial technology, and capitalist economics—along with certain themes in the Christian tradition that we will consider later—all encouraged human domination over nature.

But other voices were raised in opposition to prevailing attitudes. The *romantic literature* of the late eighteenth and early nineteenth centuries was in part a reaction to the view of nature fostered by Newtonian science and the Industrial Revolution. For Blake, Wordsworth, and Goethe, nature is not an impersonal machine but an organic process with which humanity is united. God is not the remote watchmaker but a vital force immanent in the natural world. Not rational analysis but feeling and imagination are the highest human capacities. Intuition grasps the unity of organic wholes, the interrelatedness of life. In natural settings a person may find a healing power, a sacramental presence, an experience of peace and joy. Other romantic writers extolled wild, sublime, untouched landscapes, forests, and rivers. They idealized the "noble savage" uncorrupted by civilization, and they exalted the "natural" and the "primitive."[6]

The transcendentalists in New England referred in similar terms to the presence of *the sacred in the realm of nature*. Henry Thoreau held that nature is a source of inspiration, vitality, and spiritual renewal; it can teach us humility and simplicity. "In Wildness is the preservation of the World," he wrote. Thoreau criticized the frantic pursuit of progress and affluence, the growth of technological industrialism, and the pressures of an impersonal urban life. His year and a half of living alone at Walden Pond made him more aware of the interrelationships among creatures and the natural stability upset by humans; in solitude he found serenity and peace. Unspoiled nature was for him both a symbol of qualities that he valued (freedom, courage, vitality) and a setting that would bring out these qualities in us. But he did not advocate giving up civilization. He sought a simplification of life and an alternation and balance between life with nature and civilization.[7]

Feminists have pointed to the connections between *the exploitation of nature* and *the exploitation of women*. They note the dichotomies prevalent in Western culture: spirit/nature, mind/body, reason/emotion, objectivity/subjectivity, and domination/submission. In our culture the first term in each of these pairs has been identified with men, the second term with women. The technological and economic institutions that have harmed the environment have been run almost exclusively by men. A patriarchal society values power, competition, and control more than nurture, cooperation, and reciprocity. Women are said to be more aware of the connectedness of life; they tend to reject hierarchical ordering and to welcome diversity in society and nature. These authors hold that women and environmentalists can make common cause in seeking liberation from the conceptual dualisms and the social structures of domination that have harmed both women and nature.[8]

Changes within science itself during the nineteenth and twentieth centuries can also encourage new attitudes toward nature. In the previous volume, I indicated that nature is today seen as evolutionary, emergent, and historical. Nature has a hierarchy of levels; each level must be understood in terms of larger wholes and interdependent networks as well as constituent parts. Contemporary science is not anthropocentric; humanity is part of nature in both evolutionary history and contemporary life. These ideas are found in many fields of science, but they are particularly clear in the new field of ecology.

Darwin's work was of course essential in preparing the way for ecology. In his writings, the human species is treated as a part of nature in continuity with other forms of life. No sharp discontinuities separate human from animal life, either in evolutionary history or in present morphology and behavior. The theory of evolution seemed to undermine humanity's unique status; close parallels to most human capacities could be found among other forms. In addition, Darwin's studies brought out the interconnectedness of the web of life and the complex balance of interactions in the biological world. Subsequent research in population dynamics has underscored the importance of the relation of organisms to their environment, including habitat, food sources, and predator-prey relationships. Recognition of the interdependence, diversity, and vulnerability of biological species are all central ideas in ecology.

In this century, Rachel Carson's writings combined scientific knowledge with a sense of spiritual unity with nature reminiscent of romanticism. In her influential book, *Silent Spring* (1962), the call for an ecological conscience was tied to scientific studies of the effects of pesticides on bird populations. By the early 1970s a large number of popular writings were delineating with increasing urgency the effects of a variety of pesticides, phosphates, nitrates, lead, mercury, radioactive wastes, and air pollutants. The word *ecology* had entered the public vocabulary.[9]

Four concepts have been especially important in twentieth-century ecology:[10]

1. The Ecosystem Concept. The interdependence of the forms of life in biotic communities and the complex interactions among organisms have been traced in detail by ecologists. They have described food chains linking diverse species, interlocking cycles of elements and compounds, and delicate balances that are easily upset. The interconnected web of life must be considered as a system; a change at one point can have far-reaching repercussions at other points.

2. Finite Limits. The growth of any population encounters limits; the environment has a finite carrying capacity. In nonhuman populations, growth is limited by such mechanisms as territorialism, interspecies competition, and food depletion. In the human case, neither population size nor resource consumption nor pollution generation can grow indefinitely. One of the most significant implications of ecology is its challenge to prevailing assumptions about unlimited growth.

3. Ecological Stability. Equilibrium is not a static concept but a dynamic balance of inflow and outflow. Stability is not incompatible with change and adjustment to constantly altering conditions. Balance often is achieved by feedback mechanisms that exert some control and regulation (that is, a change in one direction brings into play a counterinfluence in the opposite direction). Diversity in an ecosystem contributes to its stability and adaptability; systems with a very small number of plant or animal species are more vulnerable to disease, predators, and changing conditions.

4. Long Time Spans. The ecologist is concerned about repercussions and indirect consequences that may be distant in time as well as space. Continuity and sustainability are sought, not short-term benefits at the expense of long-term costs. Ecologists study population changes over many generations; they urge us to think about the consequences of our actions on future generations as well as on other species.

Recognition of the vulnerability of the biotic community leads ecologists to urge *restraint in human intervention*. Acknowledging human dependence on the biosphere, they advocate humility and cooperation with nature in place of attempted mastery and control. Since some consequences of our actions may be irreversible, or may be reversed only at great cost over an extended time period, we should err on the side of caution when our knowledge is so limited.

Many scientists have focused on *growth* as the crucial ecological problem. For Paul Ehrlich and Garrett Hardin, the control of population overshadows all other issues. *The Limits to Growth* and the British *Blueprint for Survival* asserted that the limits to industrial growth are set by resource reserves and pollution levels.[11] These studies advocated policies of "no growth" or "steady state" for industrial production as well as population. They held that attempts to use technology to extend the carrying capacity of the environment are running into the law of diminishing returns and the prospect of delayed side effects. While resource-conserving technologies may be more promising than these authors recognized (see chapter 7 below), we can be grateful to them—and to other scientists—for making us more aware of our dependence on a finite and vulnerable natural environment.

2. BIOCENTRIC ETHICS

First published in 1949, Aldo Leopold's *A Sand County Almanac* has become a classic of the environmental movement. These essays reflect Leopold's understanding as a scientist, his work as a forestry and wildlife manager, his first-hand experience as a naturalist, and his poetic sensitivity as a person. In a chapter entitled "The Land Ethic," he spells out some ethical implications of an ecological outlook.

The scope of one's ethics, he writes, is determined by *the inclusiveness of the community* with which one identifies oneself. Slaves once were treated as property toward which the owner had no ethical obligations. The history of ethics can be read as the extension of the boundaries of community to include the tribe, the nation, and then all people. The next stage, Leopold urges, is the inclusion of the whole land community, of which people will see themselves not as conquerors but as members and citizens.

Leopold's presentation of *the biotic community* elaborates the ideas of food chains, the biotic pyramid, the carrying capacity of the land, the interactions of predators and prey, and the destructiveness of human impacts. But it is not enough, he suggests, to appeal to enlightened self-interest. Most species have no economic value, yet they have a right to exist. What is required is a new stage in the evolution of ethics:

The "key-log" which must be moved to release the evolutionary process for an ethic is simply this: quit thinking about decent land-use as solely an economic problem. Examine each question in terms of what is ethically and esthetically right, as well as what is economically expedient. A thing is right when it tends to preserve the integrity, stability, and beauty of the biotic community. It is wrong when it tends otherwise.[12]

Right action is defined here in terms of the consequences for "the integrity, stability, and beauty of the biotic community." The criterion is the good of the total ecosystem. Integrity, stability, and beauty are holistic concepts; they are attributes of the whole system rather than aggregate functions obtained by summing up the benefits to individual members.

In Leopold's case, this holism derives from his outlook as *poet* and as *person* and not simply as *ecologist*. "That land is a community is the basic concept of ecology, but that land is to be loved and respected is an extension of ethics."[13]

He points to the importance of new attitudes and values, and therefore of educational and religious institutions, as well as scientific knowledge. He has much in common with Thoreau, Muir, and the romantic poets, though his orientation is more practical. For Leopold, the ground of unity with nature is more scientific than religious, but the personal and experiential aspects of his holism should not be overlooked. The qualities that he seeks in ecosystems—beauty, balance, and harmony—are judged by aesthetic and intuitive as well as biological criteria.

Building on Leopold's ideas, a number of authors have defended *a holistic ethics* in which the welfare of the ecosystem is the supreme value. They show that such holism is a radical departure from the individualism of other alternatives. (Even the animal rights movement, though it is not anthropocentric, still thinks in terms of the suffering of individual animals, and it has nothing to say about the value of plant life.) For the holists, by contrast, every being has value insofar as it contributes to the biotic community. But does an ethics based on ecology violate Hume's distinction of fact and value? Is this the "naturalistic fallacy" of trying to derive *ought* from *is?* Holmes Rolston says that for Leopold the *is* and the *ought* are discovered together; description and evaluation are inseparable. The very concepts of beauty, harmony, and integrity are themselves influenced by ecology.[14]

J. Baird Callicott supports Leopold's dedication to *the whole biotic community,* including nonsentient members such as trees and plants. He notes that some species, such as the honey bee, have particularly important roles in that community. "Some bacteria, for example, may be of greater value to the health or economy of nature than dogs, and thus command more respect."[15] Predators appear destructive, but in the long run they benefit their prey by keeping the population in balance with food supply. Callicott claims that ecosystem ethics is consistent with Hume's belief that even though there is no logical path from *is* to *ought,* we are all motivated by "moral sentiments" aroused by fellow feeling and a sense of kinship. Callicott says that we all tend to have a positive attitude toward the community to which we belong, and ecology simply shows us that we belong to a wider community than we had thought. "Ecology changes our values by changing our concepts of the world. It reveals new relations among the objects which, once revealed, stir our ancient centers of moral feeling."[16]

Bill Devall and George Sessions advocate biospheric equality in their book *Deep Ecology.* They propose that we can get beyond a superficial ecology (which preserves the environment only for the sake of humanity) by accepting the unity and intrinsic value of all of nature. "The intuition of biocentric equality is that all things in the biosphere have an equal right to live and blossom and reach their own individual forms of unfolding. . . ."[17] Devall and Sessions hold that human interference with the ecosystem is at present excessive and that the welfare of the biosphere requires a major reduction in the human population. Their views are partly derived from ecology, and for that reason I cite them here. But their intuition of interdependence, unity, and wholeness is also derived from Eastern meditative traditions and from Western monistic philosophies such as Spinoza's pantheism. Because their thought has philosophical and religious sources, it might equally well have been discussed later in the chapter.

Critics of biocentric ethics charge that the door is open to totalitarianism if *individual human rights* are subordinated to the welfare of the ecosystem.[18] As one example, they cite Callicott's statement that "the extent to which modern environmentalism is misanthropic is a measure of the degree to which it is biocentric." And if the welfare of the biosphere is the only criterion, are there any limits on the methods one might employ to reduce the human population? Callicott replies that ethics itself evolves, and many of the moral sensibilities of the past could be carried over within a wider biocentric ethics.[19] Yet the human values seem precarious if they can be overridden when there is a conflict. Don Marietta holds that the welfare of the ecosystem can be taken as *one* source of value and duty without assuming that it is the *only* source of value and duty.[20] Arne Naess, an early exponent of deep ecology, says that the vital needs of other species should override a peripheral human interest but should in turn be overridden by a vital human need. However, he does not show how such a conclusion follows from his premise of biospheric equality.[21]

I would argue that by subsuming the *human* under the *biological,* the ecologist neglects the distinctive features of human nature and culture.[22] Biocentric ethics provides no grounds for decision when the interests of diverse members of the biotic community conflict. I have favored a holistic approach, but I think these authors push holism too far. They often compare ecosystems to organisms, whereas I maintain that ecosystems are loosely integrated communities whose members have more independence and intrinsic value than the parts of an organism do.[23]

I suggest that the integrity of the ecosystem is important because it makes possible *the welfare of interdependent individuals,* human and nonhuman. We will see below that process philosophy acknowledges both interdependence and the intrinsic value of experience at all levels. But it holds that only individuals are capable of experience and that there is great variation among the experiences of differing types of individuals. According to this interpretation, ecological integrity is a precondition of life and therefore of other values, but it cannot serve as a definition of all value. The goals we seek must be *compatible* with the health of the ecosystem, but they need not be limited to the latter. A person or society interested only in satisfying biological needs would miss the most distinctive potentialities of human existence.

II. PHILOSOPHY AND ENVIRONMENTAL VALUES

Most of the writing by philosophers concerning the environmental impacts of technology has involved the extension of familiar ethical principles rather than the formulation of radically new principles. The utilitarians now recognize that human benefits and risks from the environment are more diverse and more important than previously acknowledged. The utilitarian framework has been expanded to include benefits and costs to future generations as well as the welfare of animals and other sentient creatures capable of pain and pleasure. Do animals have rights, and do we have duties toward them? Our answers to these questions will have implications for agricultural, industrial, and land use policies.

1. HUMAN BENEFITS FROM THE ENVIRONMENT

Many of the measures advocated by environmentalists can be justified by their contributions to human welfare. An anthropocentric position need not lead to exploitative domination if *our dependence on nonhuman nature* is recognized. The environment is important to us biologically, economically, and aesthetically. Food, health, material well-being, and personal fulfillment are all dependent on the biosphere. Enlightened human self-interest provides strong arguments for action to preserve the environment.

A number of philosophers have argued that what is needed today is not a new environmental ethic but a careful application of *traditional forms of ethics* in the light of *new ecological knowledge.* John Passmore, for example, advocates a scientifically informed utilitarianism. He starts from human interests alone, but he recognizes that if we damage the biosphere we end by injuring ourselves. Any responsibility for nature is derivative from responsibility toward humanity. The criterion of human usefulness, broadly conceived, leads him to support cautious intervention; we should "humanize and perfect nature" while respecting its vulnerability.[24] William Frankena similarly holds that commitment to the well-being of persons is sufficient justification for environmental action. The classical ethical principles are not at fault; it is we who have failed to live up to them and apply them intelligently. Frankena holds that concern for the suffering of animals should be included in the sphere of ethics, but he holds that there is no basis for duties toward nonsentient creatures except as they affect human welfare or aesthetic enjoyment.[25]

Pollution of air, water, and land by industrial technology clearly is detrimental to human health and personal fulfillment. The correlation of air pollution with respiratory diseases is amply documented. Water pollution is a health hazard, an obstacle to recreation, and a threat to the aquatic life from which people benefit. Pesticides and toxic wastes have been shown to produce cancer or injure the human nervous system, and often their concentrations are greater in the higher levels of food chains from which humans draw. Whatever harms human life-support systems harms humanity. On both ethical grounds and in the interest of a practical political strategy, environmental values can be defended because they are essential to the realization of human values.

Even the *preservation of wilderness* may be justified by its contribution to human life, though many of these benefits are less tangible. The most obvious benefit is recreation—camping, hunting, fishing, canoeing, backpacking, and so on— which has become so popular in some areas that permit systems have been instituted to prevent damage from excessive use. The enjoyment of natural beauty is another human benefit. As noted earlier, nineteenth-century romanticism appreciated beauty in wild nature (rather than in the cultivated landscapes that had been extolled from antiquity to the Enlightenment), and wilderness has been widely celebrated in Western art and literature. As urbanization and the stresses of a technological society have increased, and wilderness has become more scarce, these recreational and aesthetic values have become more important for many persons.[26]

The positive influence of wilderness on *human character* is another intangible benefit. The solitude of woods and mountains is a source of inner serenity and strength, an opportunity to live in harmony with the natural order. Wilderness can teach us moral lessons; we can learn humility and gratitude but can also gain self-reliance in facing the challenge of the wild. It has been argued that we should preserve wilderness because it has come to symbolize human qualities that we respect, such as freedom, innocence, and courage. Destroying wild rivers and forests would be an attack on the human values that these natural phenomena represent in our cultural tradition.[27] These wilderness ideals can be affirmed even if one does not accept the romantic view that human nature in a natural setting, without the corrupting influence of technology and the city, is intrinsically good.

Others have found in wilderness a *spiritual significance* that goes beyond aesthetic or moral values. Thoreau, Muir, and their successors have held that wilderness experience is a source of inspiration, an opportunity to acknowledge the divine in or beyond nature. The majesty of forests and windswept mountains evokes wonder, awe, and a sense of mystery. In quiet and solitude we learn reverence, which is an antidote to the spirit of technological omnipotence and the attempt to conquer nature. In wilderness we discover that we are spiritual beings and not just consumers; our imagination and primeval intuitive capacities, long suppressed by calculative rationality, may be awakened and expressed. Wilderness is a "sacred space" in which the experience of the "wholly other" can be powerful.[28] One survey found that the majority of Sierra Club members considered aesthetic and spiritual experience more important than recreation or resource conservation as motives for wilderness preservation.[29]

Two additional reasons for preserving wilderness are its use in *scientific research* and its role in preserving *endangered species*. Ecology is a young science and needs "natural laboratories" in which to investigate the complex interactions among the members of biotic communities. No point on earth is totally beyond the reach of human influence, but relatively untouched areas can provide a data base from which to study the effects of our intervention. We can learn from what David Brower has called "the vanishing remnant of a world that can run itself." We are extinguishing one hundred species every day, destroying libraries of genetic information that were built up during millions of years (see chapter 7). This reservoir of genetic diversity is an irreplaceable resource. Who can say what genetic strains may be useful in future plant breeding or what clues important in agricultural and medical research may be found in unexpected places? Furthermore, this diversity enhances the stability of the ecosystems on which we depend.[30]

Wilderness preservation, like pollution abatement, can thus be defended by reference to its benefits to *humanity alone*. I will suggest shortly that beyond this instrumental view of nature, nonhuman life has intrinsic value. Duties to other creatures, as well as the intangible human benefits above, cannot be quantified for inclusion in cost-benefit analyses. Yet the preservation of wilderness and of endangered species, like pollution abatement measures, involves inescapable trade-offs with jobs, economic growth, and regional development.

2. DUTIES TO FUTURE GENERATIONS

Many of the environmental effects of technology are cumulative or long-term. By depleting resources and polluting the environment, we are increasingly jeopardizing the welfare of our descendants. Population growth, soil erosion, and the use of nonrenewable energy sources will affect human life for many generations to come. After ten thousand years, the radioactive wastes from our nuclear power plants will still be dangerous to anyone exposed to them. The alteration of human genes or the depletion of the ozone layer might affect the human race permanently. What are the grounds for ethical obligations to future generations in such policy issues?

A number of *utilitarians* have made at least some reference to posterity. John Stuart Mill spoke of "the general interest of the human race," and pictured "all generations of man indissolubly united into a single image"; but he actually applied the principle of the greatest good to the greatest number only to living persons.[31] Among contemporary utilitarians, J. J. C. Smart holds that in principle all generations should count equally, but in practice the more remote effects of our actions usually are either too small (dissipating like the ripples from a stone in a pool) or too uncertain to be taken into account. But he says that if long-term effects might well be catastrophic (as in the case of nuclear war or the release of a potentially lethal virus), they should count heavily in our calculations.[32]

But can there be *obligations to people who do not exist?* Nonexistent persons do not have rights, though living persons or the state might assert claims on their behalf. Some utilitarians maintain that we do not have an obligation to be benevolent toward persons as such, but rather an obligation to produce what is good in itself. The obligation to increase the balance of good over evil does not require the specification of individuals because it is not owed to anyone. Other authors affirm that we do have obligations to future persons as potential members of our moral community, but that these obligations diminish for more distant generations because our social ideas are less relevant to their lives under conditions that may be very different from ours.[33]

Utilitarians usually *discount the future* because the more remote consequences of our actions are more uncertain. Events of low probablity or great uncertainty are assigned low weights in cost-benefit calculations. The economic system also gives priority to short-term benefits. A natural resource worth $100 ten years hence is only worth $50 to me now if I have the alternative of investing my money at 7 percent interest. The democratic process reinforces this tendency to discount the future in policy decisions. A political system with elected terms of two or four years is geared to immediate results and to current voters. Unborn citizens cast no votes. Another factor in discounting the future is the prevalent confidence that technological advances will provide solutions to environment and resource problems.

In later chapters I will urge that potential long-term effects should be more heavily weighted than in current practice. It is usually assumed that smaller weights should be assigned to consequences the further they are from us into the future because uncertainties are greater over longer time spans. But such a uniform time discount is questionable. Well-understood distant effects can be predicted with greater confidence than short-run phenomena that are poorly understood. We are as likely to underestimate uncertain risks as to overestimate them. Facing uncertainty, we should give a range of estimates and emphasize their tentativeness, rather than introducing a time discount. On the other hand, the improbability of an event is a valid reason for giving smaller weight to its consequences, regardless of its location in time (though I will propose that a small probability for a catastrophic consequence should count more heavily than in standard risk-benefit calculations).[34]

Duties to posterity can also be derived from the idea of *justice between generations*. Several authors have developed Rawls's model of contractors in a hypothetical "original position" who do not know to which generation they will belong.[35] What policies concerning technology and environmental degradation would you recommend if you did not know in which generation you would be born? What policies would be adopted by an assembly of representatives of all generations? Rawls asks us to think about the least advantaged, regardless of temporal location. Compared to future generations, the present generation may be the less advantaged with respect to economic development and technology but the more advantaged with respect to resources and the environment. Only in the poorest nations today is the present generation probably the least advantaged overall, so that in their case some transfer of resources from future to present generations would be justified.[36]

In practice it is impossible to consider an indefinitely long series of future generations in policy decisions. But in the case of *renewable* resources, such as fish, timber, and crops, the same result can be obtained by aiming for the *maximum sustainable yield*. Once the maximum sustainable yield is exceeded by excessive fishing, cutting, overgrazing, and soil erosion, the productivity of oceans, forests, grasslands, and croplands is rapidly reduced. People are then consuming productive biological "capital" rather than living on the "interest" that could continue indefinitely. They are stealing the wealth of their children and grandchildren. A fair distribution over the generations (assuming population stabilization) can be achieved by keeping within sustainable yields, which can be roughly calculated from knowledge of ecosystems today. Future sustainable yields may be somewhat higher (due to technological advances) or lower (due to environmental damage); but use levels could be readjusted in the light of new information.[37]

With respect to *nonrenewable* resources, such as minerals or fossil fuels, justice between generations would require that *the resource base* should not be depleted more rapidly than it can be extended by technology, since all generations have an equal claim on it. Technology turns previously useless raw materials such as metallic ores into useful resources. Those who deplete resources should compensate future generations by passing on improved technology and capital investment to offset the effects of depletion. If the resource base relative to technology is thus preserved from one generation to the next, it will be preserved for all generations. Brian Barry suggests that the criterion should be equality of opportunity with respect to productive potential. The present generation should pass on to the next the technological improvements that would replace the productive opportunities lost by resource depletion.[38]

Until about 1970, *technology* extended the resource base at a rate that more than offset depletion. New extraction technologies made lower-grade ores economical, discoveries expanded reserve estimates, and substitutes were found for scarce materials. As a result, the price of most minerals and raw materials (in constant dollars) stayed constant or fell from 1900 to 1970. During the 1970s and 1980s, however, shortages and rising prices affected several materials as the more accessible reserves were used up. Technological knowledge is itself a significant legacy to posterity. But the indirect costs of technology in the form of toxic wastes, pollution, and environmental degradation pass on an increasing burden of risks, although improvements in industrial processes and in pollution control technology can in some cases offset these threats to health and the environment. Some resource uses do contribute to technological development and economic growth. However, we must remember that 90 percent of natural resources go into short-lived consumer goods, and only 10 percent into capital accumulation and technological equipment from which future generations might benefit.[39]

Risks passed on to future generations are of special ethical concern for two reasons. First, they are imposed involuntarily, since future generations cannot give voluntary consent or have any voice in current decisions. Second, distributive injustice occurs if risks to future generations arise from actions in which the present generation is the main beneficiary (as in the case of radioactive wastes). We have created enormous and diverse debts that will have to be paid by our children and grandchildren: resource depletion, environmental damage, and of course huge national debts on which the interest alone will absorb a major portion of government funds needed for other purposes.

Concern for the needs of future generations must of course be combined with concern for *the needs of those now living*. Urgent current needs have priority over uncertain future ones. But the basic needs of future generations should have priority over the luxuries of the present. Barry points out that the transfer of technology and capital from industrial nations to developing nations would fulfill intergenerational and international justice simultaneously. For our obligation is not simply to our own descendants, but to future generations everywhere—and particularly to those who missed the chance to industrialize while energy and other resources were cheap. Barry proposes an international income tax, which would fall most heavily on those with highest income, combined with a severance tax on the use of nonrenewable resources, which would fall primarily on those who consume the largest quantities of such resources.[40] Such taxes may be difficult to effect, but they offer a mechanism by which justice across space and justice across time could be combined.

Anticipatory planning for long-term sustainability requires *a wider time horizon* than we have adopted in the past. As we become aware of the indirect consequences of our actions, our moral community is expanded in both space and time. The transition to a sustainable society must start now; a far greater toll in human suffering will occur if the change is forced on our descendants. We will later consider recycling, less resource-consumptive technologies, and frugal life-styles in affluent nations as important steps toward sustainability.

3. RESPECT FOR ALL FORMS OF LIFE

Are prudential considerations enough to motivate decisive action to preserve the environment? Does human self-interest, even on a long time scale, provide strong enough reasons to work for the protection of other species? Are other creatures valuable themselves, apart from their usefulness to us? Do we have duties concerning other forms of life that are not simply derivative from the consequences for humanity?

Aristotle, Aquinas, and Kant, along with most major Western philosophers, asserted that we have no direct duties toward animals because they cannot reason. *Rationality* was the criterion for a sharp separation of the human and the nonhuman and for the delimitation of those toward whom we have moral responsibility. Some utilitarians, on the other hand, have interpreted the greatest total happiness as including other forms of life. Bentham and Mill held that we do have duties to animals because they can experience suffering. They said that *sentience*, not rationality, determines whether a being can be the object of our moral obligations. Utilitarians were active in the nineteenth-century movement for the humane treatment of animals. Smart speaks for many contemporary utilitarians when he defends the goal of "maximizing the happiness of all sentient beings."[41]

Peter Singer's *Animal Liberation* is an extended defense of the duty to minimize the suffering of animals. He attacks the "human chauvinism" of restricting ethical concern to the members of our own species. Our ethical responsibilities extend to any organism that can experience pleasure or pain (he draws the line "somewhere between a shrimp and an oyster"). Singer documents in detail the suffering that is common in the raising, transporting, and slaughtering of cattle and poultry, and he ends by advocating vegetarianism. He maintains that if we have a duty not to treat animals cruelly, then animals have rights (at least the right not to be treated cruelly).[42] Such rights are, of course, not absolute but must be weighed against the competing rights of other beings. Note that in these views we have no duties in regard to the nonsentient portions of nature.

Public concern for *the humane treatment of animals* has been more widespread in recent years. The goal has been to reduce the suffering of animals on farms and in scientific research and product testing. The more extreme animal rights activists have broken into laboratories to free animals and publicize their cause. In 1985 the United States Congress passed amendments to the Animal Welfare Act, which legislated standards for the treatment of laboratory animals, and required local review committees to inspect facilities and practices. Activists have advocated greater use of alternative research methods such as tissue cultures and computer simulations. But the majority of the scientific community is convinced that animal experimentation has produced and will produce such enormous benefits for humanity (such as heart surgery and the testing of drugs for AIDS) that it should be continued, though in ways that reduce the suffering involved.[43]

Some philosophers want to broaden the sphere of ethical consideration to include *all living things*. Any living thing can be benefited or harmed by our actions. A tree or a plant does not suffer, but it maintains and heals itself, and there are conditions which are good or bad for it. Kenneth Goodpaster holds that living things do not have absolute rights but deserve ethical consideration because they have interests and needs. He advocates a "life-respecting ethics."[44] Donald Scherer says that we should promote the flourishing of individual living things. We must recognize that they always live in populations and ecosystems, but our loyalty is to individual beings rather than to ecosystems as such.[45] Robin Attfield develops an expanded utilitarianism that includes the welfare of nonsentient living things. All living things have interests and innate capacities, which can be furthered or thwarted by our actions.[46] The goal here is not the avoidance of suffering but the promotion of the welfare of living things.

I agree that the welfare of all living things should be considered, but I believe we need a *principle of discrimination* when the welfare of different forms of life conflict. If life is the only basis for assigning value, how are priorities to be established when the interests of human and nonhuman creatures are incompatible?

The *process philosophy* of Alfred North Whitehead and his followers seems to me helpful at this point. It offers a rationale for respecting all human and nonhuman creatures, along with a principle for assigning priorities. The organismic view of reality as a network of interacting events, rather than a collection of externally related objects, has much in common with the ecological perspective, as I indicated in the previous volume. The portrayal of interdependence is similar to that in ecosystem ethics, but the center of value lies in individuals, not in the whole. Process philosophy is fundamentally pluralistic rather than monistic. Reality is many leveled, and everything is part of a larger whole, but every individual is of value in itself.[47]

Process thought gives priority to *human needs* without being anthropocentric. Whitehead's stress on temporality and change and his interpretation of the world as a creative process were strongly indebted to evolutionary biology, which posits no radical gulf between the human and the nonhuman. Process philosophy describes all entities by the same set of basic categories. All beings are centers of at least rudimentary experience; no sharp lines can be drawn among creatures of varying degrees of complexity. But immense differences are present in the intensity, breadth, and originality of experience at various levels. A human being is capable of a far more varied and unified experience than an insect. But the level of experience of a chimpanzee or a porpoise is closer to that of a human than to that of an insect.

John Cobb has systematically explored the implications of process thought for environmental ethics. Every integrated being, he says, is a center of experience, which is the only *intrinsic good.* Every being also contributes to the experience of other beings, which constitutes an *instrumental good.* Our goal should be not just the reduction of suffering, but the positive enhancement of experience and the fulfillment of the potentialities of all creatures. Cobb suggests that the richness, intensity, and complexity of experience provide criteria for discrimination.[48]

Process thought portrays great *differences among beings* in both the intrinsic good and the instrumental good they can achieve. A rock is a mere aggregate, an object with no integrating center. A plant has no center of unified experience higher than that of individual cells, though the life of the cells is dependent on the total life of the plant. More complex beings are capable of both greater intrinsic good and a greater contribution to the experience of other beings. In simple forms of life, elementary feeling is present but no consciousness, much less self-consciousness. In this framework it is entirely justified to destroy cancer cells or malarial mosquitoes to save human lives. If protein for starving children could be obtained only at the price of suffering on the part of animals, such suffering would be justified. (Actually, feeding grain to cattle in feedlots is detrimental both to the human food supply and to animal well-being, so we do not have to choose priorities in that case.) All living things are valuable, but they are not equally valuable.[49]

Susan Armstrong-Buck compares the *process view* with the *animal rights* position on the one hand and *ecosystem ethics* on the other. She contends that process thought avoids both the individualism of animal rights and the collectivism of ecosystem ethics. Process philosophy values individuals in themselves and also in their connectedness. It also differs from these other views by providing criteria for discriminating among conflicting interests. In process thought we have "a greater obligation toward entities with more significant experience."[50] In a similar way, Jay McDaniel argues that differences in the richness and range of experience provide grounds for judgment among competing claims.[51]

Finally, process thought holds that all types of entities are *valued by God* and contribute to God's experience. To process thinkers, all entities derive significance from their participation in the divine life. The preservation of members of an endangered species is thus to be sought for their own sake, for the good of other members of the ecosystem, and because God's life is enriched by the diversity of the created order. In short, every creature is valuable to itself, to other beings, and to God. The process view thus leads to respect for all forms of life, commitment to the distinctive values of human life, and a general framework for judging their relative importance when they conflict.

III. RELIGION AND ENVIRONMENTAL VALUES

Let us turn from science and philosophy to religion as a source of environmental ethics. Eastern religions have frequently expressed the goal of harmony with nature. The biblical heritage includes themes supporting care for the earth. Finally, some creative writings in contemporary theology fit particularly well with the scientific and philosophical ideas presented earlier.

1. EASTERN RELIGIONS

A brief summary cannot begin to do justice to the rich diversity of Eastern thought concerning nature, but it can at least show some contrasts with traditional Western thought. *Taoism* in China portrays the world as an organic, interconnected system. Nothing exists in isolation; the parts of the whole are interpenetrating and interfused. Every particular being is a manifestation of the Tao, the nameless unity that exists before differentiation into multiplicity. Humanity is part of a wider cosmic order. To achieve a harmonious relationship to the natural world, we must respect it and adjust to its demands. The path to the recovery of harmony and wholeness is surrender, tranquillity, nonattachment, the ability to "let things be." The love of nature in traditional China is evident in its poetry and painting (especially of trees, mountains, and landscapes). But even agriculture and land management were represented as cooperation with nature rather than as conquest.[52]

The Taoist term *wu wei* does not mean inaction but rather action that is in harmony with the true nature of things. By adapting to natural processes the maximum effect is achieved with minimum effort. Humanity and nature are intimately linked, and there is an ontological equality among all the manifestations of the Tao. But reality is dynamic rather than static, and there is a creative spontaneity in the world. Yin and yang are complementary and opposing principles, yet they are united in a larger whole. The black and white portions of the famous yin/yang symbol are interlocking and flowing, but together they form a perfect circle.

Joseph Needham's monumental study, *Science and Civilization in China*, shows that the Taoists' holistic assumptions did not hinder the development of science and technology. Their goal was harmonious adaptation rather than dominion, and humans were thought of as part of nature. Yet China preceded the West in inventing the mechanical clock, the steam engine, movable type, gunpowder, and the magnetic compass. The Taoists were not against technology but said that intervention in nature must respect the totality of life. Chinese medicine was also holistic and sought the achievement of balance and harmony.[53]

There were, of course, *diverse strands* in Chinese history. The same culture that produced Taoist nature mysticism and holism also accepted Confucian principles of hierarchical order, praised emperors for massive feats of engineering, and built walled cities following strict geometrical patterns. Moreover, there are gaps between the ideals and the practices of every culture. Behavior is the product of social and economic forces and institutional structures as well as beliefs. Environmental destruction was by no means absent from classical China. For instance, wood was in great demand for building, fuel, and charcoal for metallurgy; deforestation resulted in widespread soil erosion.[54]

The changes in attitudes in China during *the twentieth century* have been mainly the result of Western influences. Chinese communism shares the Marxist assumption that nature is an object of conquest. Mao Tse-tung called for "a war against nature." In many official pronouncements nature was explicitly presented as an enemy against which the people must struggle to achieve victory. When environmental measures were implemented, the motive was primarily to improve sanitation and health or to recover industrial wastes for reuse. Major programs such as reforestation and water management have been developed in response to pressing human needs. Considering the extent of human deprivation in China, it is understandable that a revolutionary government has seen nature mainly as a resource to be exploited for the good of society. But this does represent a major shift from the characteristic cultural attitudes of the Chinese past.[55]

Zen Buddhism arose first in China from the confluence of Taoism and Mahayana Buddhism and then developed further in Japan. Here too human kinship with nature was stressed. According to Zen, the merging of self and other is known in immediate experience. Intuition and personal awareness, not analytic rationality or conceptual abstraction, reveal the unity of subject and object. In the Zen tradition, nature is to be contemplated and appreciated rather than mastered. Humankind should act on nature with restraint, bringing out the latent beauty and power of the natural world. These aesthetic elements in the Zen outlook entered many expressions of Japanese culture such as flower arrangement, the tea ceremony, and the garden. All the *wabi* arts value gracefulness, serenity, tranquillity, simplicity, and aesthetic balance. In the short poems known as *haiku,* events in nature typically provide moments of insight into the beauty, harmony, and dynamic flow of reality. The path of spiritual awareness, it is held, can liberate a person from the obsessive drives of the ego and from preoccupation with material possessions. The goal is nonattachment and freedom from control by our desires. The Buddhist ethic thus encourages humility, simplicity, and frugality, as well as compassion and service.[56]

In Japan there have been many religious traditions, including several forms of Buddhism. The indigenous Shinto heritage affirmed the sacredness of the cosmos and the continuum between humanity and nature. But the *industrialization* of modern Japan arose largely from the impact of the West. The attitudes engendered have remained in considerable tension with traditional views of nature. The postwar period saw the single-minded pursuit of rapid industrial development, and by 1965 Japan was the most polluted nation in the world. But after the media publicized the costs in human health (for example, people crippled by mercury poisoning), thousands of citizens' groups were formed, and in 1970 the government adopted the most ambitious legislation in the world for controlling air and water pollution. These measures resulted in marked improvement, though initiatives have slowed down since 1975 and conservative governments have been reluctant to antagonize industry or slow economic growth.[57]

Environmental legislation in Japan has dealt primarily with threats to human health and has done little to preserve wildlife or scenic beauty. In fact, Japan's actions have been highly destructive of endangered species, whether in killing whales at sea or in obtaining lumber from the Amazon basin, and in international negotiations Japan has strenuously opposed environmental restrictions. One might take this as evidence that economic considerations dominate modern industrial nations or as evidence that Buddhism is a minor force amid the secularism of contemporary Japan. James Whitehill suggests, however, that there is a widespread sense of rootlessness and uncertainty in Japan today, and if there is a cultural crisis and a breakdown of the prevailing confidence in technology, traditional religious ideas might play a creative role.[58] The legacy of respect for nature, poetry, and art may yet find political expression in action to preserve the environment even when human health is not at stake.

I believe that we have much to learn from *Taoism and Zen,* especially about disciplines of meditation and respect for the natural world, both of which have been neglected in the West. But most Westerners are not familiar with these religions. Furthermore, Eastern traditions have had less to say about social justice, which I take to be a crucial value today. It seems to me more promising to use the insights we can gain from the East to help us recover neglected strands of our own heritage.

2. HISTORICAL CHRISTIANITY

Biblical attitudes toward nature have been the subject of considerable controversy. The first chapter of Genesis includes the commission to "be fruitful and multiply, and fill the earth and subdue it, and have dominion over the fish of the sea and over the birds of the air and over every living thing" (1:28). Humanity alone is said to be created "in the image of God" and is set apart from all other forms of life. Moreover, nature is desacralized in biblical religion. Ancient Israel believed that God was known primarily in historical events rather than in the sphere of nature. More than any other religion of antiquity, the Hebrew tradition stressed God's transcendence of nature and the distinctiveness of the human.

A number of authors have claimed that *the biblical idea of dominion* was the main historical root of environmentally destructive attitudes in the West. In a widely quoted article, Lynn White described the separation of humanity and nature in biblical thought. Holding that ideas and attitudes are significant influences in history, White concluded that Christianity "bears a huge burden of guilt" for the environmental crisis because it has been so anthropocentric and arrogant toward nature.[59]

The replies to White have taken two forms. First, it is said that by concentrating on Christian thought, White ignored other sources of exploitative attitudes in Western history. For example, *Greek and Roman thought* also had great influence on the West. Plato and Aristotle portrayed a gulf between humans and all other beings based on the unique human capacity for reason. Aristotle stated that other creatures are devoid of the contemplative activity in which humans are most akin to God; plant and animal life exists solely for the sake of human life. Cicero, drawing upon Stoic writings, insisted that we have no obligation to respect animals because they are not rational beings. To the Neoplatonists of the early Christian era, the eternal forms are embodied only imperfectly in the world of nature. To the Gnostics and Manichaeans, nature is the realm of evil from which the human soul seeks to escape. Greek and Roman views were indeed extraordinarily diverse; some pantheistic authors were more appreciative of the natural world. But the classical sources that were taken up in the early church, the Middle Ages, and subsequent Western thought seem to have stressed the differences between humankind and nature.[60]

Other critics say that later institutions such as those of *industrial capitalism* were the main determinants of environmentally destructive behavior. It could be argued that those institutions were themselves in part the outcome of Christian assumptions, but some authors maintain that economic forces are the main determinants of social change and that White overemphasized the role of ideas in history. Moreover, extensive environmental damage occurred in non-Christian countries at many periods of history. White seems to have oversimplified a complex historical phenomenon.[61]

The second line of reply to Lynn White's thesis is to assert that *stewardship* and *celebration of nature* are also strong biblical themes. The dominion theme can be used to justify destructive practices only if other ideas that qualify or counterbalance it are ignored. Human beings are to be responsible caretakers, tilling the garden and caring for it (Gen. 3:23). Throughout the Bible, humankind does not have absolute and unlimited dominion but is responsible to God. "The earth is the Lord's" because the Lord created it. The land belongs ultimately to God; we are only trustees or stewards, responsible for the welfare of the land that is entrusted to us and accountable for our treatment of it. In the last analysis, the biblical outlook is neither anthropocentric nor biocentric but theocentric.

In the biblical view, the created world is *valued in itself,* not simply as an instrument of human purposes. In several of the Psalms (19, 89, and 104, for example) God is said to delight in the earth and the manifold variety of life, quite apart from humanity. Even in the first chapter of Genesis, each form of life is pronounced good before humankind is on the scene. "Every living thing" is included in the Covenant after the flood (Gen. 6:19), and humanity exists in kinship with other creatures. The Sabbath is a day of rest for the earth and other living things, as well as for people. Every seventh year the fields are to lie fallow; the land deserves respect and will cry out if misused. Human life and nature stand together jointly as God's creation. The idea of creation is a great unifying framework, encompassing all forms of life. The entire creation is important to God.[62]

Many biblical passages express *appreciation and wonder* in response to nature. Job is overwhelmed by the majesty of natural phenomena, including strange creatures of no use to humanity (Job 40–41). Jesus spoke of the lilies of the field and God's care for the sparrow. Value pervades all life, not just human life. The Holy Spirit is God's presence and activity in nature as well as in the human community. Further, nature is part of the drama of redemption and will share in the ultimate harmony, as portrayed in the symbolic vision of the coming Kingdom, when "the wolf shall dwell with the lamb, and the leopard shall lie down with the kid" (Isa. 11:6). Paul imagines that "the whole creation has been groaning in travail together until now," but that it will all take part in the final fulfillment (Rom. 8:2). While the focus of interest was human history, the world of nature was not neglected within the Bible, as it was in much of later Christian thought.

Let us note also the distinctive biblical grounds for *obligations to future generations.* In the biblical narrative, individuals have a strong sense of belonging to a family and a people who will continue beyond their own lives. The Bible expresses a sense of solidarity in time as well as space, a Covenant from generation to generation "to you and your descendants forever." Both Judaism and Christianity have expressed a universalistic vision of the unity of humankind embracing generations yet to come. In the biblical view, stewardship requires consideration of the future because God's purposes include the future. The land, in particular, is held as a trust for future generations. This long-range perspective derives from a sense of history and an orientation toward the future, as well as accountability to a God who spans the generations.

In later Christian history, impressive voices spoke on behalf of nature, though they were always in the minority. In the sixth century, *Saint Benedict* presented an interesting model of stewardship in practice. He advocated treating nature with care and respect. The Benedictine monasteries combined work and contemplation. They developed sound agricultural practices, such as crop rotation and care for the soil, and they drained swamps and husbanded timber all over Europe. Benedictines were creative in practical technologies related to nature.

In the thirteenth century, *Saint Francis* expressed eloquently a deep love of the natural world and a sense of union with it. He saw nature as a living whole and all creatures as objects of God's love, and hence as significant in their own right. He spoke of our sister the earth and greeted the birds as brothers, extending the family relationship and the circle of God's love to include all created beings sharing a common dignity and equality under God. Humanity, he said, is part of a wider community, and each creature has its own integrity that must be respected. Saint Francis has continued to appeal to the popular imagination, and the order that he founded has passed on both his respect for nature and his dedication to the poor.

The very mixed record of Christian thought in the medieval and Reformation periods has been amply documented. Biblical themes were combined with dualistic ideas from late Greek thought. Biblical theology was systematized, but also altered, by the philosophical categories used in Catholic scholasticism, and then in Protestant scholasticism.[63] The doctrine of creation ensured that nature was never ignored in theology or ethics, but it was considered subordinate to the doctrine of redemption. Most theologians assumed that humanity would be saved from nature, not in and with nature. The created order was too often viewed as the stage or background for the drama of redemption, not as part of that drama.[64] Only in recent decades have attempts been made to develop a more affirmative theology of nature and a biblically grounded and ecologically informed environmental ethics.

3. CONTEMPORARY THEOLOGY

Writing in 1956, when environmental awareness was rare, H. Richard Niebuhr expressed a strong sense of *loyalty to the community of life* of which we are a part. He stressed the social character of selfhood and our involvement in the natural world. Life is a gift mediated by a long history and a continuing web of interdependencies; each of us is constituted partly by our relationships to our fellow creatures. For Niebuhr, the injunction to love one's neighbor must include the whole community of present and future beings:

Who, finally is my *neighbor,* the companion whom I have been commanded to love as myself? . . . He is the near one and the far one, the one removed from me by distances in time and space . . . the unborn generations who will bear the consequences of our failures, future persons for whom we are administering the entrusted wealth of nature and other greater common gifts. He is man and he is angel and he is animal and inorganic being, all that participates in being.[65]

Since the 1970s, a number of theologians have presented variants of *the stewardship motif.* Paul Santmire, for example, maintains that both "compulsive manipulation of nature" and "romantic retreat to nature" ignore the demands of the biblical God whose purposes embrace both nature and history. Sant-mire develops the biblical conviction that God takes delight in the created realm and values it for its own sake. Humanity, in turn, has the role of care-taker, extending God's care, preserving the life of other creatures, and de-fending their rights beyond their usefulness to us. God has purposes in history and the ultimate goal is a kingdom of human and cosmic redemption. Sant-mire arrives at an ethic of responsibility and dedication to both distributive justice and care of the earth.[66] Critics maintain, however, that stewardship still assumes human superiority over nature and it can too easily revert to the atti-tude of domination.

A sacramental view of nature avoids this danger. Eastern Orthodoxy celebrates the goodness and beauty of creation and finds God's presence in it, holding that the infinite is manifest in the finite. Celtic Christianity, influenced by pre-Christian nature worship in Britain and Ireland, expresses a deep love of na-ture and a conviction that God is immanent in the natural world. The Anglican archbishop, John Habgood, says that as the water, bread and wine of the sacra-ments are means of communication with God, so all nature can be a vehicle of God's grace. These traditions offer a sense of the community of life and they seek to heal the divisions within it. They envision the redemption of the whole creation rather than the rescue of individual human beings from the world. When it is believed that the sacred is present in and under all nature, people are likely to treat it with greater respect.[67]

A more radical departure from traditional theology is Matthew Fox's *cre-ation-centered spirituality*. Fox believes in the goodness of creation and says we should talk about "original blessing" rather than original sin. He draws from some of the Christian mystics, such as Meister Eckhardt, who affirmed the im-manence of God in nature and in the human soul. He makes use of science in delineating the interdependence of all life, but he holds that we have to rely on intuition as well as logic in grasping the unity of all things. He says that we are cocreators with God in bringing to birth a world of justice and harmony with nature.[68] Fox insists that his views are a reformulation rather than a re-jection of Roman Catholic thought, but he has challenged many traditional ideas.

Here again the insights of *process thought* seem to me helpful. Process the-ologians reject the sharp separation of nature and history that has character-ized much of twentieth-century theology (especially in the existentialist, neo-orthodox, and evangelical movements). As indicated in my previous vol-ume, the God of process theology does not intervene coercively from outside but participates throughout cosmic history to bring order from chaos. God's purpose in continuing creation is the harmonious fulfillment of all beings. Process theology can represent, within a unified set of concepts, God's action as Creator and Redeemer.[69]

As I see it, traditional Christianity overemphasized divine *transcendence* and the gap between God and nature. At the opposite extreme, respect for nature has been encouraged by the idea of divine *immanence* (variously expressed in nature mysticism, pantheism, the experience of cosmic unity, or a romanticism that found spiritual qualities in nature itself). But immanence alone leaves us with an impersonal God or else identifies God with nature. According to process theology, God transcends nature but is also immanent in the temporal process. This implies that nature is not to be exploited, on the one hand, or worshiped, on the other, but is to be respected and appreciated, for it is the scene of God's continuing activity.

The process perspective helps us to expand our vision of the moral community, as both biocentric ethics and the animal rights movement do. But unlike these alternatives, it offers a principle for establishing priorities within the broader framework of a cosmic purpose. Because process thought acknowledges the unique freedom and creativity of humanity, it welcomes technology when it is consistent with respect for the community of life. It does not tell us to refrain from intervening in natural systems, but it does urge us to do so with sensitivity to the welfare of all creatures. A process theology that is indebted to both ecology and biblical religion seems to me a promising conceptual framework for environmental ethics today.

Combining the conclusions of this chapter with those of the previous one, I suggest that the biblical heritage offers significant resources for articulating an *environmental* message that can be combined with the historic Christian message on *social justice*. This would represent a distinctive synthesis, since justice has not been a strong theme in biocentric ethics or in many forms of the Eastern traditions—or, for that matter, in many of the publications of the environmental movement. A task force of the National Council of Churches has used the term *Eco-justice* to bring together environmental and social values. Other writers have argued that environmental preservation and justice are not really conflicting goals because there are common social structures and cultural attitudes that have been responsible for the exploitation of both nature and the poor. In the short run, environmental standards may sometimes lead to a loss of jobs, but new jobs can be created, including those in pollution control or nonpolluting industries. Care of the earth and justice in society are not mutually exclusive goals (see chapter 8, section III).

A major conference of the World Council of Churches (WCC) at MIT in 1979 adopted the threefold goal of *justice, participation, and sustainability*. A series of smaller WCC conferences during the seventies had explored a theology of nature, using the concept of stewardship, sacramental views of nature, and ideas from process theology. Some Third World participants had at first considered environmental preservation a luxury of affluent countries which diverted attention from justice, but they became increasingly aware of the environmental problems of their own nations (such as soil erosion, deforestation, and water pollution). The 1979 conference discussed finite resources and environmental degradation as well as equitable distribution and alternative technologies.[70] At the sixth WCC Assembly in 1983, the goal was reformulated as "Justice, Peace, and the Integrity of Creation," and the environmental commitment was reaffirmed.[71] The theme of the seventh WCC Assembly in

1991 was "Come Holy Spirit: Renew the Whole Creation." A fine statement by the American Catholic bishops in 1991 advocates both solidarity with the poor and respect for nature and explores the links between these goals.[72] But there is much work to be done before such ideas find a significant place in the life of local churches.

IV. CONCLUSIONS

I have suggested that science, philosophy, and religion each can make important contributions to the defense of environmental values, though none of them is adequate alone. The formulation of a strong Christian environmental ethics is a significant task in Western societies where Christianity is the predominant religious heritage. Within the Christian community, environmental ethics will be expressed in terms of biblical and theological ideas, such as creation and stewardship, reformulated in the light of ecology. But in the arena of public debate in pluralistic societies, these ethical convictions must be expressed in terms of environmental values that people of various philosophical and religious persuasions can support, even though there are distinctive Christian perspectives on each of them. Three interrelated values have been prominent throughout this chapter.

1. Resource Sustainability. Ecology makes us aware of the finite carrying capacity of the planet and the need to adopt a long time scale. Both utilitarianism and the concept of intergenerational justice support the claim that we have obligations to future generations. Biblical religion has an extended temporal and spatial frame, since God's purposes encompass the past, present, and future of the cosmos. I suggested that our use of renewable resources should not exceed the maximum sustainable yield and that nonrenewable resources should not be used up more rapidly than new technologies can compensate for their depletion. Population stabilization is also an essential component of sustainability. Conservation and restraint in consumption by affluent nations are necessary for the sake of posterity as well as for the sake of developing nations today, yet current economic and political institutions are oriented almost exclusively toward short-run costs and benefits.

2. Environmental Protection. Pollution of air, water, and land is a threat to human food, health, and aesthetic enjoyment. The control of pollution and the protection of the environment can be justified by their human benefits, and this is likely to be the most effective appeal in gaining public support. Environmental impacts are crucial in evaluating agricultural, industrial, and energy technologies. Even global climates and oceans are beginning to be affected. Many political battles seem to involve trade-offs of environmental protection against meaningful work, social justice, and economic development, but we cannot afford to lose sight of either set of goals. Both ethical commitment and technical and political expertise are required to devise strategies for combining them.

3. Respect for All Forms of Life. Ecology has made us more aware of the inter-connectedness of the web of life. Human actions have unintended and indirect repercussions; many ecosystems are fragile and vulnerable to large-scale human intervention. We have greatly reduced the diversity that contributes to the stability and balance of natural systems. Endangered species represent an irreplaceable genetic heritage that is of great scientific and ecological importance. But even beyond these human benefits, there are philosophical and religious grounds for loyalty to the wider community of life. I have maintained that we have duties to human and nonhuman individuals, rather than to holistic ecosystems as such, but we must always consider individuals within the larger wholes of which they are parts.

Respect for other creatures can be based on the extension of utilitarianism to include all sentient beings. It can also be based on a spiritual experience of the unity of all things or on the insights of Eastern religions. I have cited theologians who insist that the biblical concept of stewardship of the created order goes far beyond its usefulness to us. Process philosophy defends the intrinsic and instrumental value of all beings. It holds that every being is of value to itself, to other beings, and to God, but it maintains that not all kinds of beings are equally valuable since their experience differs radically in intensity and complexity.

We may now combine in one table (Fig. 2) the individual and social values of the previous chapter and the environmental values of this chapter. In subsequent chapters we will see that in decisions concerning particular technologies, differing subsets of these values are especially relevant. Chemical-intensive agriculture, nuclear power plants, and computer systems, for instance, have very different human and environmental impacts. We will find that *justice, participation, and sustainability* are the values most frequently at stake, but always in combination with other values.

INDIVIDUAL VALUES	SOCIAL VALUES	ENVIRONMENTAL VALUES
Food and Health	Social Justice	Resource Sustainability
Meaningful Work	Participatory Freedom	Environmental Protection
Personal Fulfillment	Economic Development	Respect for All Forms of Life

Fig. 2. Values Relevant to the Appraisal of Technology

Any move from ethics to *policy applications* must be made with great tentativeness. As the U.S. Catholic bishops indicated in their study of nuclear weapons, even those who agree on general principles may disagree on specific policy recommendations, which require difficult judgments of the probable consequences of alternative options.[73] Moreover, in the context of politics pragmatic considerations arise concerning effective strategies, the formation of coalitions, and realistic judgments as to prospects for public support. In some cases the initial task may be one of education or protest aimed at modifying public attitudes. In addition, each of us must make personal decisions about the life-styles we will adopt and about the kinds of technology we will seek at home or at work. This rather abstract table of values will only be helpful if it illuminates particular social and individual decisions about technology in the real world.

Part Two

CRITICAL TECHNOLOGIES

CHAPTER 4

Agriculture

In part 2 the values developed in part 1 are considered in relation to three specific kinds of technology. Chapter 4 deals with agriculture—perhaps the most crucial technology today because of the universal need for food and the continuing crises of hunger and environmental degradation. Policy choices for energy, discussed in chapter 5, involve issues of justice, participation, and sustainability, as well as economic development and the environment. Computers, the topic of chapter 6, present few problems of environmental impact or sustainability, but they raise significant questions about work and personal fulfillment, in addition to justice and participation. The diversity of these technologies—biological and physical, low-tech and high-tech, decentralized and centralized—should warn us against any simple generalizations, but we will trace some common patterns among them.

In this chapter we look first at the global problem of food and hunger. Then we take up some issues in Western agriculture: threats to the family farm and rural life; the influence of agribusiness on research priorities; and alternative practices conducive to sustainable agriculture. After that we examine agriculture in the Third World in the context of various development strategies. Finally some responses to the current global inequities in access to food are presented.

I. FOOD AND HUNGER

During the 1960s hopes were high for significant progress in the global fight against hunger. The Green Revolution (new high-yield strains of wheat, corn, and rice, together with increased fertilizer and irrigation) brought dramatic increases in production in many countries. World grain output in 1984 was 2.6 times that in 1950. Even after allowing for population increase, this represented a 40 percent rise in grain per capita.

But from 1984 to 1990, global grain production grew only 1 percent per year, while population grew at nearly 2 percent, so grain per capita fell almost 1 percent annually. Growth in rice yields virtually ceased during the period from 1985 to 1990. In Africa, grain output per person peaked in 1967, and by 1990 it had fallen by 28 percent because of population growth, land degradation, political conflict, and inadequate funding for agricultural development. U.S. grain production per capita peaked in 1981 and had fallen 12 percent by 1990, partly because the Conservation Reserve Program was introduced in 1985 to reduce crop surpluses and soil erosion. In general, the agricultural yields, which had been gaining on population growth in the sixties and seventies, were falling behind by the end of the eighties. During the nineties, world population is expected to grow by another billion—an increase of 20 percent, mostly occurring in the Third World.[1]

The great diversity among nations today should be noted at the outset. The First World includes the free market democracies of North America and Western Europe—most of which have had agricultural surpluses—plus Japan, Australia, and New Zealand. The term *Second World* was used for the communist nations of Eastern Europe and the Soviet Union; I shall say little about that area, partly because of the uncertainties of the transitions through which it is going. The First and Second Worlds together constitute the industrial nations, or "the North," with one-fourth of the world's population and 75 percent of the total gross national product. The remaining three-fourths is designated Third World, or less developed countries (LDCs), or "the South." It includes rapidly developing (or newly industrializing) nations with extensive manufacturing and world trade, such as Brazil, Mexico, Taiwan, and South Korea. China, with a fifth of humanity, has a low GNP, but social and economic benefits are more evenly distributed than in any other nation. Hunger and malnutrition are most severe in the forty poorest countries where more than a quarter of humanity lives.[2]

1. CAUSES OF HUNGER

Despite the gains from the Green Revolution, a billion people are malnourished today, half of them children for whom malnutrition frequently results in mental retardation and lowered resistance to disease.[3] Forty thousand infants die each day from malnutrition and the diseases associated with it—a loss of life equivalent to that from dropping a Hiroshima-sized bomb every two days, year in and year out.[4] Some hunger and starvation may be attributed to unexpected events such as droughts, floods, and civil wars. But most of it arises from five ongoing causes, each of which is discussed later in the chapter.

1. Environmental Degradation. Deforestation to obtain wood for cooking, heating, and building has led to severe soil erosion in many parts of Asia. Overplowing and the use of marginal land (such as steep slopes without terracing) have also produced extensive erosion. In Africa and Latin America, more intensive agriculture has replaced traditional practices of shifting cultivation and alternate-year cropping; soil fertility declines if fallow periods are reduced. The United States has lost half its topsoil since the Civil War by water and wind erosion. Overgrazing of African grasslands has hastened erosion in semiarid regions, and "desertification" is occurring in many parts of the world. Other farmlands have been harmed by salinization from continued irrigation without sufficient drainage. Overpumping has lowered the water level of underground aquifers.[5]

2. Population Growth. There are 95 million more people to feed each year, which is like adding to the globe every three years the entire population of the United States and Canada. As noted above, global population is growing faster than agricultural yields, so food per capita is falling. Ecological stress is caused by population pressures combined with destructive practices. Clearly, family planning programs must receive high priority among responses to world hunger.

3. Slower Growth in Yields. In many LDCs, more funds have been devoted to industrial than to agricultural development. High energy prices have affected fuel for tractors and irrigation pumps and raised the price of fertilizer derived from oil. Less research has been done to date on crops in tropical areas under Third World conditions than on temperate zone crops. Africa in particular has few trained scientists. Little effort has gone into improving traditional agricultural methods using labor-intensive rather than capital-intensive and energy-intensive techniques. Water is scarce in many regions and irrigation equipment is costly to install and operate, while rainfed agriculture is vulnerable to periods of drought and to climate changes. In affluent nations, additional fertilizer applications yield diminishing additional returns, and they result in pollution of surface and ground water.

4. Land Ownership. In many developing countries, a small percentage of the population owns most of the land. This is partly a legacy of colonialism under which one-crop plantations were established for export crops rather than for local food. Foreign aid policies often have strengthened ruling elites and provided little help for the peasant and small farmer. In Mexico, India, and Brazil, mechanization, tax policies, and access to credit have forced small farmers to sell out, further concentrating ownership in a few wealthy landowners. The Green Revolution in LDCs disproportionately benefited large owners who could afford the fertilizer and equipment it required. If one has neither land for subsistence farming nor a dependable source of income, one's family is unlikely to get the minimum dietary requirements. Malnutrition is largely a reflection of patterns of land ownership and rural poverty.[6]

5. Consumption by Affluent Nations. Since colonial days much of the land in LDCs has been devoted to such nonfood crops as tea, coffee, tobacco, cotton, jute, and rubber. In Central American countries where half the children are malnourished, more than half the agricultural land is used for export crops. In some parts of Colombia, carnations have replaced corn and wheat on prime agricultural land. Gambia grows peanuts, Ghana cocoa, and Senegal vegetables, mainly for Europe. The strawberry industry in Mexico is controlled by American companies who own or have contracts for some of the best land. Luxury foods for export to the well fed do bring badly needed foreign exchange but at a high cost in local health and nutrition. Prices paid for export crops are often low because a few companies dominate the market.[7]

6. Income Distribution. A more equitable distribution of existing food supplies would eliminate hunger and malnutrition. Total world food production is equal to more than twice the minimum calorie and protein requirements for every man, woman, and child.[8] The basic problem is the distribution of income. Food production is determined by those who can pay, so luxury foods for the affluent displace subsistence needs of the poor. Poverty, then, is the basic cause of hunger. Famine relief to victims of starvation is important as a short-run act of charity, but political and economic changes are required as long-run acts of justice. The economics of hunger is inseparable from the politics of international and domestic power. In the nineteenth century the colonies supplied raw materials for the Industrial Revolution in Western nations. The current international economic order hinders Third World development by perpetuating their role as exporters of commodities.

We will look at three kinds of response to the food crisis, all of which will be needed if humankind is to alleviate hunger and malnutrition. (1) *Increased production.* Improvements in crop yields and acreage can make more food available. Agricultural technologists are likely to stress this approach. (2) *Better distribution.* The global distribution of income and agricultural production capacity can be improved by development assistance and changes in international trade, land ownership, and tax and credit policies. This approach, along with increased production, is advocated by many development experts and Third World leaders. (3) *Slowing growth in consumption.* Both the growth of population in LDCs and the growth of consumption in affluent nations must be curbed. This approach is emphasized by many environmentalists.

I will suggest that problems of production, distribution, and consumption must all be confronted in any effective response to hunger and malnutrition. However, the main topic of this chapter is agricultural technology, so we will focus on production and return in the concluding section to issues of social justice in the global distribution and consumption of food.

2. ENVIRONMENTAL CONSTRAINTS

Among environmental constraints, *soil erosion* is the most serious threat to agricultural production in both the North and the South. Severe erosion by water and wind (loss of more than half of potential yield) has occurred on 7 percent of the land surface in North America, 16 percent in Asia, and 17 percent in Africa.[9] Iowa loses in a decade an inch of fertile soil that took at least a century to form.[10] In much of the Third World, population pressures and use of the best land by large landowners have driven poor farmers onto steeper slopes and more vulnerable marginal land. Overgrazing of grasslands has also led to soil erosion in semiarid regions. Massive deforestation is another major cause of soil loss. In many parts of Asia, people have cut trees and shrubs over vast areas, including steep slopes, to use as fuel for cooking because other fuels have been in such short supply.[11] Millions of acres of tropical rain forest are cleared each year in the Amazon basin and elsewhere to form cropland

and rangeland (mainly for export crops and cattle), though the soils are infertile and will usually support only two or three years of agriculture (see chapter 7). Eroded soil also results in siltation, which reduces the useful life of reservoirs and dams.

Soil conservation programs have been designed to slow this destruction of farmland. From 1986 to 1990 U.S. farmers put 35 million acres of highly erodible cropland (10 percent of total cropland) into the Conservation Reserve program; they were paid under ten-year contracts to plant this land in trees or grass. Minimum tillage and crop rotation also greatly reduce erosion rates. China has had a vast program for building terraces and planting trees and grass on slopes in the northern provinces. In Africa and India, village-based groups have been raising seedlings and planting them in deforested areas. Agroforestry is another promising approach. Food crops are planted in alleys between fruit trees or shrubs, whose roots and leaf-mulches retard erosion. Nitrogen-fixing trees such as leucaena have low nutrient requirements, and crop-tree-livestock systems have been developed to reduce dependence on chemical inputs. A wide adoption of such conservation practices is necessary if agriculture is to be sustainable.[12]

Environmental damage from *irrigation* is also widespread. The percentage of cropland that is irrigated varies from 10 in the United States to 48 in China, 77 in Pakistan, and 100 in Egypt. A common problem is salinization. Irrigated land with poor drainage becomes waterlogged, and as the water evaporates, salt is left behind—often 2 to 5 tons per acre per year. Half the cropland in Egypt is salinized enough to show diminishing yields.[13] In other parts of the world, irrigation water is obtained from underground aquifers, which have been drawn down at an alarming rate. Water tables under a quarter of U.S. irrigated land are falling at from 6 inches to 4 feet per year, resulting in higher costs for drilling wells and pumping water, and in some cases wells have run dry. The Ogallala aquifer underlying five plains states will take thousands of years to recharge. In some parts of Tamil Nadu in India, groundwater levels are falling 5 or 10 feet a year. A third of the wells in Beijing, China, have dried up.[14]

Surface water, too, is in short supply. In central Russia, most of the water in two major rivers entering the Aral Sea is used for irrigation; this sea, once the world's fourth largest freshwater body, has shrunk to a third of its former volume, and its salinity has tripled, ending its fishing industry. The water of the Jordan basin, shared by Israel, Jordan, and Syria, is insufficient for agricultural, industrial, and domestic needs. Half of the western rivers in the United States are overappropriated, and cities such as Tucson and Phoenix are buying up farmland to secure for the cities the water rights that go with the land. Many regions of the world will experience increasing competition for scarce water.[15]

Four responses to water shortages in agriculture have been proposed.[16] First, the *efficiency of water use* in existing systems can be greatly improved. Currently 60 percent of the world's irrigation water is wasted by canal seepage, poor maintanance, and runoff from overwatering. New drip systems such as those employed in Israel use much less water by putting it only where it is needed, near the roots of plants or trees. Second, realistic *water pricing* should be instituted. Users of water from U.S. government water projects pay only 7 percent of the actual costs; water subsidies average $792 per acre per year.[17] In developing countries, government revenues from irrigation average only 10 to 20 percent of the full cost. When water is free or heavily subsidized, there is little incentive to conserve; prices should more nearly reflect true cost. In the case of overpumping from aquifers it may be necessary to set quotas or sell water rights.

Third, *small-scale irrigation* at the village level is a low-cost alternative to huge government dams that are socially and environmentally disruptive. In some areas, earthworks and small reservoirs capture hillside runoff for agriculture. Water users can participate in planning, which is impossible in the top-down planning of large programs. The U.N. Fund for Agricultural Development has supported such small farmer-managed systems. Fourth, *research* on drought-resistant and salt-tolerant crops may allow additional land areas to be used productively. A new drought-resistant sorghum has two or three times the yield of traditional varieties in dry regions, and an oilseed plant has been developed with sufficient salt tolerance that it can be irrigated with seawater.

Water pollution from agriculture has also been widespread. In the United States and Western Europe, water pollution legislation in the seventies was mainly directed at municipal wastes and industrial effluents, but in the eighties it was recognized that pollution from farms is also a serious problem. Nitrates and phosphates from fertilizer have accumulated in lakes, creating algae blooms that kill off aquatic life. Nitrates, pesticides, and herbicides have all been found at dangerous levels in wells in many regions, and several major aquifers are threatened with contamination. Half of American citizens depend on aquifers for drinking water. These pollutants can be reduced somewhat by measures to control runoff from farmland, but the principal way to reduce them is to use these chemicals less intensively in the first place.

Pesticides present additional risks to food consumers and to agricultural workers. The National Research Council estimates that pesticide residues in food are responsible for 20,000 new cases of cancer in the United States each year.[18] Numerous studies have shown higher rates of lung and liver cancer and leukemia among farm workers than among the population at large, and rates among pesticide applicators are even higher. It is estimated that on Third World farms 10,000 people die each year and 400,000 suffer accutely from pesticide poisoning.[19] U.S. farm workers are neither included in occupational safety and health laws nor covered by workman's compensation laws, even though the rate of disabling injuries is the highest of any industry (even higher than construction and mining). The Environmental Protection Agency has only begun the job of testing pesticides. Those in use when regulatory laws were adopted were allowed to continue, and 80 percent of registered pesticides have not been tested by the EPA.[20]

Levels of pesticide use have risen because of their decreasing effectiveness. In 1940, losses from insect damage were 7 percent; today, despite a tenfold increase in pesticide use, losses are 13 percent. An insect can develop resistance to several chemically related pesticides or even to several classes of pesticides. Because of this history of selection, the span of effectiveness of new pesticides has been decreasing. Several superpests, including the Colorado potato beetle and the cotton bollworm, seem to be virtually uncontrollable by conventional pesticides. The excessive use of broad-spectrum pesticides has killed beneficial predator insects and soil organisms; some pests have come back with a vengeance when their natural enemies were eliminated.[21] In a later section we will look at Integrated Pest Management, which uses far lower levels of pesticides.

Monocultures of high-yield seeds lack the *genetic diversity* that provides some protection against diseases as well as pests. A million people died and a million and a half emigrated from Ireland in the potato famine of 1845 because the two varieties in use were vulnerable to blight. More recently a large fraction of the U.S. corn crop was wiped out by blight; fortunately, alternative varieties resistant to the blight were already on hand, but it usually takes five or ten years to develop new varieties with a specific resistance. Both disease pathogens and insects evolve more rapidly than plants because they go through many generations in one season. We are establishing the conditions for major crop failures.

A final consideration is *energy use.* The total U.S. food system uses 20 percent of the nation's energy, but most of this occurs in processing, packaging, and distributing farm products. Only one-sixth of food energy use is in agriculture, and of that, half is in the form of fertilizer and pesticide inputs.[22] Legumes and crop rotations are a far less energy-intensive source of nitrogen than fertilizer produced from oil or natural gas. Higher efficiency in farm machinery, irrigation, and crop drying would also save some energy. But in the food cycle the main opportunity for energy conservation is after products leave the farm, especially in the processing, cold storage, and distribution of meat.

Concerning each of these components of agriculture—soil, water, fertilizer, pesticides, seeds, and energy—the 1980s brought greater awareness of the environmental constraints within which food production must operate if it is to be sustainable. None of these constraints sets an absolute limit on achieving higher yields, but together they suggest that future efforts to increase productivity must give greater consideration to environmental consequences.

II. WESTERN AGRICULTURE

Let us look at the values affected by agricultural technology in Western nations since World War II, as farms have become larger, more capital-intensive, more specialized, and more mechanized. These changes are most pronounced in North America but are found throughout Europe. They are also evident in the Third World but under somewhat different conditions that will be examined in a later section. These changes have occurred in a few decades—a period very short compared to the history of agriculture going back several thousand years. Four characteristics of the industrialization of Western agriculture can be identified:

1. *Displacement of Labor by Machines and Chemicals.* Since the Industrial Revolution, many kinds of technology have led to the replacement of human labor by machines. Since World War II, petrochemicals have been widely used in place of natural products, further reducing labor requirements. Nowhere have the resulting impacts on patterns of human life been greater than in agriculture. More food is now produced by far fewer farm workers, resulting in much higher production per person and somewhat higher yields per acre (though less production per dollar invested, because farm equipment and inputs are expensive). In 1940, 23 percent of the U.S. population lived on farms; today less then 3 percent does.[23] (However, the food system as a whole, including farm inputs and food processing and distributing, occupies between a fourth and a third of the labor force.[24]) The agricultural work force in Europe has also shrunk, though not to the same extent. In France, for example, 7 percent of the population lives on farms, and in the former West Germany, 6 percent.[25]

Farms today are more *specialized* partly because of heavy investment in specialized equipment. Earlier in the century the typical farm was diversified, combining livestock, poultry, vegetables, and several crops. Today many farms have a single output of corn, wheat, or tomatoes or they produce only cattle in feedlots or chickens in poultry batteries. Government subsidy programs have contributed to specialization by supporting prices for a limited range of commodities.

2. *Government Subsidies.* In both the United States and Europe, farmers have had considerable political influence and public sympathy, and governments have adopted costly programs to protect farm income. Congressional subcommittees have approved subsidies for particular commodities cultivated in their constituencies or supported by farm organizations or agribusinesses. Parliaments and government ministries in Europe are similarly responsive to farm and agribusiness interests. Though enormous surpluses are already present, these subsidies tend to increase production, which lowers crop prices and leads to further subsidy costs.[26] In 1989 the European Community (EC) spent $32 billion for price supports, even though huge surpluses of butter, wine, and grain had to be sold elsewhere at a loss.[27] The Council of Agricultural Ministers sets common support prices, well above world prices, and imposes protective tariffs on imports from outside the EC. Overall, a variety of farm subsidies now cost Western governments $300 billion a year.[28]

3. *Larger Farms.* In both the United States and Europe the number of large farms has been rising. In the United States, many small farms (less than $40,000 in gross sales) remain, but they are usually operated part-time by people with some off-farm employment. The number of midsize farms has fallen dramati-

cally.[29] The change is partly the product of the particular kinds of farm machinery that have been developed. But it is also the product of government policies that have favored large farms. U.S. crop subsidies are based on acreage and production, and 1 percent of all farms receive 30 percent of all subsidies. Tax credits, depreciation allowances, and access to credits have also benefited large capital-intensive farms and encouraged corporate investment in farms as tax shelters.[30] A similar pattern is found in European countries, though with regional variations. Large farms prevail in northern Italy, but not in less-developed southern Italy.[31] In Britain, the Ministry of Agriculture has sought greater efficiency by encouraging larger farms, but smaller family farms are still common in Wales, Scotland, and northern England.[32]

4. Agribusiness Dominance. A few U.S. companies dominate the markets for each type of farm input, and they have been merging with each other (for example, chemical companies have bought up seed companies). Many kinds of farm output are also controlled by a handful of companies. Cargill has a corner on soybean products (animal and poultry feeds, flour, and cooking oil). Vertical integration occurs when food processing companies either own land or contract with landowners to buy their outputs and supply some of their inputs. One company may control seeds, farms, processing, retail sales, and fast-food restaurants. Ninety-five percent of broiler chickens are produced under contracts with firms integrating feed, processing, and marketing operations. Two companies slaughter 45 percent of beef cattle and either own feedlots or contract for them. Farmers lose both independence and income thereby.[33] Of the price of food purchased today, 68 percent goes to processing, distributing, and marketing; only 32 percent goes to the farm, and much of that is spent on inputs.[34]

The merging of farm input companies and the vertical integration of inputs and outputs by food processors is also occurring in Europe. In Britain, Imperial Chemical deals in seeds, fertilizer, pesticides, and animal feeds; Unilever integrates farm inputs and outputs. Contract managers tell farmers what seeds to use and when to plant, spray, and harvest. Food processors have a strong hold on poultry, beef, and vegetable production. Three companies have 90 percent of the British market in frozen foods. Agribusiness sales representatives and advisers promote new products and technologies on farms.[35] Almost all of these practices favor large farms with specialized equipment and a narrow range of outputs. We will look first at the human and then at the environmental impacts of these new patterns of Western agriculture.

1. FAMILY FARMS AND RURAL LIFE

These trends pose a severe threat to farms owned and operated by one family. Some commentators say this is *inevitable* and should be welcomed. They claim that technological progress has always been accompanied by the replacement of labor by machines; when workers are displaced, all we can do is help them adjust and find new jobs.[36] Such technological determinism must be questioned, however, since there are choices among technologies; the particular machines developed are the product of research priorities and of policy decisions. Again, the trend to larger farms is said to be a result of the inescapable economic laws of supply and demand. According to Social Darwinism, farms compete and the fittest ones survive; we have too many farmers, and the least efficient ones will be forced off the land. Some economists say we should treat farming like any other industry and seek only to maximize productivity.[37]

The agricultural economist Luther Tweeten defends *large farms* because of their greater efficiency (output per dollar) and productivity (output per acre). He says that the primary objective of agriculture should be an abundant supply of inexpensive food. Consumers far outnumber farmers and their interests should come first. The urban poor, in particular, benefit from cheap food. According to Tweeten, it is no violation of social justice to subsidize large farms, which account for a large fraction of farm output, because this helps to keep food prices down.[38]

By contrast, several kinds of argument have been advanced in defense of owner-operated small or midsize family farms:

1. A Cultural and Moral Tradition. Family farms are a tie to the past and they offer a distinctive cultural identity worth preserving. Two centuries ago, Thomas Jefferson saw the small independent farmer as the foundation of a democratic society. In some writings he suggested that farmers are morally superior: virtuous, honest, and hardworking. More often he claimed that dispersed ownership and self-reliant independence facilitate good citizenship and democratic politics. Similar agrarian values lie behind the Homestead Act of 1862, which gave settlers 160 acres of land in newly settled U.S. territory—favoring small landowners rather than a landed gentry or the plantation system that had developed with slave labor in the South. A 1902 law limited the use of cheap water from publicly subsidized dams and irrigation projects to 160 acres per person (though the purpose of the law has been circumvented by multiple ownership agreements).

Wendell Berry—poet, essayist, and farmer—maintains that family farming is *a way of life* and that it does embody and transmit distinctive moral values. Among these values are the dignity of meaningful work and respect for the world of nature. Neighborliness and cooperation in time of crisis (rather than individualism and competition) are typical of traditional farming communities. Berry is willing to use small-scale machines along with draft animals to help human labor but not to displace it. For him, preserving the family farm is not nostalgia for an idealized past but a defense of significant human values.[39] Sociological surveys do indeed show that, compared to urban people, rural respondents attach higher value to religion, ethical standards, work, and independence, though they are more conservative on racial and women's issues. But urban-rural cultural differences are eroding under the impact of television, off-farm jobs, and population mobility. Traditional rural values will be difficult to maintain, even if family farms are preserved.

2. Widely Dispersed Ownership. In most cultures, ownership of land has been one of the main sources of power in society. While other forms of wealth are more important in industrial societies, land ownership still represents an important source of economic and political power. People who own the farms on which they work, like owners of small businesses, own their tools of production and have a voice in the decisions that affect their lives. This social aspect of the agrarian tradition can be retained even if one does not ascribe any particular moral virtues to individual farmers. The owner-operator has greater independence and opportunity for participation in decisions than the person who works for an absentee landlord or corporate farm manager. Political influence is also spread among many persons rather than concentrated in the hands of a few individuals or corporations.

3. *Stronger Rural Communities.* In a classic sociological study in 1944, Goldschmidt compared two California towns of similar size, one surrounded by family farms, the other by corporate farms. The family farm community had greater stability and more and better services and institutions (schools, libraries, hospitals, newspapers, churches, businesses). Corporate farms made many of their purchases wholesale from distant suppliers rather than from local merchants, and they did not use local banks.[40] The same towns were reexamined in 1973 and the family farm town came out ahead in each of fifteen measures of quality of life.[41]

The Goldschmidt study has been criticized because the two areas compared had different histories and water sources, which might account for some of the observed differences.[42] But other studies support his conclusions. Community life and services to residents in 130 towns in the San Joaquin valley were compared, and family farm communities came out ahead. Tenant farmers are more transient than owner-operators and make fewer community commitments.[43] While the effects of farm ownership are difficult to distinguish from the effects of size, it appears that owner-operated farms do contribute to the vitality of rural communities. Dispersal of small industries in rural areas also strengthens rural life.

4. *Efficiency and Size.* Small farms are inefficient because they have idle equipment and underemployed people for much of the year. They are viable only when combined with off-farm employment. But according to the U.S. Department of Agriculture, moderate-size farms (sales from $41,000 to $76,000 annually) capture most of the economies of scale. Little is gained in efficiency at larger scale, and the cost of management, supervision, and coordination increases with size.[44] (Comparisons are difficult, however, because management costs can only be estimated for the smallholder, whose functions as owner, manager, and worker are interwoven). Tweeten says that a farm with $100,000 per year in sales can achieve optimal economic efficiency.[45] A study of farms in Europe concludes that the maximum efficiency is achieved on three-man farms.[46]

Why have many farms grown beyond the optimum size? Partly, of course, because the total profits to the individual owner are greater, even if there is no greater return per dollar invested. But the main reasons are government subsidy and tax policies, easier access to credit, and the desire of food processors and agribusinesses to influence or control additional stages of food production. Moreover, the calculation of economic efficiency neglects long-term social and environmental costs, the externalities borne by other parties or future generations.[47]

5. *Land Stewardship.* One would expect family farmers who plan to pass on their farms to their children to take better care of their land than corporations that have invested for the sake of short-term profits. They have a personal relation to the land and not simply an economic relation to it. They are more likely to diversify their operations, raising both livestock and crops, which allows the use of manure as fertilizer and the rotation of differing types of crops and forage. Family farmers are also said to treat their animals better than large commercial feedlot operators.

Actual data on correlations between *land ownership* and *conservation practices* have been ambiguous. One survey found that soil-conserving minimum tillage was practiced on only 44 percent of the land of owner-operators, compared to 51 percent of that of absentee landlords.[48] On the other hand, another study found soil erosion 40 percent higher on rented than on owner-operated land.[49] Major long-term conservation investments such as terracing are not made by people who rent land (especially with short leases). I believe that education for conservation is more likely to be effective with owner-operators who are personally closer to the land. But public policies providing regulations or incentives for conservation are needed because they are effective for both absentee owners and family farmers, and for both large and small landholders.

In sum, small and midsize family farms do contribute to widely dispersed ownership and stronger rural communities, offering greater opportunities for what I have called participation and personal fulfillment. But social justice requires that economic efficiency and agricultural productivity be taken seriously for the sake of low food prices for consumers, especially the urban poor. Efficiency is greatest for midsize farms, but in industrial nations with agricultural surpluses some sacrifice of productivity could occur without major increases in food prices. This suggests the need to correct current biases toward large farms by giving more help to midsize and owner-operated farms. Some states have progressive taxes on farmland to discourage land concentration, or laws restricting corporate investment in farms. Nebraska and Minnesota ban nonfarm corporations from owning farmland. In Sweden, farmland can be purchased only by people who are planning to farm it themselves, while France encourages small and midsize farm ownership.[50] We will see later that land ownership and land reform are crucial issues in the Third World.

2. AGRIBUSINESS AND RESEARCH PRIORITIES

In opposition to technological determinism, I have asserted that particular technologies and their social consequences are not inevitable but are the product of choices by individuals and organizations. Such choices are present in the selection of agricultural research priorities and technological design criteria, and in governmental, academic, and corporate institutional policies.

The *public agricultural research system* in the United States consists of the Department of Agriculture and the state land-grant universities (LGUs), agricultural experimental stations, and extension services. The LGUs were established by Congress in 1887 to further the public interest in agriculture and rural life. They have been supported by federal and state funds and more recently by grants from agrochemical and food processing corporations. (There is also extensive research done in laboratories run by corporations, but that is not our concern here.) The central goal of LGU research has been *to increase productivity*. With labor scarce, land plentiful, and energy cheap, higher yields were sought by improved seeds and increased use of machines and chemicals. Much of this research focused on higher yields of five major crops, even after there were surpluses in all of them.[51] Similarly, in Britain the Ministry of Agriculture and the agricultural research institutions have pursued efficiency and productivity as their almost exclusive goal.[52]

In a survey of agricultural scientists, Busch and Lacy asked what criteria they used in *the selection of research topics*. The respondents often mentioned the advancement of knowledge and growth in agricultural productivity as general goals. But they said that the availability of funding and the personal rewards of publication and promotion were the largest influences in actually selecting topics.[53] Disciplinary training and professional journals tend to be narrowly specialized, so the reward system does not encourage interdisciplinary research. Most scientists assume that science is value free; but it is clear that funding sources, disciplinary assumptions, and institutional goals all influence the research agenda.[54]

The philosopher Paul Thompson maintains that the dominant goal of productivity in the agricultural research establishment reflects an implicit *utilitarianism*. In a utilitarian framework, the benefits to consumers from higher yields and lower food prices outweigh the losses to some farmers who are forced off the farm. Thompson reminds us that cost-benefit analysis usually ignores social and environmental externalities, though some of these can be included in more sophisticated analyses. In addition, he says, utilitarianism neglects distributional justice, since it considers only the net balance of benefits over costs. The illness or death of farm workers harmed by pesticides might be outweighed by the benefits to society, unless we can say that people have inherent rights which are not subject to utilitarian trade-offs. Some agricultural technologies increase yields but at the same time harm community life and reinforce unjust power structures and economic inequalities. Finally, the welfare of future generations and of other creatures are not included in most utilitarian judgments.[55]

Criticisms of prevailing research priorities have come from within the land-grant universities, especially from ecologists and rural sociologists who are aware of environmental and social interactions.[56] Some LGU agronomists and economists have called for a broadening of prevalent goals to include a wider range of values and clients. Thus Vernon Ruttan wants agricultural scientists to take more responsibility for the applications of their findings and to consider impacts on health, environment, and rural life, though he says that the main policy decisions affecting these impacts must be made by society through political institutions. He urges LGUs to develop greater expertise in assessing the diverse consequences of technological developments, without giving up their commitment to higher productivity.[57]

Other critics, mainly from outside the LGUs, have charged that agricultural research has been *too much influenced by agribusiness*. It is claimed that grants and political pressure from chemical, seed, and equipment companies have determined the direction of research, including that carried out with public funds. Land-grant universities, it is said, have primarily served the interests of corporate America.[58] In some cases, scientists were unwilling to testify on the health hazards of pesticides for fear of alienating funding sources.[59] Pesticide companies are more likely to support research on pesticide-resistant plants than on pest-resistant plants, and they have opposed Integrated Pest Management that lowers pesticide use. In general, agribusiness has supported research on chemical-intensive agriculture but not on low-input sustainable agriculture.[60]

A final criticism is that publicly funded research on *farm mechanization* has benefited large farms, equipment manufacturers, and food processors but has harmed smallholders and farm workers.[61] In one controversial case, the University of California at Davis developed huge mechanical tomato harvesters and also harder tomatoes that could be picked green and ripened later with ethylene gas.[62] A suit was filed against the university on behalf of displaced workers and farmers, asking that some of the royalties from patents on these machines be used to help them. In 1987 the court ruled in their favor and directed that in the future a wider range of interests must have a voice in reviewing proposed research at public institutions.[63] But the appellate court partially reversed this ruling, arguing that the LGUs do not have a specific mandate to serve small farms.

Again, the *promotion of tomato harvesters* by Ohio State University and the state extension service helped to expand the power of food processors. Through contracts, a few companies in Ohio controlled all stages of tomato production: seeds, farming, processing, and marketing. In 1979, the Campbell soup company required all tomato contractors to use mechanical harvesters. The farm workers' union struck against the growers and then called a boycott against Campbell and Libby, the main tomato buyers. Extension service publications argued that the harvesters reduced labor problems and were economical. The machines waste part of the crop because they can go through a field only once (whereas the plants grow at varying rates), so yields are lower, but owner profits are higher. Thousands of workers lost their jobs, and smallholders lost their land, because the machines are economical only on large holdings.[64]

Defenders of tomato harvesters say that all technological change entails *job displacement,* but the losers should be helped or given unemployment compensation by society as a whole, not by the universities or the food processors introducing the change. After all, consumers are the main beneficiaries of cheaper food.[65] (In this case, the price of tomatoes did not fall, but this can be partly explained by the increased demand from the new popularity of pizza and pasta. The texture and taste of the tomatoes are inferior to most hand-picked varieties.) Moreover, the machines eliminated backbreaking work in the fields, which had been done mostly by migrant labor (including illegal Mexicans in California). As in the case of industrial automation (see chapter 6), mechanization is attractive to management because it reduces the uncertainties of dependence on labor (especially when it is unionized). Defenders of the land-grant universities also say that cooperation between universities and industries is justified because it speeds the practical applications of science and technology.[66]

I would argue that in the interest of social justice, productivity and the availability of food at low prices should indeed be central goals of agricultural research, provided the productivity is sustainable. I also agree that technological changes that result in job displacement can sometimes be justified and that it is mainly the responsibility of society to help the losers. But many forms of large-scale agricultural technology, including the tomato harvester, do little if anything to increase yields or lower food prices; they primarily benefit agribusiness. They are laborsaving and capital-intensive, which is unnecessarily damaging to workers and smallholders. (Such technologies also do not benefit the Third World, where labor is plentiful and capital scarce). Even though impacts on society and the environment should be addressed primarily through public policies, they should be considered in agricultural research institutions in the interdisciplinary assessment of new technologies before they are developed and promoted. Priorities for publicly funded research are set mainly through political processes, and direct participation of a wider range of stakeholders at both national and regional levels should be sought.

In the new field of *biotechnology,* questions concerning the relation of universities to industrial corporations have arisen with particular intensity because the financial stakes are so high. Does the secrecy required to protect potential commercial patents hinder the free exchange of information that has characterized the academic community? Do contracts with industry bias research priorities or the testimony of scientists on public issues? Who will benefit most from the research? These problems in agricultural biotechnology are discussed in chapter 7 among the ethical issues in genetic engineering.

3. SUSTAINABLE AGRICULTURE

In response to the problems of soil erosion and chemical-intensive agriculture, a movement to use conservation practices and biological resources has been gaining strength in the United States among farmers, scientists, and the public. Alternatives cover the range from *"organic farming,"* which uses no synthetic chemicals, to *"low-input"* methods that combine reduced chemical use with a variety of conservation techniques. While alternative methods are slightly more labor-intensive than conventional ones, most of these farmers use modern mechanical equipment and certified seeds, have good management skills, and own the farms on which they work.[67] A similar movement has emerged in Europe.[68]

On virtually all these farms, *crop rotation* is practiced to increase soil fertility and reduce soil erosion. Rotation also helps to control pests by interrupting their annual cycles. In successive years a field might be planted in wheat, soybeans (which fix their own nitrogen), and oats (which use fewer nutrients than wheat), followed by a couple of years of alfalfa plowed back into the land as a "green manure." Legumes, crop residues, livestock manure, and "green manure" are used in place of some or all synthetic fertilizer. When possible, livestock and crops are raised on the same farm; the animals provide fertilizer, and some of the land provides forage or pasture.[69]

Soil erosion is reduced by cover crops, shallow tillage, contour plowing, and other conservation measures.[70] By one estimate, soil erosion on alternative farms is one-third of that on conventional ones.[71] Weeds are controlled by crop rotations, tillage, intercropping, and in some cases by limited use of herbicides. These practices increase the organic matter, microbes, and earthworms in the soil, improving its aeration, water retention, and slow release of nutrients.

A variety of *pest management strategies* have been used in addition to crop rotation. An effective technique is the introduction of a predator that is the natural enemy of the target pest. In a number of cases, benign insects that prey on a specific destructive insect have been successfully imported, bred, and released to control it. Mealybugs were destroying 30 percent of Africa's cassava crop until a type of wasp that preys on them was imported from Paraguay, and yields rebounded.[72] In other cases, biodegradable botanical or microbial insecticides have been used. Integrated Pest Management (IPM) uses natural predators but allows limited use of pesticides when necessary. It requires careful study of pest life cycles, data on the pest levels at which economically significant crop damage occurs, scouting for pest populations during the growing season, and spraying only when damaging levels are approached. One review of forty-two IPM studies says that most of them reported higher crop yields than with conventional methods, and all had lower costs.[73] IPM is currently used on 8 percent of U.S. farmland. To date it has been more effective on cotton, sorghum, and peanuts than on corn and soybeans, but research is likely to expand its applicability. Sweden reduced pesticide use by 50 percent between 1985 and 1990, mainly by taxing pesticides, monitoring pest conditions, and improving application equipment.[74]

Most studies show slightly lower *crop yields* on alternative farms than on conventional ones, but *net income* is comparable because the cost of farm inputs is much lower. Lockeretz surveyed 363 fully organic farms and studied 14 in detail, matched against 14 conventional farms. The organic yields were typically 10 percent lower, but this was offset by a comparable reduction in production costs, so they were equally profitable, and the organic farms actually had more net income during dry years. Soil erosion was markedly lower, and energy consumption (including the energy used in producing farm inputs) was 60 percent less.[75] Another summary of studies of low-input farms concludes that "the use of legumes in rotation and low use of chemicals offer encouraging profitability prospects."[76] Studies in Britain and the Netherlands have reached similar conclusions.[77]

In its 1989 report, *Alternative Agriculture,* a committee of the National Academy of Sciences examined alternative farming practices and gave case studies of twelve farms (mostly low-input, though a few were organic). They concluded that reduced use of agrochemicals and greater reliance on biological processes lowers production costs and environmental damage without significantly decreasing crop yields per acre. "Farmers successfully adopting these systems generally derive significant sustained economic and environmental benefits. Wider adoption of proven alternative systems would result in even greater economic benefits to farmers and environmental gains for the nation."[78] In its policy recommendations the report advocates government programs to support alternative agriculture, and economic incentives that will reflect the social and environmental benefits to the nation. The report also calls for substantial public funding of alternative research because private industry has little incentive to produce seeds or equipment designed for reduced input use.

The response of *U.S. government agencies* to alternative agriculture has been slow. A 1980 report commissioned by the Department of Agriculture concluded that low-input farming is productive, profitable, and environmentally sound, and it recommended expanded research on crop varieties suited to such practices.[79] Under the Reagan administration, however, the department did nothing to implement these recommendations. In 1985, Congress—under pressure to reduce surpluses that were eroding crop prices, and farm practices that were eroding soil—established the Conservation Reserve Program that paid farmers to convert their most erodible cropland to grass or woodland. Congress also established a Low-Input Sustainable Agriculture program that by 1990 was providing $4.5 million annually for alternative research—a good beginning, but still a tiny fraction of the agricultural research budget.[80]

The 1990 Farm Bill for the first time provided *incentives for sustainable agriculture.* Rotation of crops no longer entails the loss of commodity support payments if it is part of a plan to reduce soil erosion and chemical use. In areas where surface and groundwater are threatened by farm runoff, farmers can also receive modest payments for conservation tillage and pest management costs.[81] The state of Iowa has introduced a tax on fertilizers and pesticides to discourage the contamination of wells and lakes and to fund research on alternatives. Norway has a similar tax.[82] But much more needs to be done, especially multidisciplinary research on integrated farm systems as a whole.[83] Reduction in current commodity subsidies in Western nations could release substantial funds for research and extension services in support of environmentally sound practices. Such public investment in the future is necessary because market forces do not take into account the long-term deterioration of land and water.

The previous chapter indicated some of the ethical grounds for supporting *long-term sustainability*. The biblical tradition in particular has an extended time horizon because it holds that God's purposes span the generations, and we therefore have duties toward future generations. Moreover, the Bible maintains that the land is a gift from God that we should cherish and treat with care.[84] Conditions today are, of course, radically different from those in ancient times, and agricultural technology opens up vast new opportunities for food production, but we are still dependent on the land. The injunction to be good stewards of the land is still valid because it derives from an underlying understanding of the place of humanity in nature and history. Many churches in farming areas today observe a land stewardship Sunday in the spring as well as an autumn harvest thanksgiving service, and these themes are explored in their educational materials for both children and adults. A religious tradition cannot tell us what practices will contribute to sustainability, or how it is to be balanced against justice in responding to immediate human needs, but it can help to motivate concern for such issues.

The welfare of farm animals has been neglected historically in the West, but it has recently been more widely discussed in religious and secular writing and in the media. One practice that has been criticized is the confinement of veal calves to very narrow stalls where they are fed an all-milk diet to make them weak and anemic so their meat will be tender and pale. The tethering of sows and overcrowding of laying hens in cages has also been attacked and has been banned in several European nations.[85] Cattle raised on rangeland or on integrated farms are probably better off than those raised in overcrowded feedlots. But it is difficult to judge how much an animal's welfare is enhanced by additional measures beyond those removing suffering and allowing some freedom of motion. The animal welfare movement has campaigned for more humane treatment and for painless slaughtering; the goal is to reduce suffering.[86] Some defenders of animal rights have objected to any killing of animals and have advocated strictly vegetarian diets that can fulfill human nutritional needs.[87]

In the previous chapter I urged concern for *animal welfare* but argued that the fulfillment of *basic human needs* should have priority if one has to choose.[88] I favor a predominantly vegetarian diet in the interest of social justice. It takes seven pounds of grain to produce a pound of beef.[89] Injustice is compounded when food crops in the Third World are replaced by feed crops to produce beef for export to affluent nations. But within developing countries, domestic use of meat from grass-fed animals grazing on land not suitable for agriculture could contribute significantly to human nutrition. Lower meat consumption in Western nations would produce healthier diets and more efficient and equitable use of land. It would also reduce animal suffering, but that objective can be more directly addressed by educational and legal measures. Farm animals should be included in the humane treatment laws now applicable only to domestic pets and laboratory animals.

III. AGRICULTURE IN THE THIRD WORLD

The same values are at stake in agricultural policy in the Third World but under conditions where food shortages are more common than food surpluses. We look first at the relation between the Green Revolution and continuing poverty and malnutrition. Then we examine agriculture in the context of alternative development strategies.

1. THE GREEN REVOLUTION AND MALNUTRITION

During the 1960s, high-yield seeds and more intensive use of fertilizer and irrigation resulted in impressive *increases in production* in several developing countries. India doubled its wheat crop in six years (1966–1971). In the seventies, several drought years, higher oil prices, and continuing population growth led to food shortages, but by the early eighties India was self-sufficient and was even a small exporter of grain. Wheat in Mexico and Pakistan and rice in the Philippines saw similar growth in production. Clearly the Green Revolution saved the lives of millions of people who would otherwise have died of malnutrition and its associated diseases. But in the late seventies some of the leading exponents of the Green Revolution were saying that technical solutions are inadequate without greater attention to their impacts on small farmers and the rural poor.[90] By the eighties, the environmental consequences of the new methods were more evident, as I indicated earlier.

Most analysts acknowledge that the Green Revolution has had harmful effects on *smallholders* and *farm laborers*. The early adopters were wealthier landowners who could afford the high inputs, and thus concentration of land ownership increased. Later adopters did include smallholders, but by then food prices had fallen so they benefited less.[91] Recent high-yield varieties are less dependent on heavy inputs of chemicals and water and are thus more attractive to smallholders. Yields per acre are higher on small farms, though yields per unit of labor are lower. But mechanization and better access to credit have tended to favor large farms. The principal impact of mechanization has been to reduce the need for labor in planting, cultivating, threshing, and milling. Herbicides and cultivation have largely replaced hand weeding.

Why has the Green Revolution, despite its accomplishments, had so little impact on the *poverty* that causes malnutrition? Lipton and Longhurst suggest that the new techniques have benefited the poor as consumers by keeping food prices down but that this has been offset by unemployment and depressed wages caused by the labor surplus. India, for example, is a net food exporter, but the poor still cannot afford to buy the food they need. The combination of population growth and farm mechanization led to unemployment and low wages in the rural areas where the vast majority of the poor live. Though employers have benefited from low wages, the poor rely on wages for income, and they lack the purchasing power to buy back what they have produced. These authors conclude that the Green Revolution as a technical solution will continue to be a mixed blessing unless its effects on the poor are directly addressed. "In the absence of redistributive social change or employment generation, the purchasing power of the poor is unchanged."[92] Increased productivity is no substitute for policies that would correct the urban biases of the past and help the rural smallholders and laborers who depend on farm income.

Another factor that has led to malnutrition in the Third World has been the substitution of *export crops* in place of *food crops* for local consumption. We noted earlier that in the colonial era cotton, rubber, cocoa, coffee, and tea were grown on large plantations in South Asia and Africa. Subsequently, crops for export were promoted by governments and multinational corporations, especially in Latin America. Lappé and Collins claim that cash crops for export perpetuate poverty and malnutrition in the South and domination by the North. Often a few transnational corporations (TNCs) control the export market (for coffee, bananas, or sugar, for instance) and a few landowners own the best land. U.S. policies of economic and military aid in Central America have largely been designed to protect U.S. investments and to prevent political change. The U.S. government has repeatedly sided with landowners and repressive governments who opposed peasant land reform movements and wage demands. Lappé and Collins advocate land reform and the production of staple foods for local populations (food self-sufficiency).[93]

Proponents of *export crops* claim that they can contribute to *economic growth* and a favorable balance of trade.[94] But much of the profit from export goes to the TNCs; terms of trade have usually been unfavorable to the South, and prices have been unstable. Many LDCs use food exports to pay the interest on their enormous debts and then have to buy food imports to meet domestic needs. The claim that various regions have distinctive "natural advantages" for the production of specialized crops neglects the role of colonial powers and TNCs in promoting their own interests. In a careful and balanced appraisal, Maxwell and Fernando reject both the "self-sufficiency" and "economic growth" positions. They maintain that governments and international agencies should promote small, diversified, efficient farms growing *both* staple crops and cash crops, allowing some exports without excessive dependence on the North.[95]

The displacement of *food crops* by *feed crops* for animals has also contributed to malnutrition. In Mexico, land formerly used for maize and beans, the staples of the national diet, are now used for export vegetables and for wheat and sorghum for cattle feed. Small farmers are displaced by vertically integrated, export-oriented enterprises. In other parts of Central and South America, cropland has been turned into pasture, which has put campesinos out of work because cattle ranching is not labor-intensive. Third World governments can obtain international loans for expanding exports of crops, feed, or beef more readily than for production for local consumption.[96]

Some aspects of the Green Revolution, such as high-yield seeds, can be combined with practices that are *more labor-intensive* and use *smaller inputs* of chemicals and fuel. At research institutes in the Third World, experts in scientific agronomy have tried to combine traditional and modern methods. In India, regional institutes have developed efficient small-scale equipment such as two-wheel power tillers—intermediate between hand or bulluck plowing and four-wheel tractors.[97] The Institute of Tropical Agriculture in Nigeria has worked out modifications of the traditional system of shifting cultivation. Mixed cropping, legumes, mulches, deep-rooted fallow vegetation, minimum tillage, and intermediate-scale equipment have been used. Research is also beginning in Africa on improving the poor people's crops: cassava, yams, sorghum, and millet.[98]

Rural development policies also can help to strengthen *small farms*. These include land distribution, communal land tenure, producer cooperatives, farmers' associations, wider access to credit, subsidies to small rather than large farms, and better agricultural extension services. There should be provisions for the participation of the farmers themselves in development planning. Small farms contribute to employment and to social equality, avoiding the polarization of large landowner and landless laborer.[99] Small and midsize farms usually have higher productivity measured in yield per acre or yield per dollar, but not yield per worker. In Taiwan, which instituted policies to encourage small farms, output per acre on farms of less than 1.25 acres was nearly twice that on farms of more than 5 acres.[100] Smaller farms also led to more equitable income distribution there. Brazil has higher average per capita income than Taiwan, but it has three times the percentage of its people in absolute poverty because it is polarized between a prosperous elite and the impoverished majority.[101]

The Green Revolution, then, can still make an important contribution to *increased food production*. Half the hungry poor live in areas that it has barely touched—in Bangladesh, West Africa, and Zambia, for example.[102] Norman Borlaug, who won the Nobel Peace Prize for his work on high-yield seeds, writes, "I am cautiously hopeful that world production can be doubled over the next 40–60 years, provided that governments give high enough priority and continuing support to the agricultural sector in their development programs." He recognizes the need for "greater equity in distribution" and urges research in semiarid and tropical areas that are "subject to severe biological constraints." He is also "apprehensive about population growth."[103] I am thankful for the continuing work of Borlaug and others on increasing production, but I believe that it must be accompanied by major policy changes affecting consumption and distribution.

First and foremost is control of population growth through effective *family planning programs*. At the current rate (1.8 percent annual growth), the world population would double in 39 years, creating a decline in per capita production, even with Borlaug's "cautiously hopeful" projection. This topic is so crucial to the global future that I will deal with it separately in chapter 7.

Second, *Third World debts* must be rescheduled. Currently, debt repayments by LDCs are higher than the new loans they receive, resulting in a net outflow of $50 billion per year, paid for largely by the export of crops, timber, and other natural resources. Germany, Canada, the United States, and Britain have forgiven $5 billion in debts owed by African countries, but enormous debts still remain.[104] In the long run all nations will benefit by more rapid development in LDCs, which is now hindered by their staggering debt load. Issues of aid and trade are taken up later in this chapter.

Third, *more equitable income distribution* must be sought by policies helpful to small landowners and wage laborers. In addition to tax, credit, and subsidy policies, land reform measures have been peacefully instituted in several countries, such as Egypt, Taiwan, South Korea, and Israel and by revolutionary governments in Cuba, Nicaragua, and elsewhere. The challenge to Third World countries today is to move by democratic political processes toward more equitable distribution, without the violence employed by movements of both the right and the left. The industrial and urban bias of past development strategies must also be replaced by policies for agricultural and rural development. Agricultural policy is inescapably part of the wider context of development policy.

2. SUSTAINABLE DEVELOPMENT

Three differing strategies for development have been advanced in recent years.

1. Industrialization and Economic Growth

During the 1960s the prevailing development strategy was economic growth. All countries, it was held, would go through stages of industrialization similar to those followed in Europe and the United States, and the poor countries would catch up with the rich ones.[105] Agriculture was considered less important than capital-intensive industry on the Western model, and huge showcase projects such as steel mills were built. Distributional measures should be avoided, it was said, because they might hinder capital accumulation and economic growth. The benefits of an expanding GNP would automatically "trickle down" to the poor. Impressive growth was indeed achieved by LDCs, averaging 5 percent a year during the 1960s, higher than the decade-long rate of any country in history. Because of population growth, GNP per capita increased at only half this rate.

However, economic growth was accompanied by *widening gaps,* both between and within nations. Far from catching up, the poor nations found themselves further behind rich nations. By the midseventies the ratio of GNP per capita in the United States to that in India, for instance, had increased to 50 to 1. The disparities within LDCs also widened. A privileged minority benefited greatly from economic growth, but there was little "trickle down" to those at the bottom.[106] As capital-intensive techniques replaced labor-intensive ones in industry as well as agriculture, unemployment increased. Policies of credit, taxation, education, and technology transfer combined to distribute the benefits of growth very unevenly. Urban populations had more political power than the larger rural populations, and this affected the allocation of funds for education, health care, and capital investment, while price controls and policies kept food prices low at the expense of farm income.[107]

Economic growth did not help *the poorest groups* in LDCs. A study commissioned by the World Bank concluded in 1974 that though per capita income increased significantly, there was little or no benefit to the bottom third of the population.[108] Wealth was concentrated in the hands of a small group. A study of 48 social, political, and economic indicators in 74 countries at that time concluded that "the position of the poorest 60 percent typically worsens, both relatively and absolutely."[109] Other authors suggested that from 20 to 40 percent of the population in LDCs had suffered an actual decline in living standards.[110]

2. Development for Basic Human Needs

In response to these trends, an alternative strategy was proposed in the 1970s. In this view, poverty and unemployment should be directly attacked. The goal was to increase the productivity and income of the poor to fulfill their needs for food, housing, and health. Since the majority of Third World populations is rural, policies that help the landless peasant and the small farmer were advocated. Social indicators (such as life expectancy, nutritional level, and literacy) were said to be better criteria of development than a nation's economic growth. Foreign investment and the domestic market had favored those with greatest purchasing power, putting luxury goods for the few ahead of necessities for the many. Instead, a more equitable distribution of the benefits of growth should be deliberately sought. Robert McNamara, president of the World Bank, advocated loans to support development plans targeted on the poorest 40 percent of the population.[111]

Such *equitable and need-oriented growth* was advocated by many development experts and organizations. Mahbub ul Haq, a World Bank economist, said that the distribution and content of the GNP are as important as its size. To him, the goal of development is not economic growth but the elimination of malnutrition, disease, and illiteracy. Production should be directed to essential commodities (food, housing, clothing) and public services (health, education, transportation). The productivity and income of the landless laborer can be improved only by greater access to land, credit, employment, and education.[112] The U.N.'s International Labor Office also endorsed the basic needs approach. It held that this required a shift from urban to rural investment, from nonessential to essential goods, and from capital-intensive to labor-intensive production. It urged land reform, institutional decentralization, mass participation in planning, and locally controlled appropriate technology.[113] But these policies were seldom implemented in practice, and with mounting debt repayments and dwindling grants and loans from the North, development slowed down in most LDCs.

Some Third World authors criticized international trade and aid and called for *national self-reliance*. They claimed that integration into a world economy dominated by the North perpetuates dependency. National self-determination and liberation from foreign domination should be sought by relying on local or regional resources and abilities. Transfer of "packages" of technology and services by transnational corporations locks the South into dependence on Northern expertise and equipment.[114] Latin American social critics called for redistribution of land and wealth, wider political empowerment, and production for the needs of the masses, not for the export market. Aid and trade within international structures of unequal power, they said, only delay the political and social reforms that would make justice and freedom possible.[115] But most Third World governments resisted such reforms and continued to seek industrialization and trade, even as they gave some support to programs of agricultural and rural development intended to meet the needs of the poor.

3. Sustainable Development

Neither the "economic growth" strategy nor the "basic human needs" strategy gave any consideration to *environmental constraints*. The U.N. Environmental Program was initiated in 1972 to promote cooperation on global environmental problems, but it was a small program independent of other U.N. agencies. By the 1980s, international development agencies and national governments were more aware of the environmental impacts arising from both industry and agriculture in LDCs. In the late eighties, the World Bank required environmental impact assessments for all projects and added sixty-five environmentalists to its staff, but it still gave priority to economic growth and export-oriented proposals.[116]

But in 1987 a commission established by the United Nations clearly recognized the links between *development* and *environment*. Its report, *Our Common Future* (known as the Brundtland Report), shows how poverty creates environmental degradation, which in turn harms productivity and perpetuates poverty. It documents the current destruction of biological resources (soil, forests, fisheries). The volume defines sustainable development as the attempt "to meet needs and aspirations of the present without compromising the ability to meet those of the future."[117] Like the "economic growth" strategy, it advocates industrialization; its goal is a five- to tenfold increase in world industrial output over a fifty-year period. But unlike the earlier strategy it insists that in-dustries must be more energy efficient, less resource-intensive, and less pol-luting. The report advocates small-scale decentralized industries, aided by electronic controls and communication networks. Like the "human needs" strategy, it gives priority to the basic needs of the poor, but it aims at equity be-tween generations as well as equity within each generation.

The Brundtland Report advocates many of the policies that I described ear-lier as *sustainable agriculture*. It recommends conservation practices for soil and water and reduced use of chemicals. It urges diversification, crop rota-tion, legumes, agroforestry, and Integrated Pest Management. It endorses ex-panded research and extension services blending traditional and modern technologies. The report describes current policies that favor urban popula-tions and large farms, and it claims that land reforms in East Asia and India have increased both equity and productivity. Grass-roots participation is held to be essential to effective development. The report concludes that environ-mentally sound development is possible but will require higher levels of aid from industrial nations, which could be paid for with funds released by the re-duction of military expenditures. It proposes an international fund to facili-tate the transfer of energy-efficient and nonpolluting technologies, perhaps financed by a tax on international trade and common resources such as ocean minerals and fisheries.[118] The report is hopeful that a new awareness of global environmental problems will bring greater cooperation between rich and poor countries.

The Brundtland Report emphasizes environmental limits but it is *optimistic about the ability of technology* to extend those limits. Many environmentalists would reply that global industrial growth by a factor of five or ten could not be sustained, even with improved efficiency and pollution control. They would advocate stronger measures to curb population growth than the report envis-ages. They would argue that growth in consumption and resource use by af-fluent nations must be more drastically curbed if catastrophe is to be avoided.

As nations prepare for the *U.N. Conference on Environment and Development* to be held in Brazil in 1992, it remains to be seen whether the rich countries, and particularly the United States, will make anything more than token contribu-tions to sustainable development in the Third World. The developing coun-tries have said that effective international agreements on forests, endangered species, and global climates require that industrial nations provide additional financial resources and access to environmentally sound technologies (see chapter 7).

IV. FOOD AND GLOBAL JUSTICE

According to UNICEF, 250,000 children die each week from malnutrition and the diseases associated with it.[119] Yet the United States and Europe have crop and dairy surpluses that depress farm prices. The U.S. imports $1 billion of agricultural products annually from Central America, where half the children are malnourished. Cats and dogs in Europe are better fed than many children in West Africa. The incomes of the richest fifth of the world population averages 50 times the incomes of the poorest fifth.[120] The affluent can outbid the poor in the global supermarket. Inequalities in access to food are more critical than any other kind of inequality because they endanger health, work, personal fulfillment, and even life itself.

1. ETHICAL PRINCIPLES

What ethical responsibilities do people in affluent nations have in relation to world hunger? Starting from the equal value of all persons (chapter 2), I would argue that all persons have a right to life and therefore *a right to the basic necessities of life,* including food enough to survive. Correspondingly, people who could prevent starvation, without sacrificing anything of comparable moral significance, have a duty to do so.[121]

It has sometimes been argued that *letting people die,* when one could help to save some of the victims, is morally equivalent to killing them.[122] But *intentions* do play a part in ethical appraisal. Whereas murderers want their victims to be dead, no one wants hunger victims to die of starvation. (We will encounter the same distinction in the case of nuclear deterrence, whose proponents claim that their *intention* is to prevent nuclear war, not to kill millions of innocent civilians). Yet acts must also be judged by their foreseeable *consequences,* whatever one's intentions—even when the links between acts and consequences are long and indirect. I bear more clear-cut responsibility for acts of commission that are mine alone and that have immediate consequences than for acts of omission in complex situations for which many people share responsibility. I can do more to affect events close to home (a starving person on my doorstep) than distant ones (famine in Ethiopia). But even distant starvation lays a claim on me because I can do something about it, both by contributing to famine relief and development programs and by advocating national and international policies to attack the causes of famine.[123]

In the name of *individual freedom* and *property rights,* some writers have maintained that people have the right to use fairly acquired wealth in any way they want (including the purchase of food), as long as there is no direct harm to others. Feeding the starving, they say, is a commendable act of charity, but it is not a duty.[124] I would reply that the wealth of affluent nations has not been fairly acquired but is in part a product of colonial exploitation.

There have been many links between *technology* and *colonialism.* Military and transportation technologies made possible the global imperialism of the nineteenth century which at its height controlled 90 percent of the non-Western world. As late as 1940, seven European nations with 200 million citizens controlled 700 million people in their empires. In the Industrial Revolution, the colonies provided cheap natural resources and labor as well as markets for manufactured goods, but the colonial powers did not transfer manufacturing technologies or expertise. I have mentioned the establishment of colonial economies built on one-crop exports. The early industrializers had enormous advantages, including cheap fuel and raw materials and the opportunity to relieve population pressure and unemployment by emigration to new colonies. Nations trying to industrialize today lack these advantages, face competition from advanced nations, and are in an international economic order that perpetuates their role as suppliers of commodities and raw materials.[125]

We resist the idea of *justice across national boundaries* partly because it would threaten our privileged position, but also because principles of justice are difficult to apply internationally. But the political scientist Charles Beitz maintains that the Rawls principle—that inequalities are acceptable only if they benefit the worst off—is applicable globally because nations today are interdependent and can cooperate for common interests. Natural resources, investment, profits, and finished products all cross boundaries in the global marketplace. The distribution of the benefits that arise from international interaction can be evaluated only by principles transcending national boundaries. Beitz maintains that a hypothetical group of persons who did not know to which country they would belong would establish procedures to ensure that all persons had access to the resources necessary to satisfy basic human needs. They would accept only those inequalities that benefit the least advantaged, in whatever country.[126]

Beitz recognizes that such a principle of *global justice* is an ideal that cannot be realized immediately, but it can set a direction for changes in national policy and for modifications in international institutions. The principle is not incompatible with concern for freedom and justice within one's own nation or within other nations. Beitz maintains that U.S. agricultural and technological assistance to LDCs is an obligation of justice, not an optional act of charity. He also discusses special drawing rights in the World Bank, preferential tariffs for LDCs, changes in the terms of trade, and other specific policies that would promote greater global justice.

In addition to considerations of justice, there are *pragmatic reasons* for concern about the growing gaps between rich and poor countries that are a threat to world trade and economic stability. In an interdependent world, the health of the global system affects the welfare of every nation. The United States, for example, imports more than half its supplies of twenty critical minerals. It is strongly affected by balance-of-payment deficits, global inflation, and fluctuations in markets and resource prices. The continued frustration of the hopes of LDCs could also lead to political instability, violence, and disruptive activities. As nuclear weapons spread around the world, industrial nations may face threats from revolutionary movements or desperate actions by impoverished nations.

Judaism and Christianity include specific teachings about *feeding the hungry*. The Deuteronomic laws upheld the right of the needy to glean in the fields. Isaiah said that God seeks social justice before religious observances: "Is not this the fast that I choose: to let the oppressed go free . . . and to share your bread with the hungry?" (Isa. 58:6). Matthew included responses to hunger in his portrayal of the last judgment: "I was hungry and you gave me food" (Matt. 25:35). Equally relevant is the biblical conviction that God is on the side of the poor and works for the liberation of the oppressed. In both the Old and New Testaments, the coexistence of extremes of wealth and poverty is attacked; the rich are indicted for their lack of concern for people in need. In the biblical view there are no absolute rights to property; the earth's resources are intended for all.

In the biblical tradition, *institutional injustice* is criticized as well as individual wrongdoing. The abuses of economic power by privileged classes represent a structural evil, an expression of group self-interest and greed. The ancient prophets spoke of judgment on their nation and called for national repentance for complicity in injustice. In a similar vein, church leaders today insist that we must move beyond famine relief to deal with the causes of hunger. They call for national as well as individual action, long-term as well as short-term policies, justice as well as charity. Local hunger task forces, denominational agencies, and interdenominational organizations have promoted political action to increase agricultural assistance in order to enable people to feed themselves.[127]

2. NATIONAL POLICIES

Emergency food aid is a commendable short-run response to human suffering. National and international grain reserves should be built up to meet such emergencies. But famine relief does nothing to correct the long-term conditions that produce hunger and starvation. It deals with symptoms rather than causes and leads to dependency rather than self-reliance. Even the repeated sale of grain at subsidized prices undercuts farm income in the Third World and hinders local agricultural development.

Development aid, however, including agricultural and technical assistance, can have a significant long-run effect. In 1989, industrial nations allocated only 0.33 percent of their gross national product to foreign aid.[128] The United States gave only 0.2 percent.[129] The largest part of U.S. foreign aid has been military assistance, and nonmilitary aid has been dictated primarily by geopolitical considerations rather than by human need. Most of it has gone to "friendly" and "strategic" nations such as Israel, Egypt, and El Salvador. Many grants or loans have been tied to the purchase of U.S. equipment. Multilateral aid avoids such direct donor-recipient relations and allows the international community to select recipients and specify conditions for assistance. The United Nations administers the International Fund for Agricultural Development, but its resources have been meager. The World Bank should imaginatively implement the "sustainable development" strategy, which in the past has been subordinated to the goal of "economic growth." The industrial nations spend ten times as much on military defense as on development aid. With the winding down of the Cold War, nations have a new opportunity to put a substantial portion of these funds into the development efforts on which true security depends in an interdependent world.

A system of *international taxation* would provide the U.N. with a dependable and equitable level of funding. Within Western nations, progressive taxes serve to mitigate the increasing gaps between rich and poor; welfare and social security measures are designed to prevent anyone from actually starving. Globally, no comparable legislation limits the extreme economic inequalities that result in dramatic differences in access to food. The institution of global income taxes on individuals would require stronger international institutions and a greater sense of world identity than now exists, but international taxes could be raised in other ways. The Brandt Report (1980) proposed taxes on arms sales and seabed minerals, and the Brundtland Report advocated a tax on international trade and ocean resources.[130]

International trade affects the economies of developing nations and thereby affects the income of citizens and their access to food. Many barriers to trade have been put up by industrial nations in the form of duties, tariffs, quotas, and outright bans. These barriers are applied to processed and manufactured products such as cloth and clothing but not to raw materials such as cotton; this obviously hinders industrial development in the Third World. Protectionism has been on the rise in the United States, Japan, and Europe. Moreover, prices for many commodities and raw materials fell during the eighties, while the price of industrial products rose, so that the balance of trade was adversely affected. The North did not respond to the pleas from the South for a New International Economic Order with lower tariffs, renegotiated debts, stronger representation in the International Monetary Fund, and international regulation of transnational corporations. Agreements are needed to prevent American and European companies from locating some of their most polluting plants in LDCs to avoid stricter environmental regulations at home.

Such international measures are of course no substitute for what can be done by *individual nations*. I have mentioned the need for family planning, land reform, rural development, and changes in tax and food pricing policies in the Third World. Western nations must take the initiative in restructuring debt, regulating transnational corporations, and encouraging the transfer of technologies appropriate to developing nations. I suggest that agriculture should have higher priority in the allocation of national research funds in the North—especially research on sustainable agriculture, which would at the same time benefit the South more than high-input agriculture does. Agricultural technology has much to contribute to both North and South, provided it is pursued with sensitivity to social and environmental issues.

Finally, global justice requires that affluent nations *reduce their own consumption,* especially of grain-fed meat and nonfood and luxury food crops grown in LDCs. In the 1980s, more grain was consumed by livestock (mainly in feedlots) in the United States and the Soviet Union than by the entire human population of the LDCs.[131] Grain-fed beef is a very inefficient source of proteins.

It takes seven pounds of grain for every pound of beef produced (whereas it takes only two pounds of grain to make a pound of chicken).[132] Ninety percent of the annual world protein deficit could be met by the grain and fish-meal fed to American cattle alone. Meat production should perhaps be confined to pasture and rangeland unsuited for crops. Strenuous individual and national efforts at energy conservation also would benefit world agriculture; our ravenous appetite for oil drives up its price and escalates the cost of fuel and fertilizer everywhere. In chapter 9 we will ask what changes in perceptions, attitudes, and value priorities might help to motivate such changes.

V. CONCLUSIONS

Because agriculture affects virtually every aspect of human life and the environment, it is not surprising that all of the values discussed in part 1 are relevant to agricultural policy decisions.

Food and health are obviously the central considerations. Consumer access to nourishing food at affordable prices comes before all else. Efficiency and productivity should be sought, but higher yields must be sustainable and the use of chemicals must not endanger health or the environment. I concluded that agricultural researchers should seek productivity but always in conjunction with a wide range of other goals.

Agriculture provides *meaningful work* that is essential for rural populations. While mechanization has often increased yields and reduced backbreaking field labor, it has been responsible for extensive rural unemployment and poverty. I suggested that Third World development policies should give higher priority to both agriculture and decentralized industry for the sake of rural employment.

Rural life offers forms of *personal fulfillment* that are worth preserving as huge urban centers around the world become more crowded and impersonal. Family farms and community-based enterprises, when they are economically viable, can encourage stable human relationships, the vitality of rural communities, and opportunities for decisions at the local level.

Social justice has been a crucial value at many points in this chapter. I argued that large machines such as tomato harvesters usually benefit food processing companies and large landholders at the expense of workers and smallholders. The Green Revolution has accelerated the concentration of land ownership in the Third World. Justice is at stake when the production of luxury foods for the rich displaces the subsistence foods of the poor or when export crops and feed crops replace staple food crops. I urged development assistance in the name of justice, going beyond famine relief in the name of charity.

The *participation* of people in the decisions that affect their lives is facilitated when land ownership is widely dispersed rather than being concentrated in the hands of a few. In industrial nations, corporate land ownership, vertically integrated food companies, and tax and subsidy policies favoring

large farms all reduce the opportunities for participation. In many Third World countries, a few wealthy families own most of the land, and land reform continues to be a critical issue.

Agriculture can contribute significantly to *economic development,* especially in the Third World. I maintained that the goal should be sustainable development, rather than economic growth as such, and that income distribution is as important as the GNP in measuring a nation's well-being. The urban bias of many Third World governments and international agencies has slowed the rural development that would help the majority of the population.

Resource sustainability has been a major theme of the chapter. I advocated crop rotation, minimum tillage, tree planting, and other ways of slowing soil erosion. More efficient irrigation and more realistic water pricing will help to conserve water resources. Other forms of land stewardship and sustainable agriculture were described. And of course sustainability requires effective family planning programs to slow population growth.

One form of *environmental protection* is the reduction of pollution caused by high levels of fertilizers and pesticides, through the use of legumes, crop rotations, Integrated Pest Management, and so forth. I noted diverse forms of agricultural damage to the environment, from salinization of irrigated cropland to desertification of overgrazed rangeland. The genetic diversity present in indigenous plants has been reduced by the promotion of monocultures from a few seed strains, which increases vulnerability to plant diseases and pests.

Finally, *respect for all forms of life* was invoked in urging the banning of farm practices that inflict suffering on animals. I did not advocate total vegetarianism, but I did propose a reduction in meat consumption in affluent nations in the interest of health and global justice. In a later chapter we will see that the clearing of tropical forests to create cropland and rangeland is condemning a large fraction of the globe's species to extinction.

Agricultural policy, in short, raises complex and difficult ethical issues because food is so essential to human life and agriculture affects so many human and environmental values.

CHAPTER 5

Energy

Until 1800, all world civilizations were based on wood, wind, water, and animal power. The Industrial Revolution in the nineteenth century was the product of technologies powered by coal. The growth of twentieth-century industrial nations was fueled by coal and cheap oil. In the 1960s it was hoped that nuclear reactors would supply much of the world's future energy needs. But by the 1980s it was clear that each of these sources entails high environmental and social costs. Yet we have done little to develop the renewable sources that might replace them or the conservation and efficiency measures that could reduce energy demand.

One-fifth of the world's population now accounts for 70 percent of the world's energy use, which is more than nine times as much per person as in the remaining four-fifths of the world. One person in Western Europe uses as much energy as eighty people in sub-Saharan Africa.[1] A U.S. citizen uses as much as 330 citizens of Bangladesh.[2] The world as a whole could not sustain the rate of energy consumption now enjoyed by industrial nations, which industrialized when coal and oil were relatively cheap. The Third World is trying to industrialize at a time when high energy prices are hindering development plans and adding to already staggering debt loads.

Values are present in energy decisions, whether recognized or not. There are inescapable trade-offs among diverse values: health, justice, participation, economic development, environmental protection, and sustainability. We will look mainly at the social and ethical issues in energy choices, but we cannot ignore technical and economic questions. We must ask both what is technically feasible and what is ethically desirable. Who benefits and who bears the costs and risks of various energy systems? Which technologies accelerate the concentration of economic and political power? Which sources are most promising for developing countries? Which policies take into account the needs of future generations? We will look successively at fossil fuels, nuclear power, renewable sources, and the possibilities for conservation.

I. FOSSIL FUELS

Oil, natural gas, and coal contain energy that arrived from the sun for millions of years, captured in prehistoric plant life and preserved in the geological strata of a few regions of the earth. Fossil fuels supply 88 percent of the world's purchased energy (excluding nonmarket fuels such as firewood). It is estimated that identified global reserves of oil would last for 44 years, natural gas for 60 years, and coal for three centuries at current depletion rates.[3] However, their geographical distributions are very uneven and their political and environmental costs are very high.

1. OIL AND GLOBAL JUSTICE

Since 1900, oil has been central in the *military and political involvement* of Britain, France, Germany, Turkey, and then the United States in the Middle East. Japan attacked Pearl Harbor in 1941 because its oil supplies had been cut off and its stores of oil were running low.[4] More recently, three oil price shocks—caused by the OPEC oil embargo in 1973, the Iranian crisis in 1979, and Iraq's invasion of Kuwait in 1990—have brought home the world's dependence on Middle Eastern oil. World oil reserves were 50 percent higher in 1990 than in 1973, but most of the new discoveries were in the Middle East, which now holds two-thirds of the world's reserves.[5]

Oil production has been declining in the U.S. since 1970, and it is now falling in the former Soviet Union, Britain, and several other non-OPEC nations. The cost of drilling for deeper deposits and in inaccessible locations (such as Alaska and the North Sea) has been growing. Higher prices encouraged some conservation measures, and world oil consumption fell during the early 1980s. But following the price drop in 1986 it rose again and by 1990 was back to the levels of the late seventies. In 1990, the United States imported almost half its oil, and its current policies will lead to increased dependence on imports in the future.[6]

The prospect that Saddam Hussein might control or influence a large part of *Middle Eastern oil production* was clearly a major factor in the 1991 military intervention by the United States (with some support from other allies). Iraq's invasion of Kuwait and efforts to acquire nuclear weapons posed a threat of further aggressive actions. In the future, nations dependent on Middle Eastern oil will continue to be vulnerable to political changes in a region of long-standing conflicts and great inequalities. Saudi Arabia's autocratic royal family will be under great pressure from Islamic fundamentalists and from impoverished Arab countries that lack oil resources. Imported oil is the largest and fastest-growing component of the U.S. trade deficit. In addition, the military costs of oil imports should really be included in their true costs. For example, keeping shipping lanes open in the Gulf in 1985 cost $200 per barrel of oil shipped, eight times its world price.[7]

Dependence on imported oil has had a devastating impact on *the Third World*. We have seen that oil is crucial for agriculture because it is the main feedstock for fertilizer and petrochemicals as well as the fuel for tractors and irrigation pumps. The profligate use of oil in industrial nations together with OPEC policies in the seventies drove up world prices, and by 1981 developing nations were spending an average 61 percent of their export earnings on oil imports.[8] The high price of kerosene, an oil derivative, accelerated the depletion of firewood and other traditional sources used for heating and cooking.

Within industrial countries, *the politics of oil* are extremely complicated because oil plays such a large part in national life. In the past, oil use has been heavily subsidized through depletion allowances, import quotas, and highway construction funds. A powerful lobby was formed by a coalition of diverse interests: oil companies and oil-producing states; auto, insurance, and highway construction companies; and labor unions. The oil lobby has made large contributions to the campaign funds of the members of Congress who are on committees dealing with energy, and it has actively opposed all attempts to raise auto fuel economy standards. Of the 20 largest U.S. companies 7 are oil companies, and many other industries, such as autos, plastics, and petrochemicals, are dependent on oil.[9] Oil companies have also reduced competition by "vertical integration" (control of oil imports, domestic production, refining, pipelines, and service stations). They have used extensive television and magazine advertising to present their interpretation of the energy situation and the policies they favor.

The use of oil also has serious *environmental consequences*. The Exxon Valdez accident off the Alaska coast in 1989 dramatized the damage to wildlife and coastal beauty that a single human error can perpetrate. Other oil spills have been caused by severe storms, and we can expect such spills to continue, though we could reduce some of the consequences by requiring double-hulled tankers and better standby equipment for more rapid containment. The fuel burned in cars and trucks is a major contributor to urban air pollution, which constitutes a serious hazard to human health. According to the American Lung Association, auto air pollution adds $40 billion to annual American medical bills.[10] The nitrogen oxides and ozone-forming hydrocarbons can be controlled by catalytic converters. But the combustion of any oil product adds carbon dioxide to the air, accelerating global warming from the greenhouse effect discussed below.

The *reduction of oil use* by industrial nations is thus essential for world peace and global justice and also for environmental protection and long-term sustainability. Since transportation accounts for two-thirds of U.S. oil use, and more than half the oil used in many other countries, auto fuel economy standards, alternative fuels, and public transportation are promising starting points, as we will see in section IV. Most conservation measures and renewable energy sources have low environmental impacts, and they rely on local or regional materials, diminishing the international competition for fossil fuels that are so unevenly distributed geographically.

2. COAL AND THE ENVIRONMENT

The known reserves of coal are larger than those of oil, but the human and environmental costs are greater.

1. Human and Environmental Costs of Mining

The long history of *mine accidents* and *black lung disease* is a classic instance of high risks falling on one group while benefits accrue to other people. The appalling conditions faced by generations of miners have gradually improved in response to labor union demands and legislated safety standards. Following the U.S. mine safety law passed in 1969, fatalities per million working hours fell from 1.2 in 1970 to 0.3 in 1977, and the incidence of lung disease also dropped sharply. Since then enforcement of safety and ventilation standards has improved somewhat, but even stricter standards and better enforcement would not impose an unreasonable economic cost.[11]

It is sometimes claimed that it is justifiable for *workers to be exposed to higher risks* than the public because the risks are voluntarily accepted. However, a coal miner usually has little geographical mobility or job mobility and may have no alternative to working in a mine. Another defense of occupational hazards is that high-risk jobs, such as the construction of tall buildings, often offer a wage premium. But rates of compensation for comparable risks vary widely, and mine workers have usually been poorly paid compared to industrial workers. Some compensation for injury has been provided under workman's compensation laws and under specific legislation providing payments to miners with black lung disease. It would serve the interests of justice and offer greater incentives to safety if the cost of these payments were paid by those who benefit, such as the consumers of coal-fired electricity.

There are also considerable *environmental impacts* from the mining of coal. Deep-mining wastes have depressed land values, polluted streams, and marred areas of great natural beauty around the world. The strip mining of surface coal has left large areas denuded and subject to rapid erosion. In the United States, a 1977 law required replanting and reclamation measures to be sustained for five years (or ten years in arid areas) with costs passed on to coal users. But reclaimed land often can only be used for pasture, and in semiarid regions, with fragile ecosystems and scarce water, it will take decades for healthy plant life to be restored and some areas will probably remain wastelands.[12]

2. Air Pollution and Acid Rain

Burning coal has even greater human and environmental costs than mining it. Coal contains sulfur compounds which form sulfur dioxide in stack emissions, and these in turn combine with moisture to produce sulfates. These sulfates in *urban air* are a major source of respiratory illness and fatality (see chapter 7). Tall smokestacks reduce local sulfate levels but at the expense of increased acid rain at distant points.

Since the late sixties the effects of *acid rain* on the forests and lakes of Europe and North America have been increasing. By 1988, 52 percent of the forests in West Germany had been affected. The U.N. Environmental Program estimates that more than a third of Europe's total forest area has been damaged. In some parts of Eastern Europe, only bare trunks and eroded hill-sides mark what were once great forests. Crop losses from sulfates and ozone are difficult to determine, but one recent study puts losses at 5 to 10 percent of production in the United States. In half of the lakes in eastern North America, many species of fish and plants have been wiped out by acidification. In Greece, public monuments and statues have deteriorated more in the last 25 years than in the previous 2400 years.[13]

Effective *pollution control technologies* for acid emissions can be installed. Passing the combustion gases through limestone scrubbers removes 85 percent of the sulfur dioxide. Scrubbers and use of low-sulfur coal account for most of the 28 percent drop in sulfur dioxide emissions in the United States between 1970 and 1987. In 1977, Congress required scrubbers on all new plants, but during the eighties the utilities and eastern coal-producing states lobbied effectively against legislation to require the installation of scrubbers on older plants. By 1990, only 30 percent of U.S. coal-fired capacity had scrubbers, compared to 85 percent in West Germany and over 90 percent in Japan. Japan and Germany are also adding catalytic reducers to remove the nitrogen oxides that produce smaller amounts of acid rain. In Japan, pollution control adds 25 percent to the cost of coal-generated power, but an air pollution disaster has been averted. Newer "clean coal technologies," such as fluidized bed combustion and multistage combustion, offer the prospect of cheaper control of both these air pollutants.[14]

The technology to control sulfur and nitrogen emissions is available, and its installation is primarily a matter of *political will*. The cost of the damage from the emissions clearly exceeds the costs of the control technology, but indirect costs do not appear on industrial balance sheets. During the eighties the U.S. administration gave economic growth priority over environmental preservation, and the conflict of regional interests undermined repeated congressional attempts to strengthen clean air laws. However, greater public interest in clean air and a more pragmatic administration finally produced tighter air emission laws, though still below the standards sought by environmentalists; the 1990 Clean Air Amendments required cutting sulfur dioxide emissions in half by the year 2000. A 1988 directive of the European Economic Community seeks to lower sulfur dioxide levels to 57 percent of the 1980 levels by 2003, and a group of nine countries has pledged a 50 percent reduction.[15]

3. Global Warming: The Greenhouse Effect

The burning of all fossil fuels entails the formation of *carbon dioxide* (CO_2). The enormous quantities released into the air by fossil fuels have increased the CO_2 content of the air by 25 percent in the past hundred years. At the present emission rate it will have doubled by 2030. Coal is the worst offender, releasing 24 kilograms of CO_2 per billion joules of heat produced; oil is next (20 kg.), and natural gas produces least (14 kg.). In the United States, coal-burning utilities account for a third of the CO_2 emitted, oil-burning vehicles another third, and fuels burned in homes and industry the final third.[16]

The average annual temperature fluctuates erratically from year to year. Nevertheless, there is a very high correlation between past CO_2 levels and the general *trend of temperature changes.* Core samples from glaciers substantiate this correlation over a period of 160,000 years (including a large temperature change during the last ice age), and it holds for the past 130 years for which there are historical records. In the latter period the change in average temperature has been small, amounting to only 1°F. But the five warmest years in the last century were all in the 1980s. It is known on theoretical grounds that CO_2 increases the reflectivity of the upper atmosphere, trapping more heat in the lower atmosphere.

The prediction of *future global warming* is uncertain because climate effects are very complex and only partially understood. Computer models for atmospheric conditions involve simplifying assumptions about winds, humidity, clouds, oceans, and so forth. One widely quoted study concludes that at present emission rates the CO_2 concentration would double by 2030, and the average surface temperature would rise between 3 and 5.5° C (5 to 10° F), with larger changes in the earth's temperate zones.[17] The melting of polar ice would raise sea levels by somewhere between 8 inches and 5 feet, flooding some coastal areas and contaminating groundwater with salt. Wind and rainfall patterns would be drastically changed, affecting agriculture in many parts of the world. (These effects would of course be worse if emission rates increased or if they continued for a longer time interval.) Climatologists recognize the uncertainties in these calculations, but they are virtually unanimous that the magnitude of the potential consequences are so great that we must not only intensify research on the problem but initiate preventive policies.[18] Such policies would be a form of insurance. The greenhouse effect may turn out to be smaller than current estimates, but it could be greater, and it would be much more difficult to reverse later.

The main way to reduce global warming is to *reduce the use of fossil fuels.* If this is accomplished by conservation measures and greater efficiency, the reduction will at the same time cut down on other environmental impacts from fossil fuels. A report by the National Academy of Sciences says that the United States could cut its carbon emissions by 40 percent with no impact on lifestyles.[19] Forests absorb CO_2 from the atmosphere as well as providing wildlife habitats and protection against soil erosion. Slowing the destruction of forests would slow global warming, and in the tropics it would help to preserve endangered species and highly erodible soil. Planting trees would have many environmental benefits, but it will have to be done very extensively to make a dent on the CO_2 accumulation.

Apart from conservation, switching from coal and oil to *natural gas* is the most effective way of reducing carbon dioxide emissions during the next decade. Natural gas is more abundant globally than oil and more widely dispersed, including extensive known deposits in the Third World (and many areas have not yet been explored). Gas-fueled combined-cycle plants are highly efficient in generating electricity because the burning gases are used to drive one turbine directly and then the heat is used to generate steam to drive a second turbine. The greater efficiency, together with the smaller amount of CO_2 released per unit of heat, means that only two-fifths as much CO_2 is emitted as a coal-burning plant of equivalent capacity, and the sulfur dioxide emissions are negligible. Several nations are experimenting with vehicles driven by cylinders of compressed natural gas. Mexico City is planning to use them on its city buses to reduce air pollution. Natural gas development is also much less environmentally destructive than coal mining. Natural gas is the most promising fuel during the transition to sustainable sources.[20]

A *carbon tax* would be an effective way for industrial nations to reduce carbon dioxide emissions, but it would have to be introduced gradually to avoid severe economic dislocations. Such a tax would favor natural gas over coal and oil, and it would strongly favor nuclear, solar, and conservation technologies that release no CO_2 once they are installed. If part of the tax receipts was invested in efficiency improvements, and part was offset by reductions in other taxes, the economy as a whole would be strengthened rather than weakened. By 1991, Finland, Sweden, and the Netherlands had introduced small carbon taxes, and the European Community was making plans for a carbon tax to be instituted in all European nations.[21] Some Third World countries, especially China, have extensive coal deposits and will find it very difficult to avoid expanding their use of coal as they seek more rapid economic growth. But the efforts of industrial nations to develop alternatives to fossil fuels would help all developing nations in making the transition to sustainable sources.

II. NUCLEAR POWER

In 1990, 421 nuclear plants were operating in twenty-six countries (with 96 more under construction), generating 17 percent of the world's electricity. In many industrial nations the percentage of electricity coming from nuclear plants was even higher: 19 in the U.S., 22 in Britain, 28 in Japan, 34 in the former West Germany, 45 in Sweden, and 75 in France.[22] But in virtually every Western nation, plans to start construction of new plants have been halted by public opposition and by escalating costs (now estimated as twice those for coal-fired plants per kilowatt-hour). All orders for new nuclear plants in the United States since 1973 were subsequently canceled. The industry blames additional safety requirements, regulatory delays, and high interest payments during long construction times (up to eighteen years) for costs of up to $5 billion per plant, ten times the original estimates. A few of the critics of nuclear power are reconsidering it in the light of the greenhouse dangers of coal and the possibility of new "inherently safe" reactor designs. Let us look successively at reactor safety, radioactive waste disposal, and prospects for the future.

1. REACTOR SAFETY AND RISK ACCEPTABILITY

The history of nuclear reactors in the United States supports *the contextualist thesis*, presented in chapter 1, that the design of a technology is not inevitable but is the product of choices made in particular social contexts. In the early 1950s, reactor designers had a choice of fuels (natural uranium, enriched uranium, or plutonium), moderators (light water, heavy water, or graphite), coolants (water, liquid metal, or gas), temperature (boiling water, pressurized water, or high temperatures), and reactor size and configuration. The U.S. Navy urgently wanted a nuclear submarine, which required a high energy density and a readily available design—most easily fulfilled by a pressurized light water reactor (LWR) using enriched uranium. The first commercial LWR at Shippingport was designed to serve also as the prototype for an aircraft carrier reactor.

These choices strongly influenced subsequent decisions. By the time the Atomic Energy Commission (AEC) and the private utilities and reactor contractors chose designs for commercialization in the late fifties, LWRs had a substantial head start from the navy program. Their decisions were based more on cost and current availability than on safety considerations (though liquid sodium coolants were ruled out partly for safety reasons). In Congress, the Joint Committee on Atomic Energy concurred that these decisions were in the national interest.[23]

The next crucial stage was *the very rapid upscaling* in size and numbers that occurred between 1965 and 1970. A hundred reactors were ordered, many of them seven times as large as the largest with which there had been any operating experience. At such sizes, containment structures could not be counted on to withstand a major accident, and plants were being built near population centers. As successive problems emerged, new safety equipment was added on (such as emergency cooling systems and redundant pumps and instruments), but the basic design was not changed. Until 1965, nuclear technology still had considerable flexibility, but by 1970 large-scale light water reactors had developed a momentum of their own, sustained by heavy financial investment and institutional commitments. The industry was locked in before it could learn from experience.[24] In Britain nuclear reactors were owned by the government and alternative designs were tried, but similar problems were present. Long lead times and huge capital investment in large plants with which there had been no previous experience produced an inflexible system, and mistakes were difficult and costly to correct.[25]

The U.S. Atomic Energy Commission's mandate to *regulate* nuclear power was compromised by its conflicting mandate to *promote* nuclear power. The AEC was secretive, suppressed information and internal dissent, and virtually excluded participation by citizens or independent experts. As often happens, a "cozy triangle" of a government agency, a legislative committee, and a private industry shared a common interest in promoting a particular technology.[26]

During the seventies the nuclear industry faced a variety of *unexpected problems*. Demand for electricity grew much more slowly than had been anticipated. As new safety measures were added, regulations became more complex and delays in construction and licensing extended the construction time to 10, 12, or in one case 18 years. High interest rates further escalated construction costs. Vocal opposition to nuclear plants was often associated with the growing environmental and antiwar movements. Communities woke up to plans for nearby plants, formulated without their involvement, and they protested vehemently and started court challenges. Some of the objections were wildly overblown, but others pointed to problems that turned out to be valid. Public confidence in both government bureaucracies and large corporations was eroding. One could no longer simply "trust the experts," for knowledgeable experts now spoke out on both sides. The Nuclear Regulatory Commission was established in 1974 with authority only for regulation, but much of its staff and many of its attitudes and procedures were taken over with little change from the AEC.

Then came two major *reactor accidents.* At Three Mile Island in 1979, a partial meltdown occurred and 90 percent of the fuel rods burst, but very little radiation escaped from the containment vessel. At Chernobyl in 1986 an explosion and fire in a graphite-moderated reactor spread large quantities of radiation across Europe. Thirty-one workers at the plant died within a few weeks. It has been estimated that the radiation will result in the death of 28,000 from delayed cancer over the next fifty years.[27] Both accidents were the result of human errors, which in the Three Mile Island case were compounded by a stuck valve and misread instruments. Previous calculations of reactor risks had failed to allow adequately for unlikely combinations of mechanical failures, system interactions, and operator misjudgments. With complex, tightly coupled systems it is impossible to anticipate all possible linkages among human errors and component failures, and events can occur too rapidly for analysis and corrective action.[28] The credibility of reassurances from the industry and the government was undermined, and public support of nuclear power dwindled in both the United States and Europe.

How should we deal with the risk of reactor accidents? Risk analysis defines risk as the *probability* of an event multiplied by the *magnitude* of its harmful consequences. Current estimates of the probability of a nuclear accident are higher than those in the midseventies, but the calculated risk remains less than many other risks that we accept, such as those in driving an auto. Nuclear proponents say that public anxiety about nuclear accidents is therefore irrational. In reply, social scientists point out that the *acceptability of a risk* involves many factors in addition to probability and consequences. In general, we are more willing to accept risks that are familiar, voluntarily accepted, personally controlled, reversible, and equitably distributed—and necessary to obtain a desired benefit (see the discussion of technological risks in chapter 8). Nuclear power scores poorly on all these characteristics.[29] Nuclear power is thus less acceptable than conventional risk analysis suggests.

In addition, there may be valid reasons for giving *low-probability catastrophes* more weight than risk-benefit analysis assigns. One accident with 50,000 victims is more socially disruptive and receives more media coverage than 50,000 one-victim accidents. A study by a national laboratory in 1982 calculated that a major nuclear accident near an urban area with unfavorable winds could

cause losses as high as $150 billion and 140,000 deaths.[30] The probability of rare events and human errors can seldom be estimated with the same accuracy as high-probability events, and underestimation of the probability of a disaster would have enormous consequences. Again, complex large-scale systems are more vulnerable to disruption than smaller and simpler systems, and they can create human overloads. Finally, risk analysis of nuclear plants assumes stable social conditions and leaves out the possibility of wars, civil conflicts, or deliberate acts of sabotage, terrorism, or threat.

Any evaluation of technological risks will be influenced by assumptions about *human nature* and *social institutions*. The biblical understanding of human fallibility would lead one to be cautious about situations in which errors can have disastrous consequences. I suggested earlier that the social dimension of sin is expressed in institutional self-interest. It should be no surprise that organizations put their own goals ahead of the public interest or underestimate risks that could be reduced by expenditures on safety. If we live in a world of stark injustice and growing gaps between rich and poor nations, it seems unrealistic to leave out the effects of social unrest and political instability. Nuclear reactors would be vulnerable in situations of conflict, and they would be tempting targets for blackmail by terrorists. Moreover, their large scale and cost represent a concentration of economic and hence political power that makes it difficult for citizens to participate in decisions about them. In short, catastrophe avoidance should be given heavier weight than traditional risk analysis assigns, but it cannot be the only criterion for choosing technologies, since benefits and alternatives must be considered along with risks.

2. RADIOACTIVE WASTES AND FUTURE GENERATIONS

The risks from high-level radioactive wastes extend over very long periods of time. Many of the waste products must be isolated for a thousand years, and plutonium itself is still dangerous after 100,000 years. A huge backlog of military and commercial wastes has already accumulated around the world. Several nations are hoping to create *permanent repositories* in which solidified wastes would be stored in concrete casks in tunnels excavated deep in salt beds or solid rock. The repositories must be isolated from groundwater, since even very slow leaching would contaminate underground aquifers. Most scientists believe that suitable geological formations can be found, but at each of the sites proposed so far, geologists have been concerned about possible changes in groundwater flow induced by fractures, earthquakes, or climate changes.

Both *scientific uncertainties* and *public opposition* have plagued all attempts to establish such repositories. In Germany, the Gorleben salt dome was selected as the first site, but it was discovered that groundwater is eroding the salt and the geological structures are not stable. Gorleben has been the scene of large public protests, and the repository opening has been put off from 1998 to 2008 at the earliest. In France, the opening of a repository has been postponed to

2010, and no site has yet been named. Japan hoped to locate a burial site in a remote and poor prefecture, but despite the offer of a large subsidy, opposition from the governor and the diet was strong. Japan has signed an agreement with China to conduct research on an underground facility in Shanxi province, in return for assistance with the Chinese nuclear program.[31]

Three ethical issues can be identified in planning for nuclear waste disposal.

1. Regional Justice

Radioactive waste disposal is an extreme case of local risks for national benefits. Utilitarians may assert that the total benefits outweigh the total risks, but justice is violated if some people benefit and others sustain the risks. People recognize that the wastes must go somewhere, but they assert, "Not in my backyard." The U.S. National Waste Policy Act of 1982 permitted an individual state to veto a site within its borders, but the veto could be overridden by a majority of both houses of Congress. In 1987 Congress instructed the Department of Energy (DOE) to proceed with detailed underground studies at Yucca Mountain, Nevada, and to abandon plans to study two other western sites concurrently. Both DOE and Congress also capitulated to political pressures to abandon plans to investigate possible sites for a second repository in several eastern states. Nevada has responded by a series of court challenges. The date for opening the first repository has been postponed from 1998 to 2010, to allow for further negotiation and scientific studies.

Fairness in *selection procedures* can help to mitigate the injustice in sacrificing local interests for the national good. But DOE did such a poor job of scientific research that it was criticized by several other government agencies. And Nevadans felt that they had been betrayed politically when the original agreement to build one western and one eastern site was violated. Another way of mitigating injustices is to require the beneficiaries to compensate those who bear the risks. The 1982 act authorizes DOE to provide *compensation to host states,* paid for by a users' fee on the electricity generated from nuclear fuel. In 1987 amendments to the act, the state hosting the first site would receive $10 million per year until the site opens and $20 million per year thereafter. This seems inadequate compensation when there are uncertain health risks, social impacts, and public fears that might stigmatize local communities. Another form of compensation would be the funding of hospitals and health care centers in the area to monitor radiation risks and to provide greater local participation in health decisions.[32] But we must resist the temptation to buy off depressed communities or Native American reservations and to deprive them of an ongoing voice in their own health and safety.

2. Intergenerational Justice

Justice is violated when the current generation benefits from electricity and passes on to future generations some of the resulting risks. According to Rawls, an intergenerational decision is fair if you would accept it when you did not know in which generation you would live. I also mentioned in chapter 3 the biblical view of a covenant between generations and the idea that God's purposes span the generations. Here is a case where the longevity of the risks forces us to think beyond the short time frame of political processes and economic calculations. The present generation has a responsibility to pay for the costs of reducing the risks to posterity. The standards of the Environmental Protection Agency (EPA) specify that radiation released from a nuclear waste depository should not cause more than 1,000 deaths in 10,000 years (which is the time it will take for the level of radiation in typical reactor wastes to fall to that of the uranium ore from which the fuel was extracted).

If we considered the welfare of future generations, would we favor storing radioactive wastes permanently or retrievably? While a *permanent* site costs more initially, it entails no future maintenance costs, and sabotage and terrorism could not occur because the wastes are buried in solid rock. But a *retrievable* site offers two advantages: it would be far easier and cheaper to get at the wastes if problems did show up; and the wastes would be accessible if future generations wanted to reprocess them (to reclaim some of the potential nuclear energy still present) or to use them in a future technology not now imagined.

Alvin Weinberg advocates *retrievable storage* with tight security measures to guard against flooding, drilling, or sabotage. He describes it as a "Faustian bargain" of energy growth at the expense of long-term vigilance. He says it would require a "nuclear priesthood" of highly disciplined technicians to maintain surveillance and guard the buried wastes.[33] But is it realistic to count on the stability of social institutions on a time scale of 10,000 years? No social order in history has lasted more than a few centuries. The occurrence of two world wars and many smaller wars and revolutionary upheavals in this century suggests that in a fast-changing world we should use methods that do not depend on human institutions over periods of many generations.

On balance it seems to me that our obligations to future generations would best be fulfilled by *permanent sealed repositories,* but that the first couple of sites should be left open and carefully monitored for leakage or other problems for several decades so that any modifications could be made before they are sealed. From current fees a substantial endowment fund should be set aside, and interest from it would cover future monitoring and modification costs and underwrite insurance against accidents. It is conceivable that a new technology might allow long-lived isotopes in plant wastes to be transmuted into isotopes with shorter half-lives, in order to keep more of the risks within our own generation. But with current technologies this would be extremely expensive, and permanent sealed repositories seem a more promising way of avoiding significant risks to future generations.

3. Public Confidence

In many nations, including France, Britain, the United States, and the Soviet Union, the *credibility of government agencies* planning nuclear waste disposal was undermined by their secrecy and their historic failure to protect the public and the environment. In many cases the same agency was responsible for nuclear weapons production, nuclear power generation, and nuclear waste disposal—and for public health and safety regulations. A 1982 report by the Office of Technology Assessment (OTA) of the U.S. Congress noted the erosion of public confidence in the Department of Energy and proposed an independent Waste Management Agency to sponsor scientific research and to encourage greater participation by citizens, state agencies, and independent experts.[34] But DOE has continued to manage the disposal of wastes, and public confidence has not been restored. While some public hearings have been held, DOE has provided only limited opportunities for access to information and for independent scientific review, and it has not been sensitive to the diversity of interests involved or the intensity of public opposition.[35] The controversies over waste disposal, like those over reactor safety, point to the importance of public involvement and the full inclusion of social and environmental costs in the planning of large-scale technological projects.

3. THE FUTURE OF NUCLEAR POWER

If waste disposal problems can be solved, should nuclear power be reconsidered now that coal seems to pose such a threat to the environment? A 1984 OTA study concluded that a revival of the nuclear industry could occur only under two conditions. First, public trust in the government agencies and in the utilities would have to be restored. Second, only simpler, smaller reactors with inherent safety features and standardized designs might achieve acceptable safety and reduce construction and licensing times.[36] By 1990 DOE and the reactor vendors were considering smaller, standardized modular reactors that could be built in factories rather than on site and that would incorporate passively stable features.[37]

Passively stable ("inherently safe") reactors rely on the laws of nature, rather than on human intervention or mechanical systems, to limit the effects of any malfunctions. In smaller reactors, emergency cooling can be obtained by gravity flow from water in elevated tanks or by the natural circulation of water, helium, or air—without the use of complex systems of controls, pumps, and valves. Reactor cores can be designed so that the reaction itself slows down as the temperature rises. The uranium fuel can be formed into pellets, each surrounded by layers of graphite and ceramic that withstand temperatures far higher than do the metal-clad fuel rods currently used. A 135-megawatt graphite-moderated helium-cooled reactor (one-seventh the size of many current reactors) and a 155-megawatt reactor cooled by liquid sodium are among the designs under consideration in the United States. Canada's 300-megawatt heavy water reactor has an excellent operating record, and Sweden is working on a light water reactor in which naturally circulating borated water is used to slow down the reaction and to cool the core.[38]

Morone and Woodhouse hold that if we learn from the mistakes of the past we would develop a second generation of nuclear reactors only *slowly and cautiously*. We would incorporate safety in the initial design rather than by complex add-on systems. We would do research for a decade on diverse types of passive design, and then for a second decade we would develop prototypes of several alternatives for testing at a remote site so that we could learn from experience and make modifications. The whole process would be open to scrutiny by the OTA, independent experts, and environmentalists. After two decades, when both the greenhouse effect and the prospects for solar energy are clearer, a decision about commercialization could be made.[39]

Such a cautious research program on smaller passively stable reactors would *keep options open*. Along with a satisfactory resolution of waste disposal, it might lead to a restoration of public confidence in the nuclear industry. But I could support such a program only if it were funded at a level considerably lower than that for research on conservation and renewable sources. Moreover, given current attitudes in the industry and DOE, the program is not likely to be cautious and open or to restore public confidence. It would probably divert funds from alternatives and encourage the belief that we do not have to take conservation seriously. For at least the next decade, measures to *conserve* electricity will almost certainly cost less than the *production* of an equivalent amount of electricity from a new generating plant of any kind. During that interval we can expect the cost of solar sources to continue to fall if we allocate more funds to research on them. Even if smaller standardized reactors do achieve an acceptable level of safety, they are not likely to be competitive with solar sources if the true costs of the whole nuclear fuel cycle are included.

In the past, nuclear power has been *heavily subsidized* by public funds. The initial research was publicly funded, and free enriched uranium was supplied to private reactors. The Price-Anderson Act of 1957 limits a utility's liability from a reactor accident to $60 million, even though recent government studies put the damage from the maximum credible accident at $150 billion. Federal subsidies to the nuclear industry were still $12 to $15 billion a year in the mideighties. Waste disposal research and management will undoubtedly cost many times the disposal fees collected to date, and the safe dismantling of reactors at the end of their useful life (thirty or forty years) could cost up to $30 billion each.[40] The British government, as part of its privatization program, tried to sell off its nuclear plants in 1989, but private investors were not interested when they discovered the estimated costs of decommissioning and waste disposal.[41] Even with simpler modular reactors, the high cost of all these processes would have to be included in any realistic appraisal.

Breeder reactors and the *reprocessing* of spent fuel would reduce the cost of nuclear fuel. Breeder reactors are highly efficient, deriving sixty times as much energy from a ton of uranium as light water reactors do and extending reserves by making low-grade ores economical. Reprocessing plants are operating in Britain, France, and India; they take spent fuel and separate out the plutonium so that it can be recycled as reactor fuel. Japan is planning to ship its used fuel to Europe and bring the reprocessed plutonium back again in convoys, which would be a tempting target for terrorists. The United States rejected both breeders and reprocessing because they involve the transportation of plutonium, of which a small quantity could easily be diverted to make nuclear weapons. A nation that wants to make nuclear weapons now has to set up breeders or reprocessing or uranium enrichment plants—all of which are complex, expensive, large-scale operations. But if plutonium were circulating in nations around the world, it would be easy for terrorist groups to steal a small quantity. It takes only 20 pounds of plutonium to make a bomb, and 100,000 pounds would be available each year if spent fuel at current levels were being reprocessed and recycled. Plutonium can be handled with relative ease, and theft during storage or shipment would be tempting to small nations or revolutionary groups.[42]

It is of course difficult to prevent *the proliferation of nuclear weapons*. There are several routes for acquiring them. Several nations have not signed the Nonproliferation Treaty of 1968, and they do not allow the inspection of their facilities by the International Atomic Energy Agency. Even with stricter and universal international inspection, some diversion of weapons-grade material could occur. But the widespread circulation of plutonium would make acquisition of a bomb much easier. Even if it were intended only for purposes of extortion, events could get out of hand, especially in the midst of tensions between major powers or Middle Eastern rivalries. If nuclear war is the largest catastrophe that threatens both humanity and the planet, anything that even slightly increases the risk of its occurrence must weigh heavily in policy choices. Effective and verifiable arms control would be easier if plutonium were not readily available (see chapter 7). I would thus be opposed to breeders and reprocessing even if they posed no safety problems. I also support proposals for international jurisdiction over the crucial portions of the current fuel cycle (uranium enrichment and waste disposal), preferably at regional international centers.[43]

I have not discussed *fusion* because the temperatures necessary for a self-sustaining reaction have proved so difficult to achieve even in the laboratory. In 1991 a European team produced a fusion reaction of 1.7 million watts for two seconds, releasing much less energy than was used to initiate the reaction but more than had been released in any previous experiments. The experimenters cautioned that commercial fusion remains at least fifty years away.[44] Fusion has the advantage that its main fuel, heavy hydrogen, is present in small quantities in ordinary water; if tritium is also needed, it could be produced in the fusion installation. Fusion technology would escape the problems of highly radioactive fuels and wastes that occur with fission, but the temperatures of millions of degrees and the local radioactivity produced would entail their own risks. Moreover, if commercial fusion can be achieved it will probably be only with very large-scale equipment that could not be scaled down.[45]

In short, I would favor a modest program of research on small passively stable reactors and on fusion, in order to keep future options open. But I believe that most of our efforts should be directed toward renewable sources and conservation because they are more promising alternatives for the next few decades.

III. RENEWABLE SOURCES

Renewable energy sources contribute to sustainability and justice, and in most cases the environmental impacts are lower than those of fossil fuels. Many forms of renewable energy involve decentralized systems amenable to local ownership and participation by individuals and communities. Their direct economic costs are only just beginning to be competitive with coal and oil, but these costs could be reduced by appropriate policy initiatives, and the indirect costs are far lower.

1. SOLAR ENERGY AND SUSTAINABILITY

An inherent feature of all renewable sources is their sustainability. Their energy will keep on arriving from the sun without depletion for millions of years. Using solar energy is living off income, whereas using fossil fuels is living off capital. One nation's use of solar energy does not affect the amount available to other nations or to future generations. While solar technologies often require considerable initial investment, the fuel itself is free; using it does not drive up the price for other people. Moreover, sunlight is available in all parts of the earth. There are regional variations in the duration and intensity of sunshine, but these are smaller than the huge geographical disparities in the distribution of fossil fuel and uranium reserves. Solar sources are thus compatible with both international and intergenerational justice. They fulfill Rawls's criterion of justice: they are the sort of technology you would try to develop if you did not know in which nation or in which generation you would live.

The main obstacle to renewable energy has been the costs, though in most cases costs have fallen dramatically and would undoubtedly fall faster with more research and with mass production. There are also some environmental problems, though these are seldom as serious as those from fossil fuels. In particular, none of these technologies produces the main greenhouse gas, carbon dioxide, and none of them except biomass produces air pollutants or soil erosion. Renewable systems cannot be turned into military weapons, and they cannot produce sudden catastrophes (except for the failure of large dams). The very diversity of solar technologies and their geographical dispersal would diversify the risks. By reducing dependence on foreign oil, they would contribute to world stability and peace and reduce balance-of-payment deficits. In terms of environmental impacts, renewable sources appear in general to be worse than nuclear plants operating normally, but better than coal or oil, for equivalent energy production. But the materials and the fabrication of the equipment for renewable technologies require energy that in many cases would come initially from fossil fuels.

1. Hydroelectricity. Electricity generated from water power has for decades been cheaper than that from any other source. Nearly a fifth of the world's electricity is now water generated. North America has developed 59 percent of its potential hydropower, Europe 36 percent, Asia 9 percent, Latin America 8 percent, and Africa only 5 percent. But large dams entail great environmental costs: inundation of forests, farmland, and homes; accumulation of sediments; and harm to fish migration and aquatic life. In some parts of the world, small dams produce electricity with only modest environmental impacts. China, for example, has 90,000 small water turbines providing electricity in rural areas.[46]

2. Water and Space Heating. In Israel, 65 percent of domestic water heating comes from rooftop solar collectors, which since 1980 have been required on all new residences up to nine stories high. In most countries, space heating with passive solar design and good insulation pays for itself in fuel savings in a decade or two, but most consumers want a more rapid payback on their investment. Building contractors are more interested in keeping the initial home price down than in reducing long-term costs. Retrofitting old houses is expensive, so solar heating does little to help low-income families. But if the cost of oil and gas rises, solar heating will become more attractive when combined with conservation measures.[47]

3. Biomass. Twelve percent of the world's primary energy today comes from organic matter, mainly wood, but a serious fuelwood crisis has occurred in the Third World from the cutting of trees with little replanting. Brazil obtains 62 percent of its automotive fuel from sugar cane converted to ethanol, which was competitive with oil in 1981; but with oil prices lower in the late eighties, government subsidies were continued largely because the program creates 450,000 jobs and helps rural development instead of sending billions of dollars overseas for oil. But ethanol from crops uses scarce agricultural land. If 40 percent of the entire U.S. corn crop were converted to ethanol it would meet only 10 percent of the nation's auto fuel demand.[48]

A more promising alternative is the planting of fast-growing trees, shrubs, or grasses on marginal land or rangeland, or the integrated cultivation of trees and crops (agroforestry). The burning or fermenting of forest and municipal wastes is another energy source. Crop and animal wastes can be fermented in tanks to create a biogas fuel, with the residue returned to the land as fertilizer. The growth of such biomass absorbs as much carbon dioxide as is released when it is burned, so it produces no net greenhouse effect, and sulfur emissions are low, but careful design and management is needed to limit soil erosion.[49]

4. Wind Power. Electricity generated by turbines at windy sites now costs 6 to 8 cents per kilowatt-hour (kwh), which is competitive with current U.S. utility purchases during peak demand periods (when utilities pay a premium in order to avoid having to build new plants) and is approaching the cost of electricity from new coal plants (5 cents per kwh). After some failures in the 1970s, the wind industry expanded in the early 1980s, especially in California, but it has slowed since 1985 when oil prices fell, federal tax credits were eliminated, and federal funds for research on renewable sources were cut by 85 percent. Turbines at windy sites could supply 20 percent of current U.S. electricity demand, and costs would fall with improved designs and mass production. They could be located in mountains or deserts or on farming and ranching lands, with little interference to other activities. India is planning for 10,000 turbines producing 1000 megawatts by 1995, and the Netherlands hopes to reach the same goal by 2000.[50]

5. *Solar Thermal Electricity.* Sunlight can be concentrated by parabolic reflectors to produce steam to drive generators. In California, 350 megawatts of solar thermal electricity are being produced at costs competitive with peak rates, and the costs are expected to fall in the nineties as larger and more efficient units are installed. Already their costs (8 cents per kilowatt-hour) are lower than that of electricity from recent nuclear plants (10 to 12 cents per kwh). Solar steam can also be used for process heat in nearby industries.[51]

6. *Photovoltaic Cells.* Electricity from PVs has the greatest potential for the future. The costs of both crystalline silicon and thin films of amorphous silicon fell dramatically in the eighties but are still prohibitive except for remote locations (communications, navigation aids, or isolated villages). In 1991 Texas Instruments announced plans to market flexible solar roof panels that look like metallic sandpaper; because they are made from metallurgical grade silicon, they are expected to cost a fifth of current PVs and would be competitive with coal-generated electricity. PVs create no carbon dioxide or air pollution (except in the fabrication of the equipment). The materials used—iron and silicon—are abundant. The land area needed for solar cells is comparable to that for coal plants of equivalent capacity if land strip-mined during thirty years is included. Panels in deserts, rocky areas, roadside rights-of-way, or on factory or home rooftops would avoid competing with other land uses. Their modular character is adaptable to small or large arrays. PVs to supply all current U.S. electricity would require a square of land 100 miles on a side—a ninth the area of Arizona, or a fifth of the national area used by the U.S. military.[52]

Wind, solar thermal, and PVs share one limitation: they all provide energy *intermittently,* varying with time of day and weather. They are useful for feeding electricity into a grid, thereby reducing the consumption of fossil fuels and avoiding the construction of new plants for daytime peak loads. Solar thermal can be combined with gas turbines in hybrid systems; gas takes over when there is no sunlight. But before wind or PVs could be used as the main source for baseloads, less expensive ways of *storing electricity* must be found. A dozen auto manufacturers, including Ford, GM, and Toyota, are planning to produce electric cars by 1995. Sodium-sulfur or nickel-cadmium batteries are an improvement on lead-acid batteries, but they are bulky and expensive.[53]

Photovoltaic-generated electricity passing through water releases *hydrogen* that can be stored and used for transportation or residential heating. Hydrogen produced from PVs in sunny desert areas could be compressed and distributed inexpensively in a pipeline system like that now used for natural gas. Autos would have to carry the hydrogen in high-compression tanks or in the form of metal hydrides. Hydrogen could also be burned in fuel cells on electric-powered cars, with twice the efficiency of internal combustion motors. If 150,000 square miles (2 percent of the world's desert areas) were used for PVs producing hydrogen, they could supply as much energy as all fossil fuels today. Hydrogen appears very promising in the long run, since it burns so cleanly; it emits no carbon dioxide, sulfur dioxide, or particulates. But extensive research is still needed on its production, storage, and combustion.[54]

Only *a combination of these renewable sources* could reduce the consumption of fossil fuels significantly. By one recent estimate, if total energy consumption remains constant, the U.S. supply in 2020 could be 7 percent nuclear, 4 percent hydro, 6 percent wind, 18 percent biomass, 24 percent solar, and 42 percent fossil fuels (as compared to 86 percent today).[55] But such a shift will take a concerted effort.

Under the influence of market forces alone the transition will be slow because market prices neglect many *environmental and human costs,* even when some of the externalities have been internalized through emission standards and environmental and safety regulations. As I have said, the market takes a short-term view and discounts costs and risks to future generations. When irreplaceable resources such as oil are used up, potential benefits are transferred from the future to the present. Neither the environment nor our grandchildren enter current economic calculations. A study in the former West Germany estimates the external costs of coal (not including global warming) at 2.4 to 5.5 cents per kilowatt-hour, and those of nuclear power at 6 to 13 cents.[56] In the United States, the state regulatory agencies often compare alternatives (including conservation) before they approve the plans of utilities for constructing new plants. The Wisconsin agency adds 15 percent to coal plant figures to reflect external costs, and in New York 25 percent is added.[57]

Moreover, fossil and nuclear fuels have been *heavily subsidized* in the past. For decades, oil was subsidized by depletion allowances, tax credits, highway construction funds, and regulated prices. Some of the subsidies to the nuclear industry were mentioned earlier. By one estimate the total annual U.S. energy subsidy in 1984 was $44 billion, mostly for coal, oil, and nuclear fuels.[58] Because the energy industry has invested heavily in these technologies, most of its research and its political influence supports continued use of these fuels. The cumulative expenditures on solar research since World War II are a minute fraction of that spent on other sources, so it is not surprising that it lags behind. Of the $7 billion that twenty-one industrial nations spent on energy research in 1989, 47 percent went to nuclear fission, 15 percent to fossil fuels, 12 percent to fusion, and only 7 percent to renewables and 5 percent to conservation.[59]

In the late 1970s under Carter, *national policies* to stimulate renewable energy technologies were adopted in the United States, but these were mostly revoked during the 1980s under Reagan. Federal funding for renewable research and development fell from $718 million in 1980 to $115 million in 1989. With the new concern for global warming and the dangers of dependence on imported oil, the time has come to reinstate tax credits for renewables and restore research and development funds at least to their former level. These could be financed by a tax on imported oil, higher taxes at the pump, or the proposed carbon tax. Or they could be paid for by a small fraction of the funds released if the space station and Star Wars research were canceled. Purchases of solar equipment and fuels for government facilities, offices, and vehicles would also help to create economies of scale. The diverse long-term benefits of renewable energy would justify such policies.

2. DECENTRALIZATION AND PARTICIPATION

In chapter 2, I defined freedom as *participation in the decisions that affect one's life.* I distinguished three forms of freedom: (1) participation in policy decisions through political processes, (2) participation in economic decisions through the marketplace, and (3) participation in work-related decisions through activities in the workplace. All three forms of participation are very limited when the individual faces a large centralized organization, such as those that produce fossil or nuclear energy. Oil companies are among the world's largest financial institutions, wielding immense economic and political power. Nuclear plants require huge capital investments, so they can be built only by large private utilities or government agencies.

By contrast, some forms of solar technology can be owned and controlled by *individuals and communities,* which would encourage local self-reliance and a plurality of centers of decision—counteracting the trend toward the concentration of economic and political power. Biomass involves many actors: growers, harvesters, processors, and distributors. A biomass system would be coordinated by the market and by government policy, but it would have room for many local initiatives. Economies of scale may rule out very small operations, but intermediate-scale activities can be efficient. Solar heating and insulation and rooftop photovoltaics offer scope for small businesses in the installation and repair of equipment, drawing on local skills and shops that would integrate well with existing building trades. Solar technology is more labor-intensive than coal-fired plants and oil refineries, which produce fewer jobs per dollar than any other major industry. The manufacturing and installation of solar components would be well dispersed geographically.[60]

Community-level systems can be run as cooperatives, small companies, or municipal utilities. The burning of municipal wastes to produce electricity helps to solve the waste disposal and landfill problems faced by many cities. Wind generators and small dams can be locally owned and operated. With any form of fuel, the *cogeneration* of heat and electricity substantially improves efficiency. Two-thirds of the energy input into most electric plants is simply wasted as heat. Instead, the heat can be used in industrial processes or in community-level systems. District heating, integrated systems, and the recovery of energy from urban sewage and agricultural residues are best carried out locally.

Decentralized systems offer *flexibility* in adapting to local conditions. Solar sources are diverse and can be matched to local resources and to differing end-use and temperature needs. They typically involve shorter construction times than large installations. In addition, they are *less vulnerable* to sabotage, terrorism, or catastrophic accidents. The stakes are smaller when errors or accidents do occur. Careful design and monitoring for environmental damage may be necessary, but not the elaborate safety precautions and security measures that nuclear plants require. Finally, the costs and the benefits of small-scale systems *tend to fall on the same people.* In the case of large power plants, nearby communities receive much of the pollution and environmental degradation, while most of the benefits go elsewhere; it is not surprising that local opposition to the siting of large plants has been rising. Decentralized systems offer a more just distribution of costs and benefits and the prospect of less vehement community opposition.[61]

Critics of decentralization say that small-scale systems are neither efficient nor socially desirable. Solar components such as photovoltaics will have to be mass-produced to be cheap enough to be widely adopted. Like the auto, solar equipment can be individually owned and decentralized in use, but it can be cheaply produced only with the economies of scale possible in large factories. Moreover, with the existing distribution of economic power it is likely that many forms of solar energy will be controlled by large companies. The social regulation of such companies may be difficult, but it is not impossible. Even decentralized systems require central regulation. A million smoky wood stoves would be no improvement on coal-burning plants when it comes to air pollution.

Even if local self-sufficiency were possible it might not be *equitable;* an urban ghetto or a rural village would receive much poorer electric service than an affluent suburb. By stressing local self-reliance, say their critics, decentralists have neglected the importance of social integration and cooperation for the common good. Most people are more interested in the cost and convenience of energy than in local control; they are willing to pay someone else to deliver energy to them. Critics also fear that renewable sources will not be adequate to meet energy demands; severe shortages would lead to slower economic growth, rationing, and government intervention to allocate scarce supplies. Authoritarian responses to scarcity might leave us with more centralization than we have now.[62]

Furthermore, the correlations between *energy systems* and *social structures* are loose and ambiguous. Solar enthusiasts and nuclear enthusiasts sometimes share the assumption that the right technology, the "technical fix," will solve our social problems. But the social and institutional context in which a technology is deployed is often crucial in determining its ultimate social consequences. Nevertheless, we must acknowledge that some technologies have distinctive potential that can be supported by deliberate social policies. The relationships between scale, efficiency, justice, and participation have to be examined separately for each part of an energy system if we are to understand the trade-offs between local participation and other values.

My own conclusion is that *a mix of small and large systems,* adapted to varied tasks and conditions, is preferable to either type alone. For example, the dispersed production of electricity reduces transmission and distribution costs, which constitute half of the consumer's electric bill. But the electric grid offers convenience and reliability in service, and it allows power to be transferred between regions with different peak-load hours or unused capacity. Local sources should therefore be tied into networks wherever possible. Large blocks of electricity for heavy industry and urban areas will probably require central generation. Some photovoltaic installations and some biomass production can be locally controlled, but competition for scarce land may make large solar arrays in remote areas or large plantations of plants or trees for fuel desirable.

But a mix of scales will require a deliberate effort to develop the *untapped potential of smaller systems.* In the past we have subsidized large-scale technologies. Past investment in such systems has created a momentum for their perpetuation, whereas the constituencies for the small are diffuse and less well organized. The centralization/decentralization debate also involves differences in value priorities and in visions of the future, to which we will return in the final chapter.

IV. CONSERVATION

Conservation is the most cost-effective response to energy shortages. Most ways of reducing energy demand are cheaper than increasing supply by an equivalent amount. It costs less to save than to produce a barrel of oil or a kilowatt-hour of electricity. But conservation also furthers every one of the values discussed in this volume. Conservation in industrial nations reduces global inequities in energy consumption. It cuts the waste of the irreplaceable resources that future generations will need. It is essential for the transition to sustainable sources. Conservation measures seldom harm the environment. A dollar spent on conservation is seven times as effective in cutting carbon dioxide emissions as a dollar spent on nuclear power. Reducing oil imports diminishes a major cause of international conflict. We will look particularly at two values that might seem incompatible with reduced energy use: economic development and personal fulfillment. Finally, we examine the distinctive needs and opportunities of the Third World.

1. ENERGY AND ECONOMIC DEVELOPMENT

In chapter 2, I said that we should seek economic development: selective economic growth that does not jeopardize equitable distribution, resource sustainability, or environmental protection. The conservation of energy and other resources can make a major contribution to economic development.

Until the oil embargo of 1973, the graphs of U.S. energy consumption and gross national product rose on similar curves, and it was assumed that demand would continue to rise in a growing economy. Restricting energy growth, it was said, would automatically restrict economic growth. But from 1973 to 1987, U.S. energy consumption actually fell, while the GNP grew by 35 percent.[63] The nations in the International Energy Agency lowered their energy use per unit of GNP by 24 percent from 1973 to 1989.[64] While higher oil prices did initially result in somewhat slower economic growth, the combined effect of various conservation measures showed that, over a longer period, economic growth can be at least partially uncoupled from energy consumption. But since 1988, with oil prices lower, energy use has started to rise again.

Forms of economic growth that are compatible with *rapid reduction in energy consumption* in industrial nations are essential to global justice and the transition to sustainable solar sources. The United States spends 10 percent of its GNP on energy, while Japan spends only 4 percent, but even Japan has barely begun to fulfill the potential for conservation. Four forms of conservation are most promising.

1. *Transportation.* Transportation fuel comprises two-thirds of U.S. oil use. In response to fuel economy standards introduced by Congress in 1975, the average fuel economy of new cars doubled from 14 miles per gallon (mpg) in 1975 to 28 mpg in 1985, but the standards were subsequently relaxed.[65] The auto industry continued to oppose higher standards, and in early 1992 Congress was still unwilling to require more than the current standard (27.5 mpg average for the new cars produced by any company). Readily available technologies could produce cars with 38 mpg, and the extra cost would be offset by the savings on fuel.[66] The Volvo LPC 2000 gets 81 mpg on highways, and the four-to-five-passenger Toyota AXV prototype achieved 98 mpg on the combined urban/highway test administered by the EPA.[67]

Mass transit and railways are far more efficient than autos and trucks, produce much less carbon dioxide and harmful air emissions, and would reduce the congestion that is paralyzing many cities. Frequent, rapid service with small buses using express lanes will attract urban riders. Electric trolleys and light railway vehicles are much less expensive than underground subway systems, and parking spaces for autos and bicycles can be provided at suburban stops. City centers should be reserved for pedestrians and bicycles. Public transportation also helps low-income families, many of whom do not own cars.[68]

2. *Buildings.* Large savings are possible in commercial and residential heating and cooling. In office buildings, better insulation, thermal reflective glass, more efficient lighting, and the recovery of waste heat from ventilated air can cut energy input in half. For new homes, insulation standards, building codes, and loan requirements could lead to much lower heat losses. Superinsulated homes in Sweden save 89 percent on heating bills and pay for their additional cost in five years. But housing stock turns over slowly; more rapid savings can

come from retrofitting existing houses with insulation and weather-stripping, saving up to 50 percent on heating bills. Yet many home owners are reluctant to invest in such improvements unless the payback time is three or four years—a shorter period than they expect in other investments. Renters have little incentive to improve their landlords' houses, and landlords and housing contractors do not pay the energy bills. Some utility companies are taking an active role in home energy audits, subcontracting and guaranteeing the installation of insulation, and arranging loan repayment from the saving in future fuel bills. By helping to control demand, utilities can avoid having to build costly new power plants. But the institutional obstacles to energy conservation in homes can be overcome only by stronger governmental leadership, including housing standards, efficiency ratings, and low-interest loans to low-income families.[69]

3. Industry. Many companies have already reduced their energy demand substantially. More efficient equipment and processes have been introduced in such energy-intensive industries as steel, aluminum, paper, glass, and chemicals. New techniques have been used for waste heat recovery and for the cascading of processes requiring successively lower temperatures. Process steam can be cogenerated with electricity, at twice the efficiency of separately generated electricity. The recycling of materials usually takes far less energy than primary extraction from raw materials. A shift is also occurring within industry from the energy-intensive processing of basic materials to fabrication, finishing, and high-tech products. Electronic systems use few materials and little energy, and they can be used to monitor and control the flow of materials and energy in other industrial processes.[70]

4. Electricity. In the past, low electricity rates were offered to large users, partly to promote sales and partly because of economies in distribution and billing. Today, "inverted rate structures" are proposed to discourage heavy use. "Lifeline rates" would start with a low price for a basic block of electricity for minimal needs per person, with rising rates thereafter. This would offer some help to low-income families and would encourage conservation. Lower off-peak rates would produce some shift in demand away from peak-load periods. For appliances, testing and labeling, and perhaps efficiency standards, would be helpful. Some air conditioners and refrigerators on the market are twice as efficient as others. Heat pumps can use the same equipment for heating in winter and cooling in summer. Lighting consumes 20 percent of U.S. electricity, and this could be greatly reduced. A 15-watt fluorescent bulb produces the same amount of light as an ordinary 75-watt bulb. It lasts thirteen times as long, more than repaying its higher initial cost; if the electricity is coal generated, each bulb saves a ton of carbon dioxide and twenty pounds of sulfur dioxide emissions during its life.[71]

Apart from conservation and efficiency measures, selective growth requires other *changes in the economy.* The energy-intensive manufacturing sector has been shrinking while the service sector has been growing (including health care, financial services, information services, and education), and this shift can be carried further. Sudden and severe energy shortages do indeed result in economic and social disruption. The 1973 oil embargo led to job layoffs and economic recession; minority groups were laid off first and low-income families were hardest hit. But gradual shifts toward efficiency and services produce a stronger economy that spends less on energy and is less vulnerable to future oil crises or price fluctuations.

Expenditures on energy conservation also create *more employment* than expenditures on energy production. A European Community study found that conservation expenditures produced more jobs than traditional energy investments.[72] A U.S. study concluded that every dollar invested in conservation produced twice as many jobs as a dollar invested in conventional energy industries.[73] Home insulation and public transportation are labor-intensive, and the jobs are well distributed geographically. Higher energy prices usually result in some substitution of labor for energy. The shift to less energy-intensive products will of course decrease jobs in some industries, but the new industries and services that replace them will more than compensate for such losses.

Issues of justice in conservation do require special attention. The paradox is that the poor cannot afford the higher initial costs for energy-efficient homes or appliances, and they end by paying more in the long run in higher utility bills. They live in poorly insulated homes and they spend a much larger fraction of their income on energy than more affluent people do. The poorest tenth of the population spends 34 percent of its income directly for energy, while the richest tenth spends only 2 percent.[74] It is misguided, however, to try to keep energy prices low to protect the poor, for this reduces incentives for conservation. Instead, both financial and energy services should be targeted especially on low-income families: fuel assistance in emergencies only, home insulation subsidy, a revised utility rate structure, greater support for public transportation, and so forth.

Conservation does involve some risks to *health and environment,* but they are relatively small. Raw materials and energy are needed to fabricate insulating materials. Some insulation gives off formaldehyde, which causes respiratory illness. Sealing a house tightly against air leakage allows a buildup of radon gas (a cause of lung cancer) from cement, bricks, or soil in some regions; this can be avoided with air ventilation and a heat exchanger, but at considerable cost.[75] Smaller cars save energy and pollute less, but the fatality rate in accidents is higher. Conservation does entail environmental and health risks, but in most cases these are less than the risks from equivalent energy production. Overall, conservation gets high marks on each of the criteria we have been using.

2. LIFE-STYLES AND PERSONAL FULFILLMENT

Improvements in efficiency allow people to obtain the same goods and service with a lower expenditure of energy. These pragmatic conservation measures would require very little sacrifice and would not harm *the quality of life.* Comparisons of 35 industrial nations have shown no correlation between energy use and a varied set of social indicators (including life expectancy, literacy, unemployment, crime, suicide rates, and environmental quality indexes).[76] Another study showed that Sweden had a GNP per capita close to that of the United States and outranked the U.S. on almost every social indicator but used 40 percent less energy per capita. In Sweden, heavy gasoline taxes, smaller cars, and excellent public transportation (aided by geographical compactness) had produced high transportation efficiency, while commercial and residential use of energy was much less wasteful. Britain came out between Sweden and the United States in energy use and on most of the social indicators.[77]

The conservation measures described above rely on technical changes, economic incentives, and legislated standards, but they would be more effective if accompanied by relatively minor *changes in individual behavior.* Turning thermostats down in winter, turning off unused lights, using public transportation, or recycling glass, paper, and aluminum cans may seem inconsequential, but many small actions add up to significant savings. Educational programs can increase awareness of the importance of conservation and of simple ways to save energy. Psychologists have shown the importance of social reinforcement and group support in behavior change.[78] But the political constituency for conservation is difficult to mobilize because the benefits are diverse, long-run, and widely diffused, whereas well-organized and powerful groups in industry, labor, and government receive immediate benefits from continued dependence on fossil and nuclear fuels.

Changes in life-styles could further reduce energy consumption by altering the goods and services people seek rather than simply providing the same goods and services more efficiently. A consumer society encourages a high demand for material goods, highly processed and packaged foods, and energy-intensive products. During the 1970s, a significant number of young people in North America and Europe were disillusioned with the prevailing materialism and individualism and the stresses of a competitive society. They sought alternative sources of satisfaction in personal growth, human relationships, smaller communities, and greater harmony with nature, and they adopted more frugal life-styles. Some of them joined communes or new religious movements. Others tried to find meaningful work, family life, spiritual growth, and fulfilling activities within traditional institutions.[79] Some of these experiments were short-lived, but others continued in the 1980s, expressing values that are neglected among the dominant goals of industrial societies.

In whatever form they are expressed, *new views of personal fulfillment* can lead to lower levels of resource consumption and energy use. I do not believe that moral exhortation or appeals to austerity or sacrifice will have much influence today. But a revitalization of Western religious traditions could bring new perceptions of the good life and a new recognition of less resource-intensive sources of satisfaction that would alter patterns of consumption in affluent nations. In chapter 9 I will discuss the shift from the prevailing industrial paradigm to *a postindustrial paradigm* more appropriate to an interdependent world with finite resources. But such paradigm shifts occur slowly except in response to severe crises. In the meantime, the practice of individual frugality must be combined with political efforts for effective legislation, economic incentives, and efficient technologies, which are the most promising social paths to energy conservation today.

3. ENERGY IN THE THIRD WORLD

The Third World is highly dependent on *fossil fuels*. The disastrous impact of higher oil prices was mentioned earlier. Oil has played a vital role in transportation, agricultural irrigation, and fertilizer production. Only a few Third World countries have coal deposits, and the release of greenhouse gases makes increased use of coal problematic. Several developing nations have significant natural gas deposits, and others have not yet been explored; these could be very helpful during the next decades, though they would require the construction of pipeline systems. *Nuclear power* has played a negligible role in the South. It is expensive and perpetuates dependence on equipment and exports from the North. A second generation of smaller, passively stable reactors might be useful in urban and industrial centers in the more advanced developing countries, but most areas lack the electrical grids necessary for distributing large blocks of electricity.

Waterpower and *biomass* account for nearly half the primary energy use of the Third World, but fuelwood, the chief source in rural areas, has been used much faster than it has been replaced. We have seen that deforestation without replanting has resulted in severe soil erosion and flooding as well as fuelwood shortages. Cheap *solar technology* would be a great boon to the sun-rich Third World. It would reduce dependence on other nations for expertise and fuel. Solar cookers, solar water pumps, methane digesters, small hydroelectric dams, windmills, and photovoltaic cells would be especially helpful. They would encourage national self-reliance and fit in with local cultures.[80]

Conservation might seem insignificant in developing countries, since the amount of energy that could be saved is so much smaller than that wasted in affluent nations. But in fact conservation is crucial for future prospects in all countries. The Brundtland commission concluded that with strong conservation programs the global energy consumption in 2030 could be only 10 percent higher than in 1980 (rather than growing by a factor of three as in some projections). The report emphasizes the fuelwood crisis, pointing out that an open fire uses eight times the energy of a gas stove of equivalent cooking capacity. It sees promise in biogas, small dams, and solar installations. It advocates incentives for tree planting, community woodlots, and the combination of food and energy production in agroforestry. With careful planning, fast-growing trees could be used efficiently, and replanting would prevent the soil erosion that attended previous fuelwood use.[81]

A report by the World Resources Institute (WRI) concludes that if *efficiency* is strenuously pursued in all countries, global energy demand in 2020 would be only 10 percent higher than in 1980, even with the expected growth in population. Energy use in the North would have to be cut in half, while that in the South would grow, and living standards in the Third World in 2020 could be comparable with those of Western Europe in the midseventies. For $4 billion, a highly efficient wood stove costing $10 could be provided to 400 million rural households in the Third World. Stoves of even higher efficiency would burn biogas from crop and animal residues or producer gas from the partial oxidation of other forms of biomass. The study urges intensive research on better technologies for the conversion of biomass to liquid fuels for transportation. It also advocates the electrification of all villages, using local resources whenever feasible.[82]

The World Resources Institute study looks at energy end use in the framework of a development strategy focused on *basic human needs*. It points out that a quarter of World Bank loans have been for energy but mostly for large projects such as dams or central power plants. "Over 90 percent of energy development aid has gone to building large systems for generating and transmitting electricity. . . . Less than 1 percent of international aid has been used to improve energy efficiency, and most of this 1 percent has gone to the industrial sector."[83] Such aid has paid for foreign consultants and equipment, benefiting the donors, but it has done little to develop indigenous technical capacities. It has favored cities and heavy industries over rural areas and small-scale technologies. The WRI report advocates self-reliant development, production for local needs, and the expansion of industries producing equipment for agriculture and rural areas rather than for export or the urban elite. It suggests that a shift from materials processing to fabrication and finishing would also reduce energy use and increase employment opportunities.

Some authors hold that developing nations will have to use large-scale electric generation (such as large dams and nuclear and coal-fired plants) if their *electricity demands* continue growing at current rates (7 percent per year, with demand increasing fourfold in twenty years).[84] They hold that developing countries are likely to give industrial growth higher priority than environmental impacts, except where the latter are already serious and obvious (urban respiratory problems, for instance). While some development of such capital-intensive technologies may be necessary, they tend to perpetuate foreign dependency and trade deficits, and the benefits will be short-lived. Third World countries should rather seek to develop their own distinctive renewable sources and seek the transfer of technologies more directly related to basic human needs.

V. CONCLUSIONS

By the criteria of evaluation outlined in earlier chapters, the "hard path" of fossil fuel and nuclear expansion is less desirable than the "soft path" of renewable energy and conservation for both industrial and developing nations. Dependence on *imported oil* has been a source of global injustice, international conflict, and environmental damage. Some of the environmental impacts of *coal* can be reduced by improved combustion and emission control technologies, but others, especially the greenhouse effect, are likely to be very serious. *Natural gas* is preferable environmentally and is crucial for the transitional period, but reserves are unevenly distributed and are not indefinitely sustainable.

Nuclear fuels are sustainable for many centuries (especially with breeder reactors) and environmental risks are low in normal operation. But we have seen that economic costs are high, accidents are potentially catastrophic, and uncertainties about radioactive waste disposal are unresolved (though probably resolvable). The risk of diversion of plutonium to nuclear weapons counts against breeders and reprocessing, and all nuclear plants are vulnerable to human errors, institutional failures, and extortion by terrorists. I expressed support for a program of careful, open research on fusion and on smaller, passively stable reactors—but only as forms of insurance in the unlikely event that promising solar sources (especially photovoltaics) and storage methods (such as hydrogen) remain costly despite intensive research and development efforts.

Renewable sources, by contrast, correlate well with all of the values I have defended. They are compatible with justice because they are available to all countries, more evenly distributed than fossil or nuclear fuels, and more readily subject to national self-determination. They reduce threats to world peace from conflicts over oil and from the proliferation of nuclear weapons. They are compatible with participation because they are diverse, predominantly small-scale and decentralized, and amenable to community ownership and local control. They are uniquely sustainable because they do not significantly deplete resources. Their environmental impact on air and water is lower than that of fossil fuels, and problems of land use and soil erosion can be controlled by careful management.

Whatever sources of energy are used, *conservation* can be justified in every nation in terms of all the values we have been discussing. Efficiency research and conservation efforts have had a small role in the budgets of most national governments and international agencies, and yet these measures are essential to economic development and selective growth. A strenuous conservation program is also required if we are to shift a large fraction of total supply from fossil to renewable fuels without highly disruptive energy shortages.

I have argued that the combination of *renewable sources* and *conservation* would not jeopardize economic development in the long run, though it would involve some increased costs and displacements in the short run. Employment opportunities are greater on the "soft path" than on the "hard." Finally, the understanding of personal fulfillment that I have defended does not require ever higher levels of individual consumption in affluent nations but encourages more frugal life-styles that would give additional support to public policies for conserving energy and other resources.

For several decades we will have to use *a mix of "hard" and "soft" technologies,* even as we try to rely increasingly on the latter. Such diversity would allow flexibility in adapting to new developments and to differing local conditions. Keeping diverse options open would hedge our bets and allow creative responses to technological advances, changing economic costs, and new knowledge of environmental impacts—all of which, along with our value priorities, will influence our future policy judgments.

CHAPTER 6

Computers

We live in a world of word processors, computerized banking, industrial robotics, and high-tech weapons. Computers are the central component in the Information Revolution that is beginning to influence almost all aspects of our lives.

In chapter 1, three broad views of technology were presented: optimistic, pessimistic, and contextualist. Views of information technology follow a similar pattern. The *optimistic views* include the popular writings of futurists and the forecasts of many computer specialists.[1] They hold that just as agricultural societies were totally transformed by the Industrial Revolution, so industrial societies will be altered from top to bottom by the Information Revolution. It is claimed that automation will bring high productivity, material abundance, the elimination of repetitive jobs, and more time for the creative use of leisure. The information society will be more egalitarian; old class divisions will be obsolete when knowledge rather than wealth is the source of power. Organizations will be less hierarchical as decision making is decentralized among smaller units connected by computer networks. Democracy will be enhanced by instant referenda and electronic voting, made possible by multi-channel interactive cable systems. Telecommunications will improve world-wide understanding in the "global village."

The *pessimists* include a number of social scientists and a few computer professionals.[2] They assert that information technology augments the power of institutions that are already powerful. It increases the gaps between the information-rich and the information-poor. Automation provides a few high-skilled jobs, but for most workers it leads to unemployment or low-skilled jobs. The new methods of electronic surveillance and computerized personal dossiers facilitate the invasion of privacy and the emergence of the computer state. A handful of companies dominate the world computer market. American and European companies control access to the channels of international communication, resulting in new forms of cultural imperialism through the global media. A large fraction of computer funding and expertise is devoted to military goals. The pessimists are often technological or economic determinists and see little prospect for altering these patterns.

The *contextualists* reject technological and economic determinism and insist that there are alternatives and choices.[3] They portray a two-way interaction between technology and society. They examine the diversity of social forces entering into the design and deployment of particular computer systems. The social consequences, in turn, vary greatly among differing contexts and with differing management strategies, worker responses, and political decisions. Automation can be used to deskill workers, or it can be accompanied by plans for reskilling, job rotation, and worker participation. Computers can centralize or decentralize managerial organization, depending on the strategies pursued. Satellites can lead to cultural imperialism, or they can serve Third World cultural and development goals.

Of these views, contextualism seems to fit best with recent empirical evidence. I will first examine computers in industry and offices and then in the lives of citizens in homes and communities. Another section is devoted to computers in national defense and in the Third World. Finally, I ask how computers affect our self-understanding and what implications recent work on artificial intelligence might have for our view of human nature. The ethical issues throughout the chapter are predominantly those of human values, especially justice and participation. The production of computer circuits requires toxic acids and solvents that must be disposed of with care, but the environmental impacts and energy requirements of computer use are negligible.

I. COMPUTERS AND WORK

In many forms of work today, computers are making enormous positive contributions. Scientists and engineers use them extensively in testing theories against laboratory data and in designing new artifacts. Airlines with computerized reservation and traffic control systems offer to millions of customers a level of convenience and safety that would be otherwise unattainable. Business and government offices handle a volume of transactions and records that would be virtually impossible even with a mountain of paperwork. But computers intensify several perennial work-related problems and create some new ones. We will start with a discussion of computer-controlled automation in manufacturing. The human consequences of the computerization of offices are then considered. Lastly, we explore the diverse ways in which computer networks can affect the locus of decision making in organizations.

1. AUTOMATION AND HUMAN SKILLS

In an automated auto plant, the body welds are carried out by computer-controlled mechanical arms. In an automated chemical plant, all temperatures and material flows are controlled by a central computer. In other industries, all inventories, orders, sales, and work assignments are coordinated by computer networks. What have been the effects of computers in manufacturing?

The main benefit has been increased *productivity* and higher *quality*. The U.S. auto industry invested $80 billion in robotic production equipment between 1980 and 1985, in response to the intense pressure of competition from Japan. In other industries, greater efficiency, smaller inventories, and better coordination of operations are made possible by computer networks. In some cases greater flexibility can be achieved. For example, small batches of metal parts can be readily produced by computer-controlled machine tools. Automation has also enabled machines to do some of the dangerous, dirty, and backbreaking work once done by human beings (in steel mills and chemical processing plants, for example).

The *short-term* impact of automation on *employment* has not been as great as expected. Moreover, automation creates new jobs even as it threatens old ones, resulting in the displacement of individuals but relatively small changes in overall employment. There are fewer jobs for laborers, machine operators, clerical personnel, and lower-level managers, but more jobs for engineers, programmers, sales and support staff, and senior managers.[4] Women and minorities, who often hold unskilled, low-paying, nonunion, and part-time jobs, are particularly vulnerable to such technological displacement.

Some studies suggest that automation will result in a net long-term gain in jobs,[5] but most forecasts anticipate *a net loss of jobs*.[6] In the past, most of the increase in productivity has gone toward wage increases for fewer workers, but in the future, a larger fraction of the gain could be used to shorten the workweek, with a less severe reduction in the number of people employed. Other workers would be employed in the growing service sector of the economy and in energy conservation, recycling, reforestation, and other activities of a sustainable society. But many persons could also devote a larger portion of their time to what James Robertson calls "own work" (self-employment) and community or cooperative enterprises meeting local needs in food, products, and services.[7] In the long run, leisure time is likely to increase, which will present a serious challenge to us to use leisure creatively.

Earlier in the century when mass production and assembly-line methods were introduced, Frederick Taylor's principles of "scientific management" were widely adopted in industry. These principles included hierarchical organization and strict *managerial control*. Sharp distinctions were drawn between mental and manual tasks and between management and labor. Some writers see such division between classes as inherent in capitalism, and they expect it to be reinforced by automation. Harry Braverman argues that power is still based on wealth (not on knowledge or expertise, as the optimists predicted). Technology is a tool of management to effect greater control over labor. Knowledge and skills are taken from workers and put into computer programs, which do not threaten to strike. The remaining workers are progressively deskilled.[8] Harley Shaiken says that social as well as technical values always enter into technological design and that automation is designed as an instrument of managerial control to undermine labor unions.[9]

The thesis that *deskilling* is used by management to control labor is defended by David Noble in a study of the automation of machine tools. In the 1960s, two systems were being developed. In one system, a skilled machinist first cuts a sample of the metal part needed, while the resulting motions of the cutting machine are recorded on magnetic tape or a computer disk. Copies of the tape or disk are then used to run a whole set of automatic cutting machines. In the second system, called Numerical Control, computer programmers create a mathematical representation of the part desired and then compute from it a set of instructions to run the cutting machines. The second system, in which machinists play no role at all, eventually won out. Noble argues that the decision was not purely economic, since the second system was more expensive and was developed only with strong Air Force subsidy until it became a standard for the industry. But machinists had been a highly skilled group with strong unions, while programmers were not unionized, so Numerical Control gave greater power to management. Small firms were forced out of business because they could not afford the programmers and equipment required.[10]

However, several authors have *questioned the deskilling thesis*. The underlying theory of owner-worker class relations, put forward originally by Marx, has been widely criticized. The lines between capital and labor are fuzzy today. Managers are closer to capital than to labor, but they really constitute a third group with diverse goals, including personal status and the growth of the organization. According to these authors, the actual human consequences of automation vary with external circumstances, managerial strategy, and worker initiatives. Upskilling occurs as well as downskilling. Skills are usually redistributed, not destroyed.[11] A study of Numerical Control in Britain, for example, found great variation among machine tool firms. Some firms recognized the value of the machinists' experience with metal and tools, and they encouraged cooperation between machinists and programmers. In some cases machinists were retrained with computer skills so they could revise programs in response to problems arising on the shop floor. Such broadened skills enhanced the flexibility of the system in innovating changes and in producing small batches of customized parts.[12] A study of 24 U.S. plants with computer-controlled manufacturing found the majority had upgraded operator skills and reported greater flexibility and commitment and improved union-management relations.[13]

Scandinavia has pioneered in provisions for *worker participation* in manufacturing. Since 1974, Volvo plants in Sweden have had autonomous work teams each assembling a major unit of the automobile. Members have had special training so they can rotate jobs and do their own quality inspections. Teamwork allowed more variety, interaction, and self-regulation; production costs were 25 percent lower than in Volvo's traditional plants. Swedish labor unions have been actively involved in the introduction of new technologies and in the training of workers in new skills to avoid massive layoffs.[14] In Norway, a 1975 law gave unions the right to negotiate new technology agreements. Management, labor, and government have collaborated both in specific local agreements and in encouraging ongoing institutions of industrial democracy.[15]

The United States has seen some new patterns of *labor-management relations* in dealing with automation. In the traditional pattern, the planning and control of work was entirely management's prerogative. Unions negotiated contracts dealing only with wages, benefits, safety, work rules, and procedures for resolving or arbitrating grievances. Both sides assumed a basic conflict of interests and an adversarial confrontation, with collective bargaining as the main form of interaction. In the 1970s, limited forms of ongoing labor-management cooperation concerning workplace issues were established.[16]

In the 1980s, more extensive *restructuring of work* occurred in many U.S. auto plants. Union membership had declined, and their power diminished during a time of recession, strong competition from overseas, and an unsympathetic administration. Japanese companies established nonunion plants in southern U.S. states and introduced their own patterns for worker participation. In this situation, both local leaders of the United Auto Workers (UAW) and plant managers of General Motors (GM) in several states were willing to cooperate in introducing and running work teams like those in Sweden and Japan. In some cases job rotation and increased pay for the acquisition of broader skills were instituted in the teams, as well as considerable self-regulation and quality control. Greater employee commitment and higher quality of work have been reported.[17]

An impressive *teamwork system* has been developed in the plant run jointly by Toyota and GM in Fremont, California. It follows many Japanese management principles, but it differs in having a union present. The plant's goals are productivity, high quality, and human development and participation. Management agreed to make no layoffs except in a financial emergency and after executive salaries and service employees have been cut. The managerial personnel went through extensive training and education programs. The system has earned the loyalty of most employees, and the quality of work has increased.[18] Some critics claim that teamwork co-opts workers and undermines unions,[19] but on the whole the Fremont experiment has worked well and has had the support of most of the national UAW and GM leadership. In GM's Saturn plant in Tennessee, which opened in 1990, the UAW took part in high-level planning of the vehicle and the plant before the plant opened, as well as in the day-to-day operation of production teams after they opened.[20] But Japanese manufacturers, with considerable help from their government, have taken a commanding lead in producing smaller, more reliable, fuel-efficient autos. The 1991 recession was a severe setback to the U.S. auto industry facing continuing competition from Japan, and its future is uncertain.

These examples suggest that the human impacts of automation and computerization vary greatly according to the policies adopted by management and the responses of workers. Traditional management practices have often led to unemployment, deskilling, and the weakening of unions, but the recent history of machine tools and auto manufacturing show that alternative practices can avoid such consequences.

2. THE ELECTRONIC OFFICE

The use of computers in offices started in the 1970s and expanded dramatically in the 1980s. Office computers serve many functions, including record keeping, word processing, and communication. Electronic work stations, connected by networks to other work stations, combine these tasks. Much of the paperwork in billing, accounting, keeping inventories, and processing orders has been replaced by computer operations. Computer networks transmit interoffice messages, electronic mail, and data from branch offices. Computers are powerful tools for market analysis, financial planning, and other aspects of management strategy. Clear benefits are evident in the reduced costs of carrying out traditional office functions and in the provision of new services. But some major problems have also been evident.

1. Employment Opportunities. The decline in the number of low-skilled clerical jobs—mainly occupied by women—is likely to continue. There will be a continued need for data entry clerks (keyboard operators), but these are dead-end jobs with almost no opportunity for career advancement. The number of supervisors will fall, since the monitoring of work can itself often be achieved by computer programs. There will be new jobs for computer specialists, support staff, managers, and of course equipment suppliers—jobs currently occupied mainly by men.[21]

2. Health. There were early reports of higher rates of miscarriage and birth defects when women worked long hours in front of computer screens (visual display terminals, or VDTs), and radiation was suspected as the cause. However, scientific studies since then have not supported such claims, and problem pregnancies seem more likely to be stress related. Other symptoms of stress have definitely occurred, including eye strain, headaches, and back and muscle problems due to poor lighting and positioning. These can be virtually eliminated by adjustable chairs and screen positions, better lighting, and attention to posture.[22] British unions have negotiated for twenty-minute breaks every two hours and medical checkups for VDT operators, and unions in the United States have lobbied for legislation governing the use of VDTs.[23]

3. Isolation. The traditional secretary had a variety of tasks that permitted social interaction: walking to files; talking with peers, supervisors, or clients; collaborating in activities; fixing coffee. Computer operators are often more isolated, especially those in separate centers for data entry or word processing. One author describes how data entry clerks in rows of individual cubicles pried a small opening between the sections of the partition so they could see each other and exchange a few words.[24] Some firms have avoided such mass production centers by decentralizing operations among multifunction offices, in which greater social interaction can occur. Many managers have succeeded in retaining personal secretaries as office managers facilitating the flow of people and information—and as symbols of their own status.[25]

4. *Deskilling.* Some authors claim that in offices, as in industry, computers are used by management to control and deskill workers. Knowledge and skills are programmed into the software so that few human decisions are required. The remaining jobs are fragmented and repetitive, boring and monotonous. Keyboard operators, like production line workers, seldom see the finished product.[26] But here again it is possible to integrate tasks and broaden job definitions, providing people with greater diversity and at least some control over their own activities. Unions and potential users of a computer system should have a voice in its design. Human resource specialists as well as computer experts can collaborate in the design of the office and in its operation.[27] One study of local government offices shows that clerical staff use computers more effectively if they have additional training and access to computer support staff.[28]

5. *Electronic Monitoring.* The output of computer workers can be continuously monitored by computer. In some firms, hourly efficiency ratings are calculated by counting key strokes. The length and destination of phone calls may be recorded, or the message itself may be monitored. Such data is used for discipline, for promotion, or for firing workers.[29] Employers do have a right to try to improve productivity, but this does not justify intrusive surveillance, which creates anxiety, resentment, and an atmosphere of distrust. Employees should have a voice in establishing a system of performance evaluation that is perceived as fair. They should have the opportunity to see and challenge the records kept on them. Such a system might include computer monitoring but should include other forms of evaluation.

6. *Gender Bias.* In the world of computers, the designers, programmers, and managers are predominantly men, while the low-paid routine jobs are occupied by women. Added to the traditional gender biases in business and industry are the new gender stereotypes associated with computers. In many schools, girls are discouraged from taking mathematics, science, and computer classes. Parental expectations and career counselors steer girls away from computer work. Only 11 percent of doctorates in computer science are held by women.[30] Even computer-based games, which are usually competitive and violent, seem to be aimed at boys and men. These forms of bias should be corrected, but in the meantime women can be helped by women's computer literacy programs and support networks.

7. *Homeworking.* Many computer tasks can be done at home, with input and output transmitted from a central office on telephone lines. This offers the benefits of independence, flexible scheduling, and savings in transportation costs and energy. These features might be particularly attractive to the disabled and to women with small children. Rank Xerox in Britain provided remote terminals for employees to work at home, but few people accepted the offer. Most people value sociability, and they want a change from home. Employers do not

like to give up face-to-face interaction. Unions fear that homework will result in long hours, low wages, and a reduction in their influence. Teleconferencing among businesspeople has grown only slowly, partly because it seems more difficult to reach decisions without the visual cues that convey subtle meanings and authority relations.[31] Thus in office work, as in manufacturing, the social context is as important as the technology in determining the human consequences of computers.

3. CENTRALIZATION AND DECENTRALIZATION

How do computers affect organizational structures and *the locus of decision making?* In the 1960s, only large mainframe computers were available and only large organizations could afford them: government bureaucracies, defense installations, banks, corporate headquarters, and a few universities. Large computers strengthened the decision-making powers of the central offices with access to them. By the 1980s, however, minicomputers and personal computers were widely distributed, and they were connected in far-flung networks that provided access to distant data bases. Branch offices and plants could keep track of local markets, inventories, production, and sales—in coordination with other branches and within guidelines set by the central office.

The extent to which *decentralization* has actually occurred varies greatly in differing settings and with differing managerial strategies. Information technology provides new opportunities for horizontal communication and greater access to information at all levels within an organization, which can reduce dependence on hierarchical authority. Policy guidelines and coordination from above are still necessary, but there is less need for detailed directives. The two-way flow of information allows local flexibility in responding to changing conditions and markets. But the power of managers at any level is partly based on their access to information, and they are often reluctant to share such access with people at lower levels.[32]

Andrew Clement shows that the use of *personal computers* (PCs) connected in networks leaves more control in the hands of individual users and middle managers, whereas large machines give central management greater control over workers and greater opportunity to monitor their performance, reducing the role of middle management. He concludes, "Because computing and telecommunications provide the potential for decentralized structures and openness of access, it has been argued that new office technologies will result in more egalitarian patterns of communication and use. However, the preexisting strongly hierarchical character of the host organization will discourage this development."[33]

Computers do seem to have affected the *career structures* of many organizations. The top part of the hierarchy typically has a vertical career structure and is characterized by promotional grades, commitment to common goals, and considerable room for personal judgment. The lower levels—the machine tenders and the clerical workers, for instance—have little autonomy or commitment and little opportunity for promotion. In between, middle management has often been squeezed out, since planning has moved upward, and computers do much of the supervision of lower levels and the transmission of information upward.[34] In an earlier day an able bank clerk could start at the bottom, serve an apprenticeship in varying jobs and branches, receive a management appointment, and end in the executive office. While this was never easy to do, it seems virtually impossible today.[35]

Shoshana Zuboff describes alternative ways in which computers can be used to distribute information in an organization. She defines *automating* as the substitution of computers for human agency within a structure of hierarchical control and authority. If the relation of management and labor is based on a military model of command and obedience, workers at lower levels will be given limited access to information and no opportunity for critical judgment. She defines *informating*, by contrast, as the use of computers to provide more information to people at all levels. It encourages the development of human capacities and intellectual skills. It assumes a cooperative relation and continual dialogue between management and labor, and the boundaries between them are blurred. It can result in a more highly motivated work force and a sense of joint responsibility. The organization can be more flexible and innovative, which would provide important advantages under competitive and rapidly changing conditions today. The goal is to fulfill the potential of people as well as that of machines.[36]

Over a six-year period, Zuboff interviewed *managers and workers* at a variety of locations in which computers were being introduced. She found great differences even among plants or offices in the same industry. For example, she studied three paper mills introducing automatic pulp processing equipment. In one of them the processing operators were given special training, participated in design decisions, had access to cost data, and were able to suggest improvements and to troubleshoot if something went wrong. They had a sense of responsibility for the enterprise. In another mill, the operators were given no training, carried out boring and routine functions in the control room, and were helpless when anything went wrong. In a third, the plant managers at first delegated considerable responsibility but ended by giving out a strict set of rules so that they would be given credit by the central office for the plant's success.[37]

Zuboff suggests that the manager's *privileged access to information* has traditionally been an important source of authority. Sharing information with lower levels of management or with workers is threatening if authority is understood in terms of command and obedience. Of course, upper-level management needs a broad range of information to coordinate operations and to formulate comprehensive policies. But it may prevent the access of lower-level managers to much of its computerized data partly in order to strengthen the chain of command. The authority structure is assumed to be justified by the requirements of efficiency and by the stockholders' property rights delegated to the top management. Zuboff elaborates an alternative conception of leadership based on dialogue, collaboration, and task-related knowledge. Participation in decision making and in broader tasks has led to greater motivation, job satisfaction, and better and more innovative products. But she acknowledges that shared information and wider participation are resisted by many managers.[38]

Elinor Lenz and Barbara Myerhoff suggest that *women will be a humanizing influence* in the postindustrial workplace. In the preindustrial age, home and work were closely linked. After the Industrial Revolution they were clearly separated; home became an exclusively female domain, and work was an impersonal male sphere of competition and hierarchy. When women were employed outside the home it was almost always at the lowest levels. But today women are more than half the workforce, and they are entering jobs formerly occupied only by men, especially in the service sector that is expanding while the industrial sector shrinks. As women enter management positions, they often express new managerial styles and concerns. They bring interpersonal skills, abilities in communication, and a sensitivity to human relationships. Women tend to prefer collaborative teamwork to hierarchical authority. Some women have welcomed the decentralization that computer networks can allow. They also see the need for child care centers, flexible schedules, parental leaves, and other provisions to enable women—and also men—to combine a career and a family again.[39]

Many institutional forces, in sum, favor the deployment of computers to strengthen hierarchical patterns of centralized control, but computers also present significant opportunities for decentralization. In a study of computers in organizations, Richard Harris concludes with a call for a coalition of the movements that have shown an interest in *the democratization of work life,* including labor unions, consumer and women's groups, religious groups, and environmentalists. He says that organizations change only slowly, but they do change—in response to internal pressures, public awareness, and legislative initiatives.[40]

I submit that the evidence concerning computers in organizations, like the evidence from manufacturing and office work, supports *the contextualist thesis* that there are alternative paths of technological development and that the human consequences of technology are the result of a two-way interaction between technical possibilities and social institutions. The creative potential of computers in work life can be realized only when attention is given to the social systems of which they are a part.

II. COMPUTERS AND CITIZENS

We look first at various ways in which citizens can have access to computer services in homes, schools, and communities. Some relationships between computer programmers and computer users are then considered. Finally, the protection of individual privacy in the use of computerized data banks is discussed.

1. ACCESS TO INFORMATION

Computers can be linked through various kinds of *communication networks*, vastly increasing their versatility. In the past, telephone lines have been the main link between computers, but the speed of communicating data is limited by the narrow band of frequencies such lines can accommodate. Coaxial cables can carry computer data and many television channels at the same time. Fiber-optic cables can carry an even larger number of telephone, television, and computer signals on the same line. Satellites in orbit relay signals from many channels at once from a transmitter to a receiver at a distant point. Direct broadcasting satellites emit a signal powerful enough to be picked up by small windowsill receivers. Taken together, computers and communications represent 40 percent of all U.S. capital investment in new equipment.[41] Within this burgeoning field of telecommunications, what ethical issues arise concerning the role of computers in the life of citizens?

Access to computers raises issues of *justice* and *participation* because information often makes possible the exercise of power in society. Computers are still too expensive for most families, though their cost has been steadily falling. Unequal access of children to computers at school affects their education and their future employment opportunities. The number of computers per pupil is far higher in white suburban schools than in inner city or predominantly minority schools. The access of citizens and organizations to data bases is also very uneven. The largest collection of data bases in the world, the Lockheed system, has a high subscription fee. By contrast, Japan has an information utility open to all citizens, and several European countries have government-operated data systems. Universities, central libraries, and government agencies could provide data bases open to citizens and consumer groups. Such systems would help to reduce the gap between the information-rich and the information-poor, a gap that widens when information is controlled by commercial interests. Computer user networks permit the informal circulation of information on topics of common interest. Through desktop publishing, computers offer creative new possibilities for individuals and small groups to disseminate information.

In many areas, *free public access to information*—essential in a democracy—has been diminishing. In the United States, the number of TV network viewers has been declining, while paid subscriptions to cable TV have been rising. In Britain, new satellite TV channels are privately owned, and even the programs on publicly sponsored BBC channels are increasingly subject to commercial criteria.[42] In several nations, public support for libraries has been falling, and in some cases user fees have been introduced. Many of the on-line data bases used by libraries were designed primarily for business and industry and are therefore stronger on financial data and the sciences than on the humanities.

Since 1981 the U.S. government has greatly reduced its services for gathering and disseminating information and has restricted access to much of the data in its agencies. Information is increasingly treated as a commodity for sale rather than as a public service. The "electronic highways" are open to information elites rather than to all citizens.[43]

But computers can also be used to strengthen *community organizations*. In a government-funded project in Scotland, proposals were solicited for the imaginative use of computers to improve community services. Among the proposals supported were a data base to help disabled people in locating services and jobs, a network for agricultural information for farmers, and an urban citizens' advice bureau. In the United States, computer networks have been established to increase the effectiveness of rural organizations, senior citizens, ecology and peace groups, and other nonprofit organizations.[44] The combination of computer networking and desktop publishing enables a variety of citizens groups to publish newsletters and to keep track of mailing lists.[45] Once again, the impact of computer systems on the distribution of social power depends on the institutional structures within which they are developed.

Computers linked by phone or cable make possible a variety of *home services*. Two-way interactive systems would allow teleshopping, library access, travel services, counseling, health care advice and medical diagnosis, as well as financial services and electronic banking. President Mitterand promised a free computer terminal in every home in France having a telephone line, and by 1990 five million had been installed. The government hopes to recover the cost of the system from the companies providing the ten thousand different services offered, from reservations and purchases to electronic mail.[46] The British Prestel system was established as a public information utility, but citizen response has been slow and its services are mainly used by businesses and professionals.

Interactive cable could be used for *direct electronic voting*. Several authors have advocated "push-button democracy" in which citizens would vote from home not only in an election but also in a referendum on proposed legislation. There could be widespread and frequent direct participation in governmental decisions at local, state, and national levels.[47] However, a study of thirteen experiments with various forms of "teledemocracy" found that competition for public attention limited the participation. In the late 1970s in Columbus, Ohio, the QUBE interactive cable system was used for a suburban traffic and zoning decision, but subscription to the system was low and it has been discontinued.[48]

I see major *problems in direct electronic voting*. It would be biased against low-income families who are less able to pay for cable services. Moreover, it would weaken the processes by which decisions are reached in representative governments, such as committee hearings, compromise proposals, the formation of coalitions, and legislative debates. Direct voting would also further weaken local and state parties and caucuses, grass-roots politics, and face-to-face contacts where minority viewpoints can be expressed. *Direct* democracy is indeed problematic, but I suggest that computer networks could strengthen *representative* democracy by facilitating communication within citizens' groups, such as those concerned about environmental, consumer, labor, peace, and social issues. They could also improve communication in both directions between citizens and elected representatives, without having a formal role in the governance structure.

2. COMPUTER PROGRAMMERS AND USERS

How are *computer designers and programmers* affected by the character of their work? One study describes work in Silicon Valley, California, as very competitive and stressful. Great commitment and long hours are demanded. Many workers complained that they had become workaholic and their human relations were strained; divorce rates were well above California averages.[49] In *The Soul of a New Machine,* Tracy Kidder follows the work of an industrial team developing a new computer—a creative but all-absorbing process that takes over the lives of those involved as they try to get a jump on their competitors. Several members of the team describe their increasing alienation from their families and from nature.[50]

Sherry Turkle conducted extensive interviews with *computer users* of all ages, from children at school to programmers at work. She encountered a variety of reactions, but among both adolescent "hackers" and professional experts she found a common pattern of addiction and isolation.[51] Was this a consequence of working with computers? Or does it reflect a tendency of people with certain types of personality to be interested in computers? Some of the hackers interviewed (an all-male group) did seem to want to escape human intimacy and the threat of rejection and said that computers are more predictable than people. Of course, many computer users lead balanced lives with diverse interests, but we should be aware of the danger of total absorption with computers, whether from commercial competitiveness or inherent interest or as an escape from personal intimacy. Diverse interests and human relationships are essential components of *personal fulfillment* as I have portrayed it.

Do computers affect *our ways of thinking?* David Bolter maintains that we think of ourselves in the image of our technologies. After the Industrial Revolution we began to think of ourselves as machines, and now we are beginning to think of ourselves as information processors.[52] Will extensive use of computers lead us to value logic and rationality at the expense of the senses, feeling, and human relationships—left-brain analytic thought at the expense of right-brain imaginative and holistic thought—as Joseph Weizenbaum believes?[53] We will return to this question when we look at recent work in artificial intelligence.

Computer models (simulations) of social and environmental phenomena are a valuable tool for policy planning at the city, state, and national levels, but

their limitations must be kept in mind. Models are based on simplifying assumptions that reflect the biases of the programmers and the selection of variables that are computable. This often leads to the neglect of unquantifiable
features such as the beauty of a scenic river, the value of an endangered
species, or the impact of a thruway on an urban neighborhood. Moreover, it is
difficult to communicate the results and limitations of a model to the public
and to the elected officials who have the final responsibility for decisions. In
the past, technical jargon has hindered public participation, and agencies
have often carried out computer studies in order to add legitimacy to decisions reached on other grounds.[54] But when the assumptions and limitations
are made clear, the results of computer models can be one important input
into the political process of decision making.

Who is responsible when use of a computer program causes injury or harm?
Legal responsibility is specified by laws applying to product liability (prepack-
aged programs can be viewed as a product), or by laws applying to negligence
(especially in individually designed programs, which are regarded as services).[55] Negligence would include inadequate testing of a program, but adequacy is a judgment relative to the seriousness of the consequences (for
example, a program to control operations in a nuclear power plant would
have to go through very stringent tests). *Moral responsibility* is a broader concept than legal responsibility. A person should be concerned about the indirect as well as the direct consequences of his or her action, even if other
agents are involved so that it is difficult to assign legal responsibility.[56]

It is common for people to *blame the computer* for errors or malfunctions. You
are overcharged on your monthly bill, and after several letters and phone calls
you are told "the computer made a mistake." But of course it was actually a
human mistake: a defective program, a command or data entry error, or neglect of a hardware failure. In all large organizations—especially when computerized—there are many anonymous inputs from people who do not see
each other, so responsibility is diffuse. In some systems, operating control is
actually turned over to a computer (in automated industrial processes,
weapons systems, automatic pilots, or air traffic control systems, for example),
but even in these cases human beings have the ultimate responsibility, since
they designed and tested the systems and recommended their use. Moreover,
an emergency override is usually provided to recover human control, which is
essential when surprises may occur and flexibility is important. If several very
complex systems are tightly coupled, unexpected events can have disastrous
consequences.[57]

The *protection of software* (computer programs) from unauthorized use raises
distinctive legal problems. While software copying (piracy) by individuals is difficult to prevent, three types of law can be invoked to prevent one company
from adopting another company's program. (1) A *copyright* can be obtained on
a particular program (like that on a literary work). But the copyright can be
evaded by using a different source program and computer language to express

the same underlying mathematical principle (algorithm). (2) A program can be treated as a *trade secret,* protected by access codes and employee contracts forbidding disclosure. But secrecy is difficult to enforce. (3) A *patent* has the advantage of protecting a program while putting it in the public domain. Other firms can use and modify the program if they pay license or royalty fees, so incentives for invention are provided without imposing a secrecy that prevents further development.[58]

The U.S. Patent Office has allowed *patents on applied programs* but not on the algorithms underlying them, since patent laws are not applicable to mathematical equations or general principles. In 1989, Xerox won its patent suit against Apple, claiming the first use of "mouse and icon" concepts, and Quarterdeck won a patent for its "windows" concept, which Microsoft had adopted.[59] Patenting is the most satisfactory form of software protection, but the limits of its applicability still have to be worked out through court rulings.

Computer crime usually involves breaking into a computer system through its telephone links by guessing a password, finding loopholes in its access system, or using an inside accomplice. Illegal access may be aimed at transferring funds or at obtaining restricted information. In other cases, computer "viruses" (instructions that erase data or destroy programs) have been introduced by hackers to demonstrate their cleverness or by employees seeking revenge on former employers. These violations are covered by criminal laws, and they have led to improvements in computer security.[60]

3. DATA BANKS AND PRIVACY

The computerization of records and the new technologies for transmitting data have raised concerns about *the invasion of privacy*. National police networks make it easier to follow suspects as they move around, but the networks disseminate local files that are often inaccurate and may include undocumented suspicions and hearsay. More than half of the millions of criminal history records circulating in the United States are incomplete, inaccurate, or ambiguous. In 40 percent of police arrests the charges are dropped, and in other cases the accused is acquitted in court. But many arrest records give no indication of the disposition of the case, and the incomplete data are used by employers and government agencies to screen applicants.[61] Erroneous credit investigation data also circulates widely. In one case a person who was repeatedly turned down for a loan discovered that ten companies had received the same mistaken data from a credit rating agency.[62]

On each of us there is *an immense variety of types of records,* such as academic grades, medical histories, insurance claims, and personnel dossiers. Various branches of government keep separate files on law enforcement, tax returns, welfare payments, census surveys, and so forth. It has been proposed that the merging of all government records into one national data bank would enhance administrative efficiency, the detection of tax and welfare fraud, and the data base for police work and national planning.[63] But it would represent a further increase in the power of the state over its citizens.

Microelectronics has also created *new methods of surveillance*. During the 1960s and 1970s, the Federal Bureau of Investigation carried out the surveillance of political dissidents and persons taking part in civil rights and antiwar demonstrations. Since then, monitoring equipment has become more sophisticated, and it is easier to retrieve data on phone calls, credit card purchases, bank withdrawals, and other transactions that enable a person's movements to be traced.[64] In 1987 the former West Germany issued machine-readable identity cards to facilitate border control and police operations. In democratic governments, regulations are needed to protect against misuse of these new methods of surveillance. But it is also sobering to realize that they would increase the powers of a totalitarian government far beyond what George Orwell could imagine in 1949 when he wrote *1984*.

Governments and private companies do indeed have a *need for information* relevant to planning and administration. In general, society benefits from the free flow of information, and such flow is hindered by secrecy on the part of either individuals or organizations. On the other hand the right to privacy can be defended as a form of *respect for persons* as unique individuals. Freedom of thought in entertaining unpopular ideas requires some emotional and intellectual space protected from social intrusion. Divulging personal information about ourselves to other people gives them power over us and makes us more vulnerable. Privacy sets limits on the power exerted over individuals by the state, by organizations, and by social groups.[65] This aspect of privacy is consistent with the biblical understanding of the value and uniqueness of each individual in the sight of God. It is also supported by the emphasis on human autonomy and self-determination since the Enlightenment. Taken alone, however, it perpetuates the individualistic view of selfhood that has been dominant in recent Western thought.

A second defense of privacy is that it makes possible distinctive kinds of *interpersonal relationships*. Each of us enters diverse personal and professional interactions, which vary in the kinds of self-disclosure they require. In situations of intimacy, trust, and friendship we voluntarily reveal aspects of ourselves that we would not want to share with everyone. Similarly, belief in the confidentiality of the information given to doctors, therapists, and clergy is a prerequisite for trust and self-disclosure.[66] In this interpretation of privacy the social character of selfhood is recognized. It is consistent with the biblical understanding of persons-in-community.

Both *defenses of privacy* appear valid, and together they must be balanced against the public good served by legitimate social uses of information. The defense of privacy is supported both by the negative concept of freedom (absence of interference) and by the positive concept of freedom (participation in the decisions that affect one's life) that I emphasized in chapter 2. Privacy can be more readily defended in terms of individual rights than in terms of utilitarian calculations of costs and benefits. Any view of privacy reflects an interpretation of personal fulfillment, which in turn expresses one's understanding of human nature.

In the United States, some *privacy-protection laws* have been passed, but their scope has been very limited and their enforcement ineffective. A 1971 law requires that a credit agency must, upon request, send to any individual a copy of its records and a list of the places to which copies have been sent in the last six months. However, few people request such copies, and there are no restrictions on the gathering or exchange of information. The 1974 Privacy Act gives citizens access to the records of government agencies (except those in intelligence or law enforcement) and provides procedures for challenging errors. The Office of Management and Budget has responsibility for oversight, but it makes no inspections to check on compliance.[67] The 1984 British Data Protection Act is also very inadequate.[68]

Privacy-protection agencies have, however, been established with considerable success in several European countries. Since 1972, the Swedish Data Inspection Board has to license all personal data systems, both governmental and private. It carries out inspections covering the collection, storage, transfer, and use of information about citizens. West Germany had an independent data-protection agency that could challenge proposed data management procedures and audit actual practices.[69] The technical measures to prevent unauthorized access to personal data are similar to those used to protect sensitive national security information or to prevent computer theft of funds or trade secrets: passwords, fingerprints, or voice signatures. But attention must also be given to the collection of data in order to be sure that only information directly relevant to the organization's purposes is obtained and that the least invasive method is used whenever possible.

Computers in the life of citizens thus raise a variety of value issues. Concerning communication networks and computers in homes, schools, and communities, I raised two questions. Does the deployment of the technology provide for *justice* in equitable access to both the sending and receiving of information? Does it increase or decrease the freedom of individuals to *participate* in the decisions that affect their lives? The institutional structures within which these technologies are used affects the balance of power between individuals and large organizations (governmental or corporate). Economic forces seem usually to favor the control of information flow by large organizations, but the institutions of political democracy provide opportunities for the regulation and use of these technologies (see chapter 8). Personal information in data banks will enhance the power of organizations over individuals unless their privacy is protected—in the interest of justice and participation, but also to enhance individual autonomy and interpersonal relationships, which are important dimensions of *personal fulfillment*.

III. COMPUTERS FOR WAR AND PEACE

Two additional uses of computers deserve particular attention because of their relation to national goals. The first is the military use of computers. The second is the potential contribution of computers to the development goals of the Third World.

1. MILITARY COMPUTERS

The transistor was a product of pure research and industrial application at AT&T's Bell Lab in 1947. Integrated circuits etched on silicon chips were developed at Texas Instruments a decade later. But *military purchases* created a large market for these industrial products, which helped to bring their prices down. The Department of Defense funded research on electronic communications and computers for weapons systems, and in these early stages such research undoubtedly aided the growth of the civilian computer industry. During the 1960s and 1970s, both ground-based and on-board computers were extensively used in ballistic missile guidance systems, air defense systems, and the NASA space program.

By the 1980s, however, military electronics was harming *the civilian electronics industry* more than it was helping it. By then, home and commercial computers and communications equipment had a thriving market and needed no market-creating purchases. Defense contractors absorbed a large fraction of scientists and engineers, especially in the physical sciences. Fifty percent of the world's physicists and engineers were working on military research and development. In the United States, 70 percent of all government R & D funds were going to defense and space programs.[70] Japan and Germany, which had no large defense programs, were overtaking the United States in many fields of technology, including electronics. The Japanese dominated the world's markets in televisions, stereos, videocassette recorders, and computer chips. American trade deficits, especially in relation to Japan, increased dramatically. In Britain, the Maddock Report (1983) concluded that civilian electronics was being harmed by defense research, and a more recent study concludes, "Cuts in military expenditure may be what is required for the long-term health of the British electronics industry."[71]

Some *civilian spin-offs* have occurred from defense spending, including advances in aerospace and robotics, but spin-offs have been rare in the later stages of development. Most military applications are highly specialized and have few commercial parallels. Who can use a Stealth bomber or its distinctive technologies? Military equipment is designed for high performance under extreme conditions to achieve specific military objectives, regardless of cost. Civilian technologies have very different goals. Defense contracts virtually guarantee a profit, and cost overruns are very common, so there is little incentive for efficiency and competitiveness. Procurement practices favor large established firms at the expense of smaller, newer, more innovative firms.[72] However, the Defense Advanced Research Project Agency seems to have been an exception within the Pentagon. With a small budget, minimal bureaucracy, and outstanding leadership, it has sponsored some imaginative research, including advanced computer and artificial intelligence projects in several universities.

During the 1980s and 1990s, miniaturized computers have been at the heart of the *high-tech weapons* that have been employed with such dramatic effect. In the Falklands the British lost 114 planes and a warship, mostly to Argentina's French-built "smart" missiles.[73] The U.S. Patriot missile, deployed in 1991 to protect Israel from Iraq's SCUD missiles, is initially guided by radar signals and a ground-based computer until it homes in on its target, using its own computer system. The cruise missile, whose effectiveness was evident in the Gulf War, uses an on-board computer to match radar altitude readings against its prerecorded terrain maps; near the end of its path it matches images from its TV camera against prerecorded pictures of the target itself. In other precision-guided systems a pilot or soldier projects a laser beam onto the target to guide a bomb, a missile, or an antitank rocket. The greater accuracy of such targeting reduces the civilian damage, but it reinforces the impersonal and detached character of modern warfare. Both military personnel and the public can follow the action like a video game, but we are all far removed from the victims of our actions. Moreover, only 9 percent of bombs dropped on Iraq were precision guided, so civilian casualties were in fact widespread.

A final example of the military use of computers is the *Strategic Defense Initiative* (SDI) or "Star Wars" program initiated by President Reagan in 1983. Unlike the Patriot system, SDI would be comprised of long-range and space-based weapons designed to destroy intercontinental missiles in flight. SDI is subject to criticism on many grounds: enormous cost ($1 trillion or more to complete), vulnerability to countermeasures, violation of the ABM Treaty, and its threatening effect on an opponent (especially when coupled with first-strike weapons like the MX missile). Even proponents of SDI now concede that it would be ineffective in defending cities or in countering cruise missiles; at most, it might provide some protection for missile sites.[74] The criticism relevant to this chapter, however, is that SDI coordination would require the most complex computer in history, with no possibility of testing it under war conditions.

Large software programs are difficult to test and correct under the best of conditions. Correcting one error can introduce new errors. A programmer cannot try out all possible combinations of unexpected input events. Heathrow airport had five protracted software failures in 1988 in which planes were diverted and backed up all over Europe. A minor malfunction in the U.S. telephone system in 1990 started a chain reaction that crippled half the switching facilities nationwide.[75] An OTA study of SDI in 1988 concluded:

Existing large software systems, such as the long-distance telephone system, have become highly dependable only after extensive operational use and modification. In OTA's judgment, there would be a significant probability (i.e., one large enough to take seriously) that the first (and presumably only) time the [SDI] system was used in a real war, it would suffer a catastrophic failure. The complexity of the software, the changing nature of system requirements, and the novelty of the technology to be controlled raise the possibility that the system may not even be able to pass the more realistic of the peacetime tests that could be devised for it.[76]

Moreover, SDI design allows only a very short reaction time (less than thirty seconds for the boost phase), so it has to be essentially automatic. Human responsibility would be minimal once the system was in operation. Even in more limited proposals for computerized "battlefield management" systems human judgment must not be excluded, especially when decisions would be affected by political goals and by interpretation of the intentions of an adversary in an unforeseen situation. Apart from the role of computers, military technology raises other ethical issues that are taken up in the next chapter.

2. COMPUTERS IN THE THIRD WORLD

In discussing agriculture I outlined some strategies of development in the Third World. I indicated that prevailing practices in technology transfer have perpetuated dependence on Western exports and equipment. Information technologies have further increased the power of *transnational corporations* (TNCs). The chips are stacked in their favor. They can move information and funds across boundaries, seeking lower labor costs and global markets. Some of their activities elude the jurisdiction of any country. Environmental and safety standards adhered to at home are seldom observed abroad.[77]

A handful of companies dominates *the world market in computers.* The top five companies are all American, led by IBM, which in 1988 had 60 percent of world sales. AT&T combines communications with computer marketing worldwide.[78] The Japanese are strong in TV equipment, and several European firms have secure positions based on specialized products or protected domestic markets. A study of computers and computer-aided manufacturing in the Third World finds them highly dependent on this small group of industrial nations.[79] Several American firms have established plants for circuit assembly in Southeast Asia and Mexico, where labor costs are low. Women have been particularly vulnerable to exploitation, and substantial eyesight damage has been reported. As these processes become automated, production will return to the United States; the low-skilled jobs may only be temporary, and little technical expertise will be left behind.

Several newly industrializing countries, however, have made substantial progress in *the production of computers.* In 1984 Brazil banned computer imports, and 270 Brazilian companies had been formed by 1986. U.S. designs can be used under licensing agreements, and IBM has invested in joint ventures, but 70 percent of the ownership must be in Brazilian hands.[80] Spectacular high-tech growth has occurred, but not without its costs; Brazil has a huge debt load, and resources have been diverted from agricultural and housing needs, especially in rural areas. South Korea is prominent in the production of computer chips and computer-controlled machine tools. India has many scientists and engineers and good research facilities, and it produces most of its own integrated circuits. It has been selective in importing both hardware and software, and many of its plants operate under joint ventures or licensing agreements with European or American firms.

In most of the Third World, however, there are no computer production facilities, and growth in *the use of computers* has been slow. The World Bank and other aid agencies have made some efforts to foster the use of information systems in agricultural planning, water management, health care, and banking. Grants for such systems were included in several development projects: agricultural extension in Peru, energy efficiency in Malaysia, railroad expansion in Pakistan, and so forth. In some cases, citizens have been given training in electronics and programming.[81] Such attention to indigenous expertise is important if continued dependence on Western experts is to be avoided. Sri Lanka has instituted a program to expand computer training in schools, colleges, banks, and industries, and it has set up regional centers to create or modify software to meet local needs.[82]

The Brundtland Report (which I discussed in chapter 4) suggests that microelectronics could facilitate *sustainable development.* "New technologies in communication, information, and process control allow the establishment of small-scale, decentralized, widely dispersed industries, thus reducing the levels of pollution and other impacts on the local environment."[83] Such small-scale industries would provide nonfarming jobs outside the crowded cities. Electronic process controls could help reduce energy and resource use as well as pollution. The report acknowledges that this will require the training of local experts, along with technical and financial assistance from advanced countries. Another study gives examples of computer-aided manufacturing in small-scale industries in Brazil, Hong Kong, and Singapore. Computerized machine tools can be easily reprogrammed for small batch production and are adaptable to decentralized rural plants or small firms subcontracting from larger ones. But Third World authors have also warned that useful electronic technologies cannot be a substitute for needed political and economic reforms.[84]

A report for the International Labor Office advocates the integration of *computers* with *traditional techniques* so that rural and low-income populations may benefit. Software applications suitable for small businesses, local government offices, and small-scale industrial processes have been developed. Electronic load controls have been designed to improve the efficiency of small hydro turbines in Thailand. Local computers and telecommunications have improved rural health services and facilitated agricultural planning in India.[85]

Third World nations have also been concerned about *the control of satellites* used for the long-distance transmission of telephone, television, and computer signals. In 1962, government communication agencies around the world formed Intelsat, a consortium that now has 118 member nations and 4000 channels. In addition, satellites are owned by several private companies, individual industrial nations, and a few developing nations (India, Brazil,

Mexico, and an Arab consortium). Other Third World countries have wanted to be sure that orbital positions and transmission frequencies will be available to them when they are able to have their own satellites. Industrial countries advocated a policy of "first come, first served," but in a series of conferences called by the U.N.'s International Communications Union, the developing nations insisted that space is "the common heritage of mankind." In 1988, a regional allotment plan was approved in which every nation will be guaranteed an orbital slot and a frequency. However, most developing countries do not have the funds or equipment for their own satellites, and they will be dependent on industrial nations for many years to come.[86]

Third World critics hold that unregulated satellites open the door to *cultural imperialism*. They see the influx of foreign TV programs as a threat to indigenous cultures. In Latin America, four-fifths of TV programming originates in the United States.[87] Western entertainment indirectly conveys Western values, such as materialism, individualism, and competition. American advertising is found worldwide; global "Dallas" is accompanied by global Coca-Cola. In the Philippines, nine of the largest ten TV sponsors are U.S. companies seeking to enlarge their markets. These media encourage a consumption mentality and a Western life-style; indigenous culture is seldom represented. Most of the messages are aimed at urban rather than at rural populations.[88]

In a series of UNESCO meetings starting in 1978, developing nations called for a *new world information and communication order.* They asserted that communications are crucial in the formation of their national identity but are increasingly dominated by industrial nations. Global wire services are virtually a Western monopoly, and they give a Western slant to stories about events in both developing and industrial nations. In the UNESCO debates, Third World governments asserted their right to control their media to serve their national interests and development goals. Western delegates defended the privatization of communication in the name of freedom of the press, and they defended its commercialization in the name of competitive efficiency. In 1983, the United States and Britain withdrew from UNESCO partly in protest over this issue.[89]

A few developing nations have experimented with satellite broadcasting designed particularly to meet *the needs of rural citizens.* In 1975, the U.S. space agency loaned four hours a day of satellite time to India for a year for the Satellite Instruction Television Experiment. Indian-built community TV sets with chicken wire dish antennae were placed in 2400 villages. Morning transmissions in four languages were aimed at rural schools, while evening adult programs dealt with agriculture, health, nutrition, and family planning, together with entertainment based partly on indigenous cultural forms. In 1983, India's own satellite was launched, with twenty-four TV channels and channels for radio, telephone, and data transmission. The interest in rural education continues, though most of the channels are devoted to commercial programs and business services.[90] Indonesia, Brazil, and Peru have also developed rural satellite programs. Since 1984, sixty-five nations in Project Share have cooperated in the use of satellites to serve health, education, and development needs.[91]

Surveillance satellites can capture and transmit high-resolution images of the earth. They are used extensively for military intelligence and are also being used to verify compliance with arms control treaties. A satellite system established to transmit infrared images of land areas has proved useful in monitoring agricultural and natural resource conditions around the globe. Originally a government service, it has now been commercialized and is more readily available to transnational corporations than to Third World governments.

There are, in short, significant opportunities for the use of both computers and satellites to further the development goals of Third World nations, but the achievement of these goals is often hindered by the global reach of TNCs and by the commercialization of space. Around the world, government funding of computers for development has amounted to a tiny fraction of the funding of military computers.

IV. ARTIFICIAL INTELLIGENCE

Computers already surpass humans in some intellectual abilities, such as carrying out complex mathematical calculations. Recent parallel processing computers may allow other human abilities to be duplicated. Are we moving closer to designing intelligent robots or androids like those portrayed in science fiction? Work in artificial intelligence (AI) will undoubtedly produce some impressive applications, but it also raises basic questions about human nature. Since Darwin, human dignity has been threatened by our resemblance to animals. Now human uniqueness seems to be threatened by our resemblance to computers.

We must first examine what has been accomplished so far in AI research. Then we look at the implicit assumptions about human nature underlying various approaches to AI and try to distinguish science fiction from reasonable expectation in looking to the future.

1. PROGRESS IN ARTIFICIAL INTELLIGENCE

In the 1960s, the speed and memory capacity of *digital computers* with *sequential processing* increased rapidly. *Digital* means that the basic components have two states (on or off), in terms of which all information is encoded. *Sequential* means that the processing follows a well-defined succession of steps, specified by explicit formal rules. One step has to be completed before the next step can be initiated. The information in a well-defined verbal or numerical input can be encoded, stored, manipulated, and transformed in various ways to yield a determinate verbal or numerical output.

During the 1960s, some research was pursued on *the simulation of networks*. Certain features of neural networks in organisms were represented in computational programs. But in an influential book in 1969, Minsky and Papert showed the limitations of one-layer networks and mistakenly assumed that these limitations would apply also to multilayer systems.[92] Moreover, various types of computer research were competing for funding, and proponents of other systems claimed greater prospects of short-term success, especially for military and space applications. Networks were virtually ignored in the 1970s, illustrating again the thesis that the design as well as the applications of a technology are social constructions.

The 1970s saw a search for programs to deal with *ordinary human language*. Millions of dollars were spent on programs to translate texts from one language to another, but with little success. One problem was that the same word may have different meanings depending on the context in which it occurs. Language also includes grammatical ambiguities that can be resolved only by drawing from a background of common sense (for example, "I saw the Grand Canyon flying to New York"). Roger Schank provided a context for sentences in stories by writing a set of "script" programs portraying typical human situations, such as "eating in a restaurant." The computer was then given a story, and it could answer questions about the story by making use of the relationships set forth in the scripts. Schank claimed that "the computer understands the story" since it can answer novel questions about it.[93] But of course the story must be set in one of the scripted situations, and only a limited range of questions can be anticipated by the programmer. In interpreting a sentence or a story, human beings draw from an enormous and varied memory based on real-life experience. The selection of relevant interpretive clues does not seem to follow formal rules.

Other programs in the 1970s were designed for the *perception* and *manipulation* of objects in the laboratory. Terry Winograd hooked two TV cameras and a robot arm to a computer and programmed it to manipulate a set of blocks on a table.[94] The system can carry out typed instructions and take part in a dialogue about the blocks. "Put the pyramid on top of the red cube," you write. "Which pyramid?" it asks. "The largest one," you reply, and it carries out the order. The system operates in a very limited domain, cut off from the rich variety of human life. In most situations human perception, like language, involves ambiguities that we resolve by drawing from our store of previous knowledge and expectations. What we see depends on what we know and what we expect. Visual pattern recognition has also proved very difficult to formalize. With faster circuits and larger memories, more background information can be stored in a computer; but the search process and the selection of items that might be relevant among the immense number that are irrelevant are no easier.[95]

In the 1980s, considerable effort was devoted to the development of *expert systems*. Here the goal is to write a program that duplicates the decisions of human experts in a particular field. A "knowledge engineer" interviews an expert and tries to formalize the explicit or implicit procedures he or she uses. A series of if-then rules and inferences are formulated, and these are reviewed and revised by other experts. A program called MYCIN was written to diagnose bacterial infections and prescribe appropriate antibiotics. Other programs were developed from the practice of experts in geological prospecting and industrial troubleshooting. Some expert systems have been a commercial success, and they often work well in narrow technical domains that can be isolated from other considerations. But the systems are blind to larger contexts, and they have difficulty deciding where the boundary of the domain lies and when something outside it might be significant. They are more useful in rule-governed situations where occasional errors can be tolerated, such as locating breakdowns in telephone or industrial equipment, than in complex situations where human lives are at stake.[96]

The history of *medical diagnostic programs* is instructive. INTERNIST, a program designed to diagnose more than five hundred diseases and disorders, was based on the statistical correlation of symptoms in past confirmed cases.[97] But the program was not adopted in clinical practice. It did not do as well as experienced clinicians in diagnosing patients with two or more maladies at once, or with diseases whose symptoms change significantly as the disease develops. Moreover, the program lacked the physiological and biochemical knowledge that enables a physician to understand causal (as distinct from statistical) connections between symptoms. Finally, it often took an hour or more for a physician to enter data on a patient, using special codes, and this could not compete with a phone call to a specialist using ordinary medical language.

A *revised version* of INTERNIST does not attempt to make a diagnosis, but acts as a consultant to facilitate access to data and to suggest hypotheses to explore and additional tests to consider.[98] Perhaps future programs can include models of organ systems, biochemical reactions, and disease processes and also explain the reasoning on which its conclusions are based.[99] Even then, computers should be used as auxiliary tools to assist rather than replace the judgment of human medical experts.[100] Legal and moral responsibility for erroneous diagnoses must remain with the physician.

The 1980s also saw the development of *parallel processing* in which many separate units can carry on operations simultaneously and interact with each other. This offers greater speed and flexibility in computation because a problem can be broken down into component problems that can be pursued at the same time rather than serially.

In addition, work on *network simulation programs* was resumed after a decade of neglect. In some designs the strength ("activation probability") of the connections between the nodes of the network are variable and can be modified with experience. If successful patterns are reinforced, a kind of "learning" occurs. Information is stored in a distributed pattern as a holistic representation, not by a one-to-one correspondence between separate data items and separate memory locations.[101] A voice synthesizer program can be "trained" to turn a written input into speech output. Initially the connections between letter combinations and audible sounds are random, but every time a given input produces the correct sound, the connections used are reinforced.

Slowly the system improves until it can pronounce new words with fair accuracy.[102] But the system requires considerable initial preparation and task analysis, and the final patterns yield few general principles of pronunciation, and most of these have exceptions. Progress on learning systems has been modest to date, but they are likely to be an important direction of future research.

2. ARTIFICIAL INTELLIGENCE AND HUMAN NATURE

Many AI researchers defend *the formalist thesis* that all intelligence (natural or artificial) consists in the manipulation of abstract symbols. According to Allen Newell and Herbert Simon, a world of discrete facts can be represented by a corresponding set of well-defined symbols. They claim that the relationships among symbols are abstract, formal, and rule governed; symbols can therefore be processed by differing physical systems (natural or artificial, protein based or silicon based) with identical results. Newell and Simon assert that the brain and the computer are two examples of devices that generate intelligent behavior by manipulating symbols. Human thought is simply one form of information processing. Mind is to brain as software programs are to computer hardware. These authors say that a machine is intelligent if in performing tasks it exhibits behavior that we would call intelligent if performed by human beings.[103]

But critics of formalism have said that human language and perception are *context dependent*. Hubert Dreyfus has portrayed the importance of commonsense understanding, background knowledge, and nonlinguistic experience in the interpretation of human language. He has also argued that we perceive patterns (in recognizing the face of a friend, for example) as gestalts or wholes, not as aggregations of discrete facts. Linguistic and perceptual understanding, he insists, are active processes, strongly influenced by our expectations, purposes, and interests. Knowing typically involves contextual interpretation rather than the acquisition of separate unambiguous facts.[104]

Dreyfus has also emphasized *the role of the body* in human learning. Much of our knowledge is acquired actively through interaction with our physical environment and other people. We learn to ride bicycles not by studying physics or by acquiring a set of rules, but by practice. We use the skills of "knowing how" rather than the propositions of "knowing that." Such "tacit knowledge" cannot be fully formalized. In a child's development, growth in perception is linked to action and bodily movement. Dreyfus sees in the formalist thesis a legacy of rationalism: the assumption that knowledge consists of formal rational relationships existing independently of the body and the material world. He claims that formalism assumes a dualistic view of human beings, in which mind and body can be considered separately. In his more recent writings, Dreyfus acknowledges that robots do have mechanical bodies and do interact with the world (through auditory and visual inputs and motor outputs), but he says that their bodies are so different from ours that their forms of understanding will differ radically from ours.[105]

Terry Winograd, whose block-manipulating programs described above were hailed as early successes in artificial intelligence, has more recently repudiated formalism and stressed the importance of *individual and social life* in human understanding. He now accepts the view of the philosopher Heidegger that our access to the world is primarily through practical involvement rather than detached analysis. According to Heidegger, understanding is aimed not at abstract representation but at the achievement of our goals and interests. Our speech is communication for particular purposes, a form of action. Winograd also draws from the philosopher Wittgenstein, who insists that there is no private language or individual representation of the world, but only communication in contexts of social interaction. Language reflects our social practices, cultural assumptions, and "forms of life" in a public interpersonal world. Unlike computers, people also make commitments for which we hold them responsible. Winograd has redirected his own research and is working on the design and use of computers to facilitate human communication and social interaction, rather than to simulate individual behavior in isolated domains.[106]

How do formalist assumptions about human nature compare with other views in *Western thought?* Consider first some parallels with the Greek dualism of mind and body. Plato saw human rationality as our distinctive characteristic and said that the human mind participates in the realm of eternal forms, which are only imperfectly embodied in the realm of matter. AI formalists grant that programs must be run on material hardware (human or artificial), but they claim that the programs themselves express abstract logical relationships—rather like Plato's eternal forms—independent of any physical system. Classical Christian thought, influenced by Neoplatonism, defended a dualism of body and soul, which is a different way of minimizing the importance of the body. Many Christians today would affirm that God gives each human being an immortal soul, which distinguishes humans absolutely from any form of animal life and from any conceivable form of artificial intelligence. No machine, they would say, could ever have a soul.

In the previous volume, however, I argued that the Bible itself, prior to the influence of Greek dualism, presents a more holistic and social view of human nature:[107]

1. A Unitary Person, Not a Body-Soul Dualism. The bodiliness of persons is assumed in the Bible. We are thinking, feeling, acting beings. The person is an embodied agent—a psychosomatic unity, as we would say today. God is concerned about the whole person and all aspects of human life. Emotions and actions are as important as concepts. This view of human beings as psychosomatic unities is closer to the ideas of Dreyfus and Winograd than to the assumptions of AI formalists.

2. The Social Self. Each of us is constituted by our relationships and the covenants into which we enter. We are called to create a just and caring society, not to save our individual souls. In the biblical view, we are always individuals-in-community. The self is not a separate entity, but the highest level of a multilevel integrated person within an interdependent social and physical environment.

3. Responsibility and Rationality. The biblical assertion that humanity is created "in the image of God" refers not primarily to rationality but to our capacity for responsibility and relationships to God and other persons. Moral choice requires reflection on goals and values, and some degree of freedom and individual agency, even if we are constrained by our genes and by cultural conditioning. Reinhold Niebuhr has described human existence as combining freedom and finitude and as expressing both self-transcendence and bodiliness. [108]

Clearly this biblical view of human beings as *bodily, social, and responsible* persons differs from the assumptions of the AI formalists concerning human nature, and it clearly distinguishes humans from any forms of AI constructed so far. But can we say that no future form of artificial intelligence could share these characteristics?

3. PROSPECTS FOR THE FUTURE

There are still enormous differences between *computers* and *brains*. A brain has a thousand trillion synapses and ten million times the storage capacity of a large magnetic disk today. The brain's speed is a million times that projected for computers a decade hence.[109] Except for visual processing, we know little about how the levels of neuron nets are interconnected. We do not understand the neural basis of memory and learning. Neuron operation is analog rather than digital; the frequency of neuron output pulses varies continuously. During embryonic development, nerve cells connect (by selective adhesion) to particular types of cell, but there is no exact prewiring such as computers require. New knowledge from neuroscience will undoubtedly affect future computer design, but we should not underestimate the differences or the difficulties.

AI research in the 1990s and beyond is likely to narrow the gap between humans and intelligent machines. Parallel processing will lead to more flexible hardware and software, with greater *capacity to learn from experience.* At the same time, robots with sensor and motor capabilities will be able to interact with their environments in more varied ways. Their mechanical bodies will be different from our biological bodies (even if, like androids, they resemble humans on the outside). What they learn from experience will differ from what we learn, but they will have more in common with us than any previous artifacts. Like human beings, each such robot will have a distinctive history and memory.

The *process of socialization* in humans occurs over a span of many years. In computers, information processing is very rapid, but interaction with the environment takes considerable time. Robots might be socialized partly by being fed vast quantities of information, but if the critics of formalism are correct, participation in human culture and forms of life would require active interaction over a longer period of time. Dreyfus maintains that only computer systems nearly identical with the human brain and endowed with human motives, cultural goals, and bodily form could fully model human intelligence. That may be too strong a claim, but it points to the importance of culture as well as body in human understanding and in any attempt to duplicate such understanding in machines.

Biological organisms start life with complex structures that are the product of millions of years of *evolutionary history*. The most successful patterns of past interaction with the environment were selected, encoded in the genes, and expressed in the corresponding physiological structures. The perceptual systems of organisms do not provide exact representations of the world; they extract information relevant to the needs and the actions of the organism. A frog's visual system, for example, is sensitive to the visual patterns that contribute to its survival; some optic fibers in frogs respond only to small dark spots—undoubtedly an advantage in catching flies. So, too, human neurophysiology evolved in parallel with distinctive human goals and interests. Scientists can study human physiological structures without having to repeat evolutionary history, but they must still consider how these structures function in human life and provide for similar active functions in artificial systems.

Emotions seem to be a distinctive feature of human experience. Most AI researchers claim only to simulate cognitive processes, and they hold that cognition is quite independent of emotions. Schank writes, "It would seem that questions such as 'Can a computer feel love?' are not of much consequence. Certainly we do not understand less about human knowledge if the answer is one way or the other. And more importantly, the ability to feel love does not affect its ability to understand."[110] Other authors hold that we can analyze the function of an emotion in evolutionary history and then try to construct an AI program that fulfills the same function. For example, the main behavioral function of fear is danger avoidance, which might be programmed directly. Margaret Boden and Aaron Sloman say that computers could not experience feelings but could represent the cognitive components of emotions (for example, the external causes of anger and its relation to one's beliefs and ensuing actions).[111] P. C. Rogers replies that conceptual analysis is not enough; the ability to experience emotions is a necessary condition for meaningful language about them.[112]

AI experts have postulated that *consciousness* is a high-level central control system that focuses attention and selectively activates memory and action programs. *Self-consciousness* is said to be the ability of a system to form a symbolic representation of itself. With such a self-referential model, the control system could compare possible actions in terms of its own goals.[113] But I do not think that the idea of self-reference captures the essential subjectivity of self-consciousness. We do not understand very much about the amazing phenomenon of human personality, but it does not seem likely that it can be duplicated in an AI system. *Human freedom* is indeed limited by the constraints of genes and upbringing, but it cannot be entirely dismissed. Computers are designed to be predictable and rule governed; random variables can be introduced, but randomness cannot be equated with free choice. Human creativity and novelty go far beyond what has been achieved in computer-generated music and poetry, and human experience has many dimensions that are hard to even imagine in computers.

Because human beings are creative as well as intelligent, they can bring genuine novelty into the world—including new kinds of computers. We do not know in what directions artificial intelligence might go in the more distant future, especially if analog computers or protein components were used. We can expect improvements in the ability of AI systems to simulate and surpass a variety of human capacities in the performance of specific tasks, which could help us immensely, as long as we are aware of their limitations. Human beings and computers together will be able to do things that neither can do alone.

V. CONCLUSIONS

When used appropriately, computers can contribute to each of the values listed earlier. Computers are relevant to *food and health* when they are used in agricultural research and planning and in medical research and hospital management. I suggested that expert systems can aid medical diagnosis but should not replace the judgment of physicians.

Meaningful work is available to computer designers and programmers (mainly men), and to anyone who uses computers creatively on the job. But data entry (mainly by women) is usually boring and isolating. Automation causes job dislocation, but the net loss of jobs seems to be small. We saw that in some plants automation has resulted in the deskilling of workers and the demise of unions, but in others both workers and unions have a voice in work-related decisions.

Many people undoubtedly find *personal fulfillment* in working with computers. A few become so absorbed in it that human relationships suffer, or their range of experience is narrowed so that a problem does not seem significant unless it is computable. I suggested that some uses of personal information in computerized data banks are an invasion of privacy, which should be protected in the name of individual autonomy and the importance of personal relationships. I argued that artificial intelligence is not a threat to human dignity unless we start to think of ourselves as only information processors and symbol manipulators. I defended a view of bodily, social, responsible selfhood and human consciousness that contrasts with the formalist assumptions of some AI researchers.

Social justice requires concern about the gaps between the information-rich and the information-poor, since information is a source of social power. Public or low-cost access to computers, communication channels, and data bases should be sought in schools, homes, libraries, and community organizations. The access of women to jobs requiring computer expertise has been hindered by both cultural assumptions and discriminatory practices. Again, the global dominance of transnational corporations in communication and information systems restricts the opportunities for Third World nations to seek their own development goals.

Participation in the decisions that affect our lives can be either increased or decreased by computers, depending on how they are deployed. Some organizations use computer systems to centralize control, whereas others allow more decisions to be made in branch offices. The desktop computer enhances the capacities of small organizations in office operations, publication, and communication. Computers and networks can also strengthen citizens' movements and enable elected representatives to be more responsive to the electorate.

Economic development is furthered by the efficiency and productivity that computers and automation make possible, but I have maintained that selective economic growth should be sought. The computer and communications industries are not themselves resource intensive or highly polluting. The decline in heavy industry and manufacturing and the growth of the service and information sectors are thus environmentally desirable trends. Intensive work on military computers, however, has diverted funds and personnel from research and development more directly related to human needs.

Resource sustainability is not endangered by computers since they do not make heavy demands on either materials or energy in their production and use. They can also facilitate greater efficiency in other systems, such as computerized controls for auto engines, wind turbines, hydroelectric generators, or industrial processes. Surveillance satellites are used to monitor global agricultural and natural resource conditions.

Environmental protection is not jeopardized because computer production and use are not polluting if the disposal of toxic wastes from manufacturing electronic equipment is carefully managed. On the positive side, computer models advance our understanding of ecosystems, and computerized instruments assist the monitoring of industrial pollutants.

Like most other technologies, then, the potential effects of computers are very mixed, and the consequences for humanity and for the environment depend on the choices made in their design and deployment. The choice of options that fulfill human and environmental values requires reflection on the goals of human life, and wisdom that goes beyond technical knowledge. T. S. Eliot's words are a reminder that information alone is not enough:

> Where is the life we have lost in living?
> Where is the wisdom we have lost in knowledge?
> Where is the knowledge we have lost in information?[114]

Part Three

TECHNOLOGY AND
THE FUTURE

CHAPTER 7

Unprecedented Powers

In part 3 we turn from analysis of particular technologies to more general questions about the future of technology. In this chapter we try to come to terms with some of the unprecedented powers of recent technology, some of which exceed anything in human history in their scope and duration. The first section deals with environmental threats that are now global and long-term. The second section examines genetic engineering, which gives us the ability to alter plants, animals, and human beings—and even to make germline modifications that are passed on to future generations. The third section discusses the destructive power of nuclear weapons. How will we use these new powers of life and death: the capacity to harm the global environment, to redesign living things, and to destroy civilization?

I. ENVIRONMENTAL DEGRADATION

For two decades we have been aware of damage to local and regional environments. But we are only now beginning to know the extent of our environmental impacts. We live on an endangered planet. Let us consider air and water pollution and then the global threats to climates, species, and renewable and nonrenewable resources. These are not mainly the result of dramatic accidents like the Exxon Valdez oil spill, the Chernobyl nuclear accident, or the Bhopal chemical disaster. They are the cumulative effect of the normal operations of our technologies and our prevailing agricultural and industrial practices and consumer habits.

1. AIR AND WATER POLLUTION

Despite extensive efforts to control pollutants, many forms of air and water pollution have grown worse since 1970.

1. Air Pollution

In the chapter on energy we saw that sulfur dioxide from *coal-burning plants* causes acid rain and is a major contributor to respiratory diseases and property damage in urban areas. In the formerly East German industrial center of Leipzig, life expectancy was six years less than the national average. Acid rain from steel mills is dissolving statues in Krakow, Poland.[1] Scrubbers can effectively remove sulfur dioxide from smokestacks, but they add about 25 percent to the cost of coal-generated electricity.

Auto emissions include carbon monoxide, nitrogen oxides, and hydrocarbons (the last two react in sunlight to form ozone and urban smog). The U.S. Clean Air Act of 1970 set a series of deadlines for gradual reductions in the tailpipe emissions of these pollutants (with lead added to the list in 1978). But Congress, under pressure from auto manufacturers, repeatedly postponed the deadlines for meeting the stricter standards. By 1990, ambient air levels for all these pollutants had fallen significantly, but 150 million Americans still lived in areas that exceeded safe levels of ozone for a significant portion of each year. In 1990, Congress and the Bush administration finally agreed on amendments requiring reductions in nitrogen oxide and hydrocarbon emissions for new cars after 1994. But they have not agreed on stricter fuel economy standards that would reduce urban smog, global warming, and oil imports at the same time. Los Angeles, the most severely polluted city, may have to resort to more drastic measures, such as travel restrictions, electric cars, or the use of methanol fuel.[2]

Air pollution in *Japan* and *Britain* has had a similar history. By the late 1960s, Tokyo had the worst air of any city in the world; some people had to wear cloth masks when they went out on the street. In 1970 Japan adopted the most ambitious pollution control program in the world and enforced the strictest sulfur dioxide and auto emissions standards. But during the 1980s a conservative government was reluctant to antagonize industry, and the environmental budget was sharply reduced. By 1990 Japan lagged behind many other industrial nations. Britain made impressive progress in the 1960s when coal burning was restricted in urban areas, and sulfur dioxide levels fell sharply. During the 1970s the average emissions of the standard air pollutants remained essentially unchanged. In the 1980s Margaret Thatcher's administration was more sympathetic to industrialists than to environmentalists. In response to growing public concern and an active environmental movement, Thatcher finally endorsed a number of environmental goals in 1988. But creative ideas in the environment ministry were vetoed by the ministries of energy, transport, and treasury, and the White Paper issued in 1990 proposed few new initiatives.[3]

2. Water Pollution

For *surface waters* in the United States, the Clean Water Act of 1972 set ambitious goals but left great latitude to the Environmental Protection Agency (EPA) to establish effluent standards and to grant exemptions. Municipal treatment plants were built in many cities with the help of federal funds. Later amendments focused on toxic pollutants from industry and on agricultural runoff (especially fertilizers, pesticides, and sediments). But little progress has occurred in the reduction of runoff, and EPA's budget was drastically cut during the 1980s. Overall water quality in 1990 was about what it was two decades earlier; a few rivers had improved while a quarter of the lakes had deteriorated, many of them clogged by fertilizer-fed algae.[4] Britain saw similar water improvements in the 1960s and 1970s, and fish reappeared in the Thames above London. But British rivers were more polluted in 1990 than they were in 1980.[5]

The contamination of *underground water* (aquifers) occurs only slowly, but once it happens it can last for decades. Half of U.S. citizens depend on municipal or private wells for drinking water, and reports of contaminated wells are widespread. Some of the pollution of aquifers, especially by pesticides and fertilizers, arises from the percolation of surface waters. Other chemicals leach from toxic waste sites or from injection wells (which receive half of all toxic wastes). Brines from oil extraction, leaks in abandoned oil wells, and leaks from petroleum storage tanks are all sources of organic pollutants. When sludges from air and water treatment processes are disposed of on land, the toxic metals in them will leach into ground and surface waters. The toxics have been moved from one medium (air, for example), to a second medium (land), and reappear in a third (water).

3. Hazardous Wastes

Chemical industries today generate enormous quantities of hazardous wastes (which include flammable, corrosive, and toxic wastes). The petrochemical industries proliferated after World War II, bringing us detergents, plastics, synthetic fabrics, pesticides, and other new products. Many of these materials are not biodegradable when discarded, and some—especially among the chlorinated compounds such as dioxin—are highly carcinogenic. American industries with more than ten employees are required to report the toxic chemicals they release to the environment; in 1987, 20 billion pounds were reported. But OTA says that if underreporting, small firms, distributed sources, and unregulated chemicals were taken into account, the total toxics released would be more like 400 billion pounds annually.[6]

Many *industrial wastes* go to landfills, lagoons, and injection wells, from which they can seep into surface waters or aquifers. Love Canal was the most famous of many such cases. The 1980 Superfund law set up a $1.6 billion clean-up trust fund, financed largely by a tax on chemical production, and an additional $8.6 billion was added in 1986. But effective waste site cleanup is expected to total $500 billion.[7] By 1990 EPA had identified 30,000 abandoned hazardous waste sites, of which 1200 were on a priority list, but only 34 had been cleaned up.[8] The laws regulating hazardous waste disposal have not been strictly enforced, so new hazards are accumulating faster than old ones are being cleaned up. New sites are usually in low-income areas where citizen opposition is not well organized.

4. Prevention Strategies

The main response to such air and water pollution around the world has been to introduce *end-of-pipe devices* to remove pollutants from emissions and effluents. Without such abatement technologies, air and water quality would be much worse than they are today. In the next chapter some regulatory strategies are discussed: emission standards, "best available technology" requirements, emission taxes, and marketable licenses. But over the past twenty years pollution abatement measures have barely kept pace with increases in population, consumption, and industrial growth, and new risks keep coming to light. Cleanup efforts are costly so they have been resisted, and enforcement has often been lax. Pollutants removed at one point often show up somewhere else, including transfers between media (air, water, and land). Widely distributed sources in agriculture and small industries are hard to control. Clearly, we need to avoid creating so many pollutants in the first place, rather than just trying to get rid of them after we have created them; an ounce of prevention may be worth a pound of cure. We must deal with causes as well as symptoms.[9]

Changes in *industrial processes and products* can greatly reduce pollutants. On-site recovery and recycling can turn wastes into useful inputs, and alternative techniques can be introduced. In place of volatile solvents, water, compressed air, or sonic vibration have been used to clean metal parts. A printing company replaced solvent-based inks with water-based inks and saved $35,000 a year in waste disposal costs.[10] General Dynamics reduced its hazardous waste output from 28,000 tons to 8,000 tons in four years. 3M cut its wastes in half in a decade, saving $300 million.[11] Such changes require commitment on the part of management and programs to educate and reward employees for waste reduction. In the mideighties, Japan generated only 20 pounds of industrial hazardous waste per capita annually, compared to 80 pounds in the Netherlands, 200 pounds in Britain, and 4,000 pounds in the United States.[12] (Even allowing for the fact that nations differ in their definitions of hazardous waste, it appears that we can learn from Japan's example.)

In addition to such voluntary action by industry, *government policies* should be oriented to prevention as well as treatment of pollution. The most serious toxic chemicals can of course be banned completely (as we have done with PCBs and DDT). The use of pesticides, herbicides, and fertilizer can be greatly reduced by the practices of sustainable agriculture discussed earlier. Taxes on wastes would provide financial incentives, and the revenue could be used for waste-prevention research and development, public education, and assistance and retraining for workers displaced when plants have to close. Higher fees for disposal of industrial wastes and for municipal waste from homes would have similar effects.

Finally, *consumers* can reduce waste by insisting on less packaging and on biodegradable products and by their purchasing, conservation, and recycling habits. In chapter 9 we will return to the role of the consumer, including ways in which less consumptive life-styles might reduce both pollution and the depletion of scarce resources.

2. GLOBAL THREATS: AN ENDANGERED PLANET

Pollution problems were familiar in the seventies, but a group of new issues that are global in scope came into prominence in the eighties.

1. Deforestation

An acre and a half of forest is destroyed every second, or 41 million acres a year, an area the size of Austria.[13] Massive deforestation has repercussions on *almost every aspect of the environment.* In discussing agriculture, I mentioned that the loss of forests has led to rapid soil erosion in many parts of the world. Tree cutting, especially on steep slopes, also results in rapid runoff and flooding. In dry areas, loss of trees has accelerated the expansion of deserts. In discussing energy I described the fuelwood crisis, the additional global warming attributable to deforestation (because trees absorb carbon dioxide), and the damage to forests caused by acid rain in Europe and North America. Deforestation destroys the habitats of plants and animals, including endangered species, as we shall see.

Reforestation policies could reverse these trends. Integrated land use planning would encourage sustainable and selective timber harvesting. Land tenure laws should give greater security to rural farmers and to forest dwellers who usually have no land rights. In many parts of the world, planting trees and crops together (agroforestry) brings several benefits. The trees provide windbreaks and reduce soil erosion, the fallen leaves improve crop yields, while selective pruning provides fuelwood. In Nepal, villagers are planting fodder grasses and fast-growing trees on denuded slopes. In Kenya, the Greenbelt Movement has enlisted children in 670 school or community gardens in growing millions of seedlings for replanting. In the state of Kerala in India, church groups, cooperatives, and women's groups support a major replanting project.[14] Both India and China have government replanting programs working through forestry and agricultural extension agents. Plantations of fast-growing trees have up to ten times the annual yield of natural stands.[15] In the U.S., starting in 1986, farmers were given federal payments for erodible cropland that was replanted in trees and grasses in the Conservation Reserve, which by 1990 included 35 million acres.

Tropical deforestation is especially rapid and serious. In much of Central America, forests have been cleared to make cattle ranches, mainly for beef exported to U.S. fast-food chains. In the Amazon basin, rain forests are being cleared for timber (exported to Japan and Europe), for cattle, or for agriculture, mainly for export crops. But as E. O. Wilson writes, "The irony of cutting down tropical forests in order to grow crops or graze cattle is that after two or three years the nutrient-poor soil can no longer support the agricultural activity for which it was cleared in the first place."[16] Half of the land in the Amazon basin which has been cleared for pasture or crops is now abandoned.[17] A program in Indonesia to resettle families from the overcrowded main islands to the forested outer islands has encountered the same problem: the soils are too thin and fragile to support sustained agriculture.

2. Endangered Species

Species are endangered in many parts of the world, but the problem is most urgent in *tropical rain forests,* which cover only 7 percent of the earth's surface but are home to 50 to 80 percent of the earth's species. It is estimated that one hundred species of plants and animals become extinct each day from the loss of these forests.[18] This is an irreversible process; extinct species can never be restored. Millions of years of evolutionary information are being destroyed in a few years; this is like burning libraries of unread books. Loss of biodiversity leads to biotic impoverishment and reduced adaptability. It also destroys resources valuable for scientific research and technological development. A quarter of all prescription drugs originated in wild plants and animals. The rosy periwinkle, found in Madagascar, yields two alkaloids effective against Hodgkins disease and some types of leukemia; the income from these alkaloids exceeds $100 million a year.[19]

Action must be directed toward *preserving habitats* since species live in ecological communities. But many of the countries in which tropical forests are located are heavily debt laden. They are under pressure to earn foreign currency by exporting timber and clearing land for export crops, and they have very limited funds for conservation programs. These areas of exceptional biodiversity are a global asset, and rich nations have a responsibility to contribute to their preservation rather than to their destruction. Several private foundations have bought up a total of $100 million in Third World debts (at heavily discounted rates) in exchange for local expenditures on conservation.[20] Such debt-for-nature swaps should be greatly expanded, though they will make only a small contribution to the problems of debt reduction. The World Bank, national banks, and aid agencies should review the impacts of their loans and grants on habitat destruction. Some endangered mammals, reptiles, and birds can be preserved in zoos, and rare plants can be kept in botanical gardens and seed banks, but such off-site programs have limited capacity and are no substitute for on-site habitat protection.[21]

3. Ozone Depletion

The rapid *depletion of stratospheric ozone* by chlorine atoms was not anticipated. The worst offenders are the chlorofluorocarbons (CFCs), which have been widely used (in refrigerators, air conditioners, aerosol propellants, and foam production) because they are not toxic or flammable and are cheap and versatile. In the late seventies, environmental groups organized an effective boycott of CFC spray cans, which were banned in the U.S. in 1978. By 1987, research in Southern latitudes showed convincingly that chlorine is the main culprit; by initiating a chain reaction, one chlorine atom can destroy 100,000 ozone molecules. The ozone layer provides a shield for ultraviolet radiation from the sun. Satellite data in 1991 showed that a 5 percent loss in ozone has already occurred over the U.S., and an additional 12 million cases of skin cancer and 200,000 deaths can be expected in the next 50 years, according to the EPA.[22]

Global action to *phase out CFC production* has been an example of effective international and interdisciplinary negotiation. Starting in 1985, the United Nations Environmental Program sponsored meetings in which scientists, environmental groups, industrialists, and government officials all took part. Environmentalists wanted a complete ban on CFC production. Industrialists initially said that even a partial reduction would be too costly and that scientific uncertainties should be resolved before taking any action. But scientific evidence was mounting, and the threat of a ban spurred industry to step up research on substitutes. In the Montreal Protocol (1987) the industrial nations accepted a 50 percent reduction in CFC production by the year 2000. The London Agreement (1990) went much further: a total phaseout by 2000 and approval of a $240 million international fund to help developing nations adapt to it. In 1992 a more rapid phaseout was being sought. One possible replacement (CFC-12) is three times as expensive and has a small residual ozone effect, but the benefits far outweigh the costs, and other substitutes are being developed. Some other less prevalent ozone eaters remain for future negotiations.[23] Chlorofluorocarbons also contribute to global warming, but recent data suggest that ozone depletion itself has a cooling effect, so phasing out CFCs probably will not reduce global warming.[24]

4. Global Warming

International action on *global warming* will be much more difficult because the short-term economic impacts will be greater, though the long-term consequences of not acting could be as catastrophic. As indicated in chapter 5, current levels of carbon dioxide release would probably raise sea levels between eight inches and five feet and disrupt climate and agricultural patterns worldwide by 2030.[25] Climate models are complex, and scientific uncertainties remain, but continued delay would make the potential harm more serious. As a minimum, all nations should cooperate on measures that have other benefits apart from slowing global warming. The most promising action is stringent energy conservation. Improvements of 3 percent per year in energy efficiency (which many nations achieved between 1973 and 1985) would in twenty years cut in half the energy needed to produce the same goods and services. Other multiple-benefit policies discussed earlier include greater use of natural gas in place of coal, development of solar sources, reforestation, and a tax on carbon emissions.[26]

Third World nations will need technical and financial assistance in making such changes. The negotiation of national carbon dioxide targets will not be easy. Justice clearly requires an increase of Third World energy use, so the brunt of curbing carbon dioxide emissions will have to fall on the largest current emitters, the industrial nations. No nation is motivated to act alone, but common action could avert a tragedy that would affect us all. Once again the United Nations has a crucial role to play in dealing with a global environmental threat. Forty nations have endorsed the Hague Declaration (1989) calling for a global environmental protection agency with the power to levy fines and to submit disputes to the International Court of Justice for adjudication.[27] In preparing for the 1992 U.N. Conference on Environment and Development, developing countries have said they will sign agreements on forests, biodiversity, and global climate only if they receive financial assistance and favorable terms for the transfer of energy-saving and other relevant technologies. The United States and the European Community have resisted such conditions, and the U.S. has tried to weaken the global warming treaty.[28]

5. Population Growth

Larger populations exacerbate almost every environmental problem: pollution, soil erosion, deforestation, and so forth. The world population of 5.3 billion is growing at 1.8 percent per year. Each day there are 260,000 more people to feed. Women in Nigeria have an average of 6.5 children during their lifetime; if current birth rates continue, Nigeria will grow from 119 million in 1990 to 273 million in 2020, exceeding the present U.S. population. If the global average can be brought down to the replacement rate (2 children per couple) by 2020, the world population will stabilize at 10 billion. But if it does not reach the replacement rate until 2080, the total would be 15 billion, though soaring death rates would probably stop growth long before world population had tripled.[29]

Active *family planning programs* have produced significant reductions in birth rates. The drop has been most rapid in nations in which substantial and widely distributed economic development has also occurred (such as South Korea, where the average fell from 6 children per woman in 1960 to fewer than 2 in 1990). But birth rates have also declined in nations with slower economic development. In Indonesia the birth rate fell by 42 percent between 1970 and 1985, largely because of educational programs and a choice of contraceptive methods provided free through 40,000 village centers.[30] Thailand cut its growth rate from 3.2 to 1.6 percent in sixteen years through village programs tied to health care services. Mexico and India have used the media to promote smaller families and family planning services, with some success. For every peso that Mexico puts into family planning, nine pesos will be saved in maternal and child care alone. Improvements in the status of women and in literacy levels also contribute significantly to the reduction of birth rates.[31]

Funding for family planning is minuscule in relation to the importance of the task. Today it totals $3.2 billion annually. If this were increased to $10 billion, global population could stabilize at 10 billion people.[32] This would be a bargain if it averted the poverty and environmental disaster that would result from continued growth. Of current development aid, less than 1 percent goes to population assistance. The U.N. Fund for Population Activities is grossly underfunded. The Reagan administration cut off U.S. payments to the fund because some of the organizations it supports allow abortion as one option. Funding of research on new methods of contraception has been falling just when it is most needed. (One promising method is Norplant, a removable hormonal implant that lasts for five years; it currently costs $2.80 per year, which will fall as a mass market develops.) No other investment promises greater benefits for the world's future than family planning. It would be a real global bargain if the North would help to finance family planning, forest protection, and efficient energy technology in the South (and curb its own energy and resource consumption), while the South agreed to curb population growth and greenhouse emissions.

In sum, technology has increased our ability to feed and house the world's population, but at the price of *environmental damage* that has become increasingly serious and global in scope. Environmental problems pose a greater threat to the future of most nations than does military aggression. Yet the United States is typical of industrial nations: the total 1989 research and development budget of the Environmental Protection Agency was 1 percent of R & D in the Department of Defense, and less than the cost of one MX missile. Of every tax dollar, 1 cent was spent on the environment, 24 cents on defense, and 14 cents for interest on debt (largely created by military spending).[33] We must seek reordered priorities and a broader definition of national security that includes environmental protection. In industrial nations, efforts should be directed primarily at redesigning technology and reducing consumption rather than removing pollutants from effluents. Because the new impacts are global in scope, the U.N. is now crucial to the preservation of the planet.

3. THE "LIMITS TO GROWTH" DEBATE

Will environmental degradation and resource depletion limit future economic and industrial growth? Three positions have been presented in this ongoing debate: no growth, progrowth, and selective growth.

1. Critics of Growth

In *The Limits to Growth* (1972) a team at MIT used computer simulations to show that if population and industrial production continue to grow exponentially, global limits will be exceeded within a few decades. The main limiting factors in their study were agricultural production (land area and yield per acre), nonrenewable resources (minerals and fossil fuels), and the capacity of the environment to absorb pollutants. In these simulations, the interaction of cumulative effects brought rapid environmental deterioration. The constraints in these projections arise not from literally running out but from diminishing returns in the use of scarce resources. After the richer and more accessible ores are used up, for example, the mining and extraction costs of other deposits will be higher. Even if population growth is halted, the authors concluded, industrial production cannot continue to grow.[34]

The *Global 2000* report, prepared by the Carter administration in 1980 using data from various U.S. government agencies, also took a dim view of prevailing global trends. "If present trends continue, the world in 2000 will be more crowded, more polluted, less stable ecologically, and more vulnerable to disruption than the world we live in now. Serious stresses involving population, resources, and environment are clearly visible ahead."[35] The study underscored the degradation of living systems and the interrelated problems of agriculture, water, soil erosion, and deforestation.

Several social scientists have argued that in a world of severe resource scarcities *democracy is doomed*. William Ophuls holds that only governments with strong powers to regulate individual behavior can deal with the destruction of the environmental "commons." In crises, people accept authoritarian governments with powers to compel obedience. In the name of survival, Ophuls says, we will have to accept drastic restrictions of personal liberty. Faced with catastrophe, the sacrifice of freedom is the lesser evil. An even gloomier picture is offered by Robert Heilbroner, who anticipates intensified competition for dwindling resources and increasing regimentation by all governments. Democratic institutions, he says, will be unable to cope with internal strife, international conflict, and threats to survival. Both of these authors predict a dismal future if past trends in resource use continue.[36]

2. Advocates of Growth

The Limits to Growth evoked sharp rebuttals from some economists, who said that *market mechanisms* will provide an automatic adjustment to resource scarcities. When a resource becomes scarce its price rises, which discourages its use, encourages the search for new sources and more efficient extraction technologies, and provides an incentive to use substitute materials. They said that the market is an efficient feedback system for responding to resource scarcities.[37] Other economists were less sanguine. They pointed out that the market discounts the future and is not an effective instrument for conservation. Free market responses to the degradation of biological resources are delayed and seldom affect the causes of degradation. Moreover, the world market is far from free. Monopolies, cartels, tariffs, and subsidies are present, and differences in political power among nations distort prices.[38]

Advocates of growth have usually expressed great *confidence in technology* to extend existing limits. In the past, the price of most minerals has stayed fairly constant, and estimated reserves have often grown, because improvements in exploring, mining, and extraction techniques have compensated for the need to use less accessible and lower-grade ores. New technologies can turn previously useless substances into useful resources. Virtually inexhaustible minerals (iron, aluminum, and magnesium) can be substituted for scarcer ones such as copper (though at the cost of additional energy).[39]

In a 1984 volume, Julian Simon and Herman Kahn challenged each of the conclusions of the *Global 2000* study. "If present trends continue, the world by 2000 will be less crowded (though more populated), less polluted, more stable ecologically, and less vulnerable to resource-supply disruption than the world we live in." The price of oil was falling, and the authors supported nuclear power as the hope of the future. Arable land and yields per acre will grow, they said, and the outlook for food will improve. People will have funds to control pollution as they get richer. These authors were not worried about population growth, for they expected birth rates to fall with economic growth and rising living standards. Market mechanisms would adjust to any temporary shortages.[40] This was the philosophy of the Reagan administration, which relied on the free market and economic growth and opposed almost all new resource regulations.

How can the pessimists and the optimists reach such *divergent conclusions,* starting from the same data? I suggest that the optimists often extrapolate long-term historical trends, whereas the pessimists believe that we face radically new situations when ecological limits are approached. Most of the authors in the Simon and Kahn volume look at one problem at a time in isolation, whereas the more pessimistic studies emphasize interconnections and competing uses for the same resources, such as water and forests. Above all, the optimists have great faith in the ability of the free market and technology to respond to scarcities, and they give little attention to indirect costs and externalities. For them, nature has only an instrumental value for human use and no inherent value worth preserving in itself. I would particularly question their expectation that the free market, left to itself, will produce in the Third World the rapid economic growth on which they count to halt population growth. Previous chapters have shown that the gaps between rich and poor have been increasing, both within nations and among nations.

3. Selective Growth

I believe that technology can extend resource limits more than the pessimists expect but that environmental constraints are more severe than the optimists acknowledge. Selective growth calls for deliberate policy choices and will not occur from market forces alone. It will be the result of both individual and social decisions to encourage some kinds of technology rather than others. Scarcities are a product of maldistribution, institutional deficiencies, and shortsighted technological design, as well as finite resources. The important questions about economic growth are: Whose growth? What kind of growth?

In *industrial nations,* considerable economic growth can occur without growth in energy use or in the use of materials. We should encourage the shift that has already started from the resource-intensive manufacturing sector toward the postindustrial service sector, including health, education, communications, and human services. The shift produces a net gain in employment since the service sector is labor-intensive. Other new jobs would be related to energy conservation, renewable resources, and environmental protection.

New technologies suitable for a world of resource constraints would be promoted. Resource use would be a major criterion in technology assessment and in government research and development policy. Engineering design criteria would give greater emphasis to initial cost in energy and materials, operating energy efficiency, and product durability. The technologies of recycling and waste reduction would be fostered more vigorously. Information technologies, including computers and communication systems, are not resource-intensive. A sustainable society, in short, need not be antitechnological, but it must be selective in its choice of technologies.[41]

Sustainable development in the Third World would entail some growth of heavy industry and manufacturing, but economic growth would occur mainly through appropriate technologies of modest scale, including agricultural and renewable energy technologies. This will require both internal reforms and very substantial financial and technical assistance from affluent nations if widespread starvation and social conflict are to be averted. In some cases Third World countries may be able to leapfrog over older industrial technologies to more efficient processes and high-tech alternatives, but a more adequate infrastructure of manufacturing and transportation will still be needed.[42] Justice clearly demands that most of the world's growth in resource use should occur in the Third World. But we should not underestimate the difficulty in altering the prevailing patterns of growth, nor the catastrophic impacts on the environment if they are not altered. We will return to these questions in chapter 9.

II. GENETIC ENGINEERING

Recent knowledge in molecular biology has brought an unprecedented increase in human power over living things. In the past, new forms of plants and animals were created by selective breeding over many generations. Today, more radical modifications can be rapidly introduced by directly altering the genetic information controlling the growth of individual organisms. Gene-splicing techniques hold out great promise for producing new kinds of drugs, plants, and animals and for treating human genetic diseases. But they also raise significant issues of environmental risk, social justice, animal welfare, and human dignity. The more distant prospect of altering human germ-line cells raises disturbing questions about our power to redesign human beings.

1. MODIFYING MICROBES, PLANTS, AND ANIMALS

In the early 1970s it was discovered that strands of DNA can be cut by restriction enzymes and then *recombined* in new ways. The gene (DNA sequence) corresponding to a particular protein can be spliced into a plasmid ring in a bacterium. When the bacterium divides, a new copy of the gene is also formed. Such genes, replicated many times, can then be used to control the assembly of large quantities of the corresponding protein. Insulin mass-produced by this method was first marketed in 1982, and hepatitis vaccine and human growth hormone were available in 1986. Gene-splicing can produce drugs, hormones, and vaccines of lower cost and greater purity than similar products prepared from animal serum or natural sources.

Extensive research is under way in the *genetic engineering of plants* with particular properties. Increased tolerance to salinity would allow plants to grow in regions where they do not now thrive. If a crop has been engineereed for tolerance to a particular herbicide, farmers could use that herbicide to destroy weeds without harming the crop. The gene for the protein trypsin, which is toxic to insects but does not harm humans, might be incorporated into plant seeds. Plants with the ability to fix nitrogen from the air would greatly reduce the need for fertilizer. The bacteria present on the root nodules of soybeans, clover, and other legumes have this ability, and it is hoped that wheat, corn, and other crops can be similarly designed to make their own nitrogen compounds. Bacteria on strawberries have been altered so they do not act as nuclei for the formation of ice crystals, making the fruit less vulnerable to frost. Other microorganisms have been developed to aid the industrial fermentation of wood and crop wastes (to make liquid fuels) or to speed the degradation of oil spills or toxic wastes.[43]

Genetic engineering could greatly benefit agriculture, but it must be used with care because of its environmental and social impacts.

1. Environmental Risks. When gene-splicing was discovered in 1973, molecular biologists were concerned that genetically altered bacteria might accidentally escape from laboratory experiments and multiply, with unpredictable and potentially harmful consequences. Following a six-month voluntary moratorium on research, a set of strict guidelines for government-sponsored research was established. But by 1979 the initial fears seemed unjustified and the guidelines were relaxed. Most experiments have been conducted with a strain of *E. coli* bacteria that has been so weakened that it has difficulty establishing itself in the human body, much less outside it. No health incident has been reported among those working with recombinant DNA.[44]

During the 1980s the debate centered on the *deliberate release* of genetically altered organisms for field testing. Government guidelines in 1982 required prior approval for such deliberate release, and by the late eighties several field tests had been approved. An OTA study in 1988 concludes that harmful impacts from large-scale commercial use are very unlikely but cannot be ruled out. The study maintains that small-scale field tests are valuable for assessing such risks, provided they are carefully monitored, for example, by using genetic markers to trace the spread of the new genes.[45] One unresolved question is the possibility that genes introduced in one species might be transferred to other species. If the gene for herbicide resistance were somehow transferred from a crop to its weedy relatives, it might produce a "superweed" that would be hard to control.

Some critics have called for a moratorium on *field testing* until we have better ecological models for assessing the risks. They point out that foreign species introduced in new environments (such as Kudzu vines, gypsy moths, and Dutch elm disease) have sometimes been difficult to control.[46] But most scientists hold that field tests of minor genetic alterations are justified if they are carefully monitored and the results are reviewed before approval of commercial-scale use.

2. Impacts on the Farm. In 1980 the U.S. Supreme Court ruled that genetically altered microorganisms could be patented. Patenting has accelerated the trend toward seed production by a few large companies (whereas traditional plant breeding was distributed over many firms). Such concentration reduces both genetic diversity and economic competition. Some chemical companies have bought up seed companies in order to integrate the sale of chemicals and seeds. Monsanto is working on gene-splicing to produce crops resistant to its broad-spectrum herbicide, Roundup. Farmers would then buy modified seeds and herbicide as a package from the same company. The increased use of the herbicide might reduce tillage for weed control, but most environmentalists deplore such an increase. A few companies would dominate seed production, and genetic diversity would suffer. Companies seeking a large market usually favor large farms; the further industrialization of agriculture and the requirement of expensive inputs would make it harder for small farmers to compete.[47] In the more distant future, the greatest threat to the farm might be the growth of food substances in plant cell cultures in large vats in industrial plants.

3. *Impacts on the Third World.* Most of the world's genetic diversity in plants is found in developing nations. It would be unjust for First World companies to secure seeds from the Third World, alter them genetically, patent them, and sell them back at high prices. Indigenous genetic diversity would be reduced, and farmers would be dependent on seeds and inputs designed for First World conditions. Whereas research on Green Revolution seeds was carried out mainly at international centers or public universities, biotechnology research is conducted mainly by corporations or by universities under industrial contracts requiring secrecy and patent rights, so Third World access will be more difficult. Some Third World countries can develop their own capacities in biotechnology, geared to their own needs. Others can try to conserve their own genetic resources in living plants and in seed banks. The U.N. has proposed regional and international seed banks, with free access (as a common global heritage), or with payment of royalties to the nation from which the seeds came.[48]

Current research on genetically engineered drugs and vaccines is mainly directed to the needs of high-income populations (such as prescription drugs, on which profits are high), rather than *needs in the developing world* (such as vaccines against intestinal diseases). Frost-free strawberries have received more attention than improvements in cassava or other Third World subsistence crops.[49] Another concern is the production of bioindustrial substitutes for Third World agricultural products. For example, genetically engineered enzymes might convert vegetable oil or palm oil into substitutes for vanilla beans from Madagascar or cocoa from West Africa, on which their local economies are highly dependent. In Southeast Asia, 22 million jobs would be lost if a good synthetic substitute for rubber is found

Genetic engineering related to animals is also under way. The most controversial case to date has been the production of a protein almost identical to natural *bovine growth hormone* (BGH). When injected daily in cows, milk production increases 10 to 15 percent. But the European Community and two U.S. dairy states (Minnesota and Wisconsin) have declared a moratorium on its use. Several issues are involved:

1. Human Health. Scientific panels in both Britain and the United States have reported no harmful effects.[50] In 1990 the U.S. Food and Drug Administration published industry studies showing that BGH breaks down in the human digestive tract. But critics decry the agency's dependence on industry data (part of which has been kept secret from the public), and they claim the breakdown may not be complete. Up to 9 of the 191 amino acids in engineered BGH differ from those in natural BGH, so there might be long-term effects that warrant further study.[51]

2. Humane Treament of Cows. Bovine growth hormone at low doses, carefully administered, seems to cause no harm to cows. Reports of metabolic stress and greater susceptibility to diseases have not been substantiated. But the use of BGH will further encourage the industrial model of "factory farming" in overcrowded feedlots.

3. Impact on Small Farms. BGH will be adopted first in large farms because it requires management skills and increased consumption of high-quality feed. This will accelerate the demise of small and medium-sized family-owned farms and further undermine the quality of rural life (see chapter 4). BGH would be of little use in the Third World because it requires high-bred cattle and nutritional concentrates, and it causes cows to overheat in hot climates.[52]

4. The Milk Surplus. BGH might lead to cheaper milk for consumers, since labor costs are lower. But past productivity increases in agriculture have tended to benefit food processors more than consumers. Moreover, it is ironic to seek higher productivity when both the United States and Europe have been plagued by dairy surpluses and have poured billions into price-support subsidies to dairy farmers. In 1986, the U. S. government paid farmers to slaughter a million cows to reduce the milk surplus. The decisions of farmers, consumers, and government agencies concerning BGH should take such social impacts into account, along with questions of health.

Some additional questions arise concerning gene-splicing in animals. First, is *the patenting of genetically modified animals* justified? In 1987 the U.S. Patent Office allowed the patenting of animals, and in 1988 it granted a patent for the "Harvard Mouse," specially modified for cancer research. The World Council of Churches has opposed animal patents because they will concentrate the control of livestock breeding in the hands of a few industrial firms rather than traditional breeders. Farmers might be required to pay a royalty for each newborn animal. The WCC also holds that exclusive rights to genetically altered animals are not justified since they represent relatively minor modifications in creatures that were not produced in the laboratory. It maintains that adequate financial incentives for research can be found without the concentration of power that exclusive rights will create.[53]

Second, is *the transfer of genes between animal species* justifiable? It occurs in nature, but it is very rare. Reproductive isolation and other biological barriers keep gene pools distinct in evolutionary history and enable new species to form. Some theologians and defenders of animal rights have opposed all deliberate transfer of genes between species as a violation of "the intrinsic nature" of each creature.[54] This seems to me a dubious position, since species are continually changing anyway, though of course very slowly. Moreover, the proposed modifications are relatively small; we are not talking about pigs that fly. However, I do hold that we must not treat other creatures as mere commodities to alter and use for our own benefit. Beyond avoiding inflicting suffering on animals, we must consider their welfare and their harmonious interaction with their environments. In place of the anthropocentric and technocratic assumptions expressed in our domination of nature, we should encourage a greater respect for all living beings.

2. HUMAN GENETIC ENGINEERING

Genetic screening discloses the presence of human genetic diseases, and in some cases it can lead to therapy. Genetic engineering of human somatic or germ-line cells offers further prospects of therapy, but it could also be used for eugenic goals that are more questionable.

1. Screening for Genetic Diseases

Genetic diseases afflict twenty million U.S. citizens and are responsible for 25 to 30 percent of admissions to children's hospitals.[55] Some inherited diseases produce great suffering, mental retardation, physical deformity, or early death. Some diseases appear only when a child inherits a defective gene from each parent. A genetic counselor can calculate the probability that a child of gene-carrying parents will have the disease; they may be advised to adopt a child rather than risk conceiving a child of their own. After a fetus is conceived, prenatal tests (amniocentesis) can determine if it has any of a growing list of genetic diseases. The presence of some genetic disorders can be detected in childhood or early adulthood before any symptoms have appeared (Huntington's disease, for example, which is devastating and incurable).

In some cases such *genetic information* can lead to corrective therapy. When the gene for the enzyme PKU is missing, a special diet is required for the child to survive. If a very serious and incurable genetic disease is detected early in pregnancy, a woman faces the agonizing decision of whether or not to have an abortion to prevent the development of a child who will suffer intensely. In the future, genetic engineering may provide cures for some of the diseases that are now incurable.

Genetic screening programs raise several ethical issues.[56] Should genetic screening be voluntary (to protect the rights of patients or parents) or mandatory (in the interest of reducing the incidence of genetic diseases and their social costs)? Who should have access to the results? If insurance companies had access to test results they would probably raise rates for those at risk. If employers had access, they might discriminate against potential victims, even though they had no symptoms of disease. Do persons who might carry the gene for Huntington's disease have the right *not* to know whether they have it and will die early? I would side with a voluntary and confidential program, favoring individual rights rather than social benefits, so that people can participate more fully in the decisions that affect their lives, though I think individuals should also be encouraged to consider the social consequences of their decisions.

2. Somatic-Cell Therapy

Therapy on somatic cells affects only *the individual treated* and not the germline cells that influence future generations. Research is currently directed at diseases that involve a single gene and that are fatal, since an experimental therapy can be more readily justified in a terminal condition that has not responded to any other treatment. In 1991, the first clinical trials of genetically altered cells on cancer patients were approved. Ten patients were to receive white blood cells armed with the gene for a toxin that destroys tumor tissue.[57] Tay-Sachs disease and sickle-cell anemia are promising candidates for future genetic therapy research. The hope is that as genetic knowledge and genetic techniques improve it will be possible to add a missing gene or to remove or replace a defective one.

Proposals for *clinical trials* of somatic-cell therapies in the United States must go through four stages of review, from local ethics committees to a series of committees at the National Institutes of Health. These reviews examine experimental trials on animals, provisions for confidentiality, the informed consent of patients (or their parents), and comparison with alternative therapies.[58] An OTA study concluded that somatic cell therapy is essentially similar to other accepted medical practices and raises no distinctive ethical issues as long as regulations for the protection of patients are strictly adhered to. But the study found no agreement on germ-line therapy and urged further public discussion and congressional hearings before it is attempted.[59]

3. Germ-line Therapy for Genetic Defects

Germ-line therapy is a more distant prospect, but it offers several *possible benefits*. Instead of having to repeat somatic cell therapy on individuals in successive generations, germ-line changes would be passed on to future generations. This would be a more efficient use of medical resources, and the prevalence of harmful genes that cause intense suffering would be permanently reduced. Moreover, some diseases affect inaccessible somatic cells (such as those in the brain) or widely distributed body tissues (such as those in cystic fibrosis). In 1989, the defect responsible for cystic fibrosis was located (an error in a single DNA triplet in the string of 1480 triplets that produces an important protein). If this defect could be corrected in a fertilized egg, it would affect many different body tissues and also benefit future generations.[60]

The prospect of germ-line therapy has evoked considerable controversy. Clearly it has *long-term and widespread consequences* since it affects future generations. Correcting even single-gene defects is a complex problem because it involves the mechanisms that regulate the expression of the gene at the right time and place in the developing organism. Some of the indirect consequences of intervention might be harmful, delayed, and irreversible. In 1983 Jeremy Rifkin secured the signatures of fifty-eight scientific and religious leaders on a statement calling for the banning of all germ-line research, on the grounds that such research is an unjustified tampering with the natural order and that it would open the door to eugenic uses of germ-line intervention.[61] A World Council of Churches report was more moderate; it proposed a ban on human germ-line experiments "at the present time" while ethical reflection is pursued and guidelines are developed.[62]

4. The Selection of Desirable Genes

All of the examples so far are attempts to diagnose or treat harmful genes. But the same techniques could be used to *select desirable genes*. Amniocentesis allows the sex of a fetus to be ascertained, and if followed by selective abortion, parents could choose the sex of their child. If a woman's eggs are fertilized by her husband's sperm in a test tube (*in vitro*), the egg to be reimplanted could be selected for sex and perhaps for other characteristics. The combination of in vitro fertilization and germ-line intervention could be used to seek positive genetic improvements, though this is a more distant prospect. In each of these cases, parents would be interested in selecting the characteristics of their children. But the social consequences might be disastrous; for example, parents in a patriarchal society might choose to have many more sons than daughters.

The selection of desirable genes could also be aimed at *the improvement of society*. Eugenic proposals in past history advocated selection of the individuals who would be allowed to reproduce (paralleling the breeding of animals selected for particular characteristics). In Nazi Germany, a group of young women exemplifying "ideal Aryan characteristics" were chosen to be the mothers of an elite group of children. In the United States, a deep-freeze bank has been established containing the sperm of men of outstanding mental or physical abilities, from among which women can choose to be artificially inseminated.[63] In the future, genetic engineering could perhaps be used to improve society, but its technical possibilities seem to be very limited. Most human characteristics are the product of many genes in complex interaction. There is no one gene for intelligence, much less for artistic ability or love and compassion.

Let us examine some of the ethical and theological issues raised by these various levels of human genetic intervention.

1. Attitudes Toward Disabilities. The alleviation of suffering by the diagnosis and therapy of genetic diseases is a commendable goal. But we must take care that it does not lead to resentment or condescension toward people with disabilities. If parents think of an unborn child as a product for whose quality they are responsible, they may feel a burden of guilt or be unable to deal with the child's limitations, many of which are not genetic. The worth of an individual does not depend on the absence of defect or disease. Unconditional love and acceptance within the family and respect for persons in society must not be compromised by our efforts to eradicate genetic defects. We will always need the courage to live with our own limitations and the compassion to accept those of others.[64]

2. Dangers in Eugenic Programs. Proposals for social eugenics—whether by selective breeding or by germ-line intervention—appear very dubious. By what criteria would selection be made? All cultures would probably want to avoid the suffering and death of severe genetic diseases, but less severe "defects" are culturally defined, and the specification of positive traits reflects even stronger cultural biases. Ideals of beauty, physical ability, and mental accomplishment vary historically. We must be cautious in imposing on future generations our images of perfection. We do inevitably impose many of our expectations on our children, and all education and social change imposes our values on the future, but genetic changes are more irreversible and long-range—and also more uncertain in their unforeseen consequences. If human traits could be improved genetically, who would be competent to make such decisions? Parents project their unrealized ambitions on their children, scientific elites have their own biases, and governments seek to enhance their own powers. Eugenic programs would also reduce genetic diversity and encourage intolerance toward differences.

3. The "Slippery Slope" Argument. Rifkin and others have opposed even germ-line therapy for genetic defects because it could be the first step toward germ-line eugenics. It is easier to draw an absolute line between somatic and germ-line cells than between "correcting a defect" and "making an improvement." Rifkin says that it is hard to stop once you start down this slippery slope. In some societies, color of skin might be considered a "defect," and genetic powers could be abused by tyrannical governments. Rifkin expresses a pessimistic view of technology and human nature in general.[65] I reply that it is possible to make moral distinctions in both theory and practice. Social regulation can allow valid uses of a technique while limiting the abuses. In principle I would draw the line between therapy and eugenics, even though this requires more careful discrimination than distinguishing somatic from germ-line cells.

But I would approve germ-line therapy only under three conditions. First, extensive studies of human *somatic-cell* therapies similar to the proposed germ-line therapy must have been conducted over a period of many years to acquire data on the indirect effects of the genetic changes. Second, the effects of similar germ-line therapy in *animals* must have been followed over a period of several generations to ensure the reliability and long-term safety of the techniques used. Third, widespread *public approval* must have been secured, since the therapy will affect unborn generations who cannot themselves give informed consent to treatment. None of these conditions are currently even close to being met, so for the present only somatic-cell therapy is justifiable.

4. Intervention in Nature. Some critics oppose all human genetic engineering as "tampering with nature." In one version, such alterations would be a violation of ecological integrity and a denial of the maxim that "nature knows best." Another version is based on the idea of natural law and the conviction that the world has permanent structures in accord with divine intentions. Human nature, in particular, is said to be fixed and inviolable. But this criterion is too broad. All medicine (and much of civilization) is an intervention in nature, even if it is aimed at fulfilling what one takes to be the true functioning of nature. Typhoid, after all, is part of nature. Moreover, in an evolutionary perspective, nothing—not even human nature—is fixed; all structures are changing. But we can at least meet these critics partway. Let us be grateful for the amazing human genetic heritage, aware of its complexity and fragility, and cautious about changing it when our knowledge is so limited. We should be wary of talk about redesigning humanity.

5. "Playing God." A religious version of the previous argument holds that we are usurping divine prerogatives when we tamper with God's creation. I would reply that creation was not completed once for all. I suggested in the previous volume that God works through the continuing evolutionary process and through our lives today. Human beings are endowed with intelligence and creativity; we can be coworkers with God in the fulfillment of God's purposes. We can cooperate with God in the continuing work of creation in nature and history.[66] At the same time, the biblical tradition speaks of human sinfulness and our tendency to use power to advance self-interest at the expense of others. This tradition is critical of the unbridled drive for mastery and control, and it rejects all attempts to seek technical fixes as a substitute for changes in human relationships and social institutions.

3. THE SOCIAL CONTEXT OF RESEARCH

Science and technology, I have said, are always the products of particular social contexts. The main institutions affecting genetic engineering research are universities, industrial corporations, and government agencies and legislatures.

In the 1970s, a prominent issue was the balance between *freedom of inquiry* and *government regulation* in research on recombinant DNA. When gene-splicing was discovered, a six-month moratorium was voluntarily initiated by the leading scientists in the field. Some of them wanted to avoid any outside regulation of DNA research, which, they said, would compromise freedom of inquiry and would bring in lay persons who do not understand the technical details. They asserted that the scientific community is self-regulating, with its own professional associations and rewards and punishments. Outside interference might unduly restrict the progress of research.

Most scientists, however, recognized that some *government regulation* of genetic engineering was justified. They acknowledged the right of the National Institutes of Health (NIH), the main funding agency, to impose guidelines on its grant recipients (private industry has voluntarily accepted the NIH guidelines). Public funds and public safety are involved in this research, so the public has a right to set conditions. Moreover, scientists have personal stakes in their field of research (recognition, promotion, funding, and sometimes stock ownership) and they cannot be expected to remain unbiased—even though their initial reaction in the DNA case was on the side of caution. In addition, freedom of inquiry refers only to the pursuit of knowledge and the communication of ideas, not to actions. Molecular biology was rapidly entering areas of application in biotechnology, which had important social consequences. Policies and priorities for such applications are legitimate matters for public concern.[67]

Some of the regulations under debate in the 1980s were about *environmental risks,* especially in the field testing of genetically altered microbes. Prior review and approval of proposed field tests were required to guard against environmentally harmful consequences. Other regulations governed *research on human subjects.* Clinical trials required advance approval to protect the welfare of patients. When a scientist in California violated his NIH-approved protocol for research on thalassemia patients, his research funding was cut off. Congress has debated legislation to extend the NIH guidelines and also to make them mandatory for industry.

A second issue that intensified in the 1980s was the relation between *universities* and *industrial corporations.* The Hoescht company agreed to contribute $50 million to molecular biology research at Harvard and Massachusetts General Hospital in return for exclusive licenses on any products resulting from the research. Similar contracts were signed by Washington University with Monsanto and by other universities, medical schools, and agricultural schools with various firms. Moreover, many of the top molecular biologists in universities are consultants or serve on the boards of biotechnology companies and often hold stock in them. These academic-corporate ties have several benefits for universities. They bring in income for salaries and expensive equipment at a time of cutbacks in federal funding; they lead to career opportunities in industry for graduating students; and they facilitate the two-way flow of ideas between academia and industry and the rapid application of promising discoveries. But they also raise some important questions—not unique to molecular biology, but more acute because of the huge financial stakes.[68]

Can the protection of industrial *trade secrets* be reconciled with the tradition of *open discussion* in the university and in the scientific community? In theory the publication of research results need only be delayed long enough for the sponsoring company to review it and file for patents, but in practice long delays have occurred, sometimes hindering the publication of Ph.D. dissertations.[69] Students are often under pressure to do research beneficial to the company with which their professor has ties. A survey of molecular biologists revealed considerable concern about the effects of secrecy on collegiality and cooperation among faculty members, and on the exchange of information within the scientific community. Many respondents said that industrial contracts resulted in a distortion of research priorities toward applied research—and within applied research, toward short-term payoff.[70] The commercialization of university biology weakens the effectiveness of peer review as an instrument by which the scientific community governs itself. Another complaint is that industry too often gets the royalties from research partly paid for by public funds.

Another criticism of academic-corporate ties is that they compromise the role of scientists as *impartial advisers in public decisions*. The university has always been a source of independent expert advice in review committees, regulatory and funding agencies, and legislative hearings. A recent volume by two scientists asks, "With many if not all of the world's top molecular biologists on the boards of directors or consulting for biotechnology companies, will we receive the impartial advice we need?"[71] Since policy decisions may affect regulations, conditions of research, and the value of company stock, a conflict of interest makes it difficult for such scientists to speak out on behalf of the public. These problems have no simple solutions, but more adequate public funding of research would give universities more independence. Universities could also insist on stronger rules to limit the outside activities of their faculty and to require prompt publication of research findings.

A final group of issues concerns *social justice* in the allocation of scarce medical resources. Access to genetic counseling and genetic therapy should be open to the widest possible range of patients and not become a special treatment open only to those who can afford to pay for it. We should perhaps question efforts to cure extremely rare genetic illnesses (in some instances, with only a hundred cases worldwide), unless they are of exceptional scientific interest. An even more basic decision is the division of funds between medical treatment and preventive medicine (in which many more lives could often be saved for the same expenditure). Biotechnology companies invest heavily in research offering high returns in affluent countries, and they neglect major health needs, both at home and abroad, that offer smaller returns.

In conclusion, genetic engineering can make important contributions to the future, but its environmental and human consequences require careful scrutiny and protective regulation. The directions in which it develops should be the concern of public policy and not simply the product of scientific expertise and market forces. The new power to modify living things must be used with caution because its effects are so far-reaching.

III. NUCLEAR WEAPONS

Of all the unprecedented powers in our hands, none is potentially more destructive than nuclear weapons. For forty years we lived with the threat of a nuclear holocaust that could wipe out a large part of humanity and other forms of life. With the ending of the Cold War and the disintegration of the Soviet Union, a major East-West confrontation now seems highly unlikely, but as nuclear weapons spread to other nations the temptation to use them on a smaller scale is increasing. We still have not faced the possibility that the most impressive monument to our technological abilities will be nuclear destruction. But the end of the Cold War also opens up new opportunities for the world to move decisively toward the abolition of nuclear weapons. We will examine (1) the arms race, (2) ethical and theological issues, (3) arms control treaties, and (4) the prospects for common security.

1. THE ARMS RACE

Nuclear weapons technology can be understood only within the context of the arms race since World War II. Let us consider in turn the causes of the arms race, its direct and indirect costs, and the consequences of nuclear war.

1. Causes of the Arms Race

Among the many causes of the continuing arms race, six have been predominant:

1. The East-West Conflict. Until recently, the United States and the Soviet Union each saw the other as the main justification for enormous military expenditures. The ideological differences separating communism from economic capitalism and political democracy appeared irreconcilable. American armaments were built up to deter Soviet aggression against other nations or in "proxy wars" around the world. Nuclear weapons for NATO forces in Europe were intended to offset Soviet superiority in conventional forces.

2. Action-Reaction in Weapons Development. The Soviets acquired an early lead in intercontinental ballistic missiles (ICBMs) and deployed the first antiballistic missile system. Every other advance in strategic weapons occurred first in the U.S.: the A-bomb, the H-bomb, submarine-launched missiles, multiple-headed missiles, cruise missiles, and so forth. Some technological improvements were highly destabilizing. For example, new guidance systems upgraded the accuracy of ICBMs so they could be targeted on silos housing enemy missiles. A secure second-strike capability was supposed to be the basis of stable deterrence, but both sides kept improving their first-strike capabilities. Each new system gave its owner a temporary advantage, but the other side soon acquired similar weapons.[72]

3. Defense Lobbies. Lobbying by military and industrial interests has strongly promoted the funding of new weapons systems. Military technologies developed a constituency and a momentum of their own when jobs and local economies were at stake. Members of the armed services committees in Congress formed alliances with the Department of Defense and defense industries in supporting particular proposals.[73]

4. Competition for Raw Materials. The Americans and the Soviets intervened frequently in the Third World, partly to support political movements with which they were ideologically sympathetic, but also to secure economic resources. U.S. intervention in Central America has had economic as much as political motivation, and it led to the support of leaders cooperating with U.S. interests even if they were military dictators with no concern for human rights. In the

1980s, the U.S. conflict with Iran and arms sales to Saudi Arabia were influenced by the fear that the Soviets might manipulate the politics of oil in the Middle East. After Iraq invaded Kuwait, the possibility that Saddam Hussein might threaten other oil areas and effectively control half the world's oil was one reason for the military response of the United States and its allies.

5. *Regional Conflicts.* From 1981 to 1988, $337 billion in arms was delivered to the Third World, half of it to the Middle East. Iraq spent $43 billion on advanced military technology from the Soviets, France, and other countries.[74] It also acquired equipment for making chemical, biological, and nuclear weapons from companies in the former West Germany, the United States, and Britain. Many nations have sought nuclear weapons because hostile neighbors had them or were working on them; among such rival pairs were Israel and Iraq, India and Pakistan, North and South Korea, and Brazil and Argentina. With the demise of communist governments in Eastern Europe and the Soviet Union, the greatest threats to peace in the 1990s will arise from regional and ethnic conflicts. Several nations are trying to acquire an "Islamic bomb." Added to the risk is the danger that the United States will further project its military power globally without having to consider possible Soviet reactions. The United States is currently the world's largest arms exporter, and the sale of high-tech weapons is booming since the Gulf War.[75]

6. *Male Dominance.* A more subtle cause of war is male dominance in political and military decisions, and the association of men with violence. Most violent crime is male, and the perpetrators of sexual and family violence are 98 percent male.[76] In many cultures, combat and war have been seen as tests of masculinity. Such gender differences may have some biological basis, since women have more experience in nurturing life, but the differences are primarily the product of socialization. Boys are taught to be tough, to compete, and not to express their feelings. "Macho" male images are common in films and on television. Feminists do not, of course, want women to perpetuate the passive roles associated with their victimization; they advocate the self-defense and empowerment of women, but by nonviolent methods. Perhaps as men share in the nurture of children, and as women have a larger role in public life, both men and women will seek less violent ways of settling conflicts between nations.[77]

2. The Costs of the Arms Race

The arms race has exacted a high cost in relation to each of the following values that I have defended in this volume.

1. *Economic Development.* By the end of the 1980s, global military expenditures were costing nearly $1 trillion annually.[78] Unlike investment in industrial equipment or agricultural development, money put into weapons generates no further wealth. Furthermore, it produces relatively few jobs. $1 billion spent in producing guided missiles creates 9,000 jobs; the same amount would create 21,000 jobs in local transit or 63,000 jobs in educational services.[79] The indirect spin-offs from military to civilian technologies could have been

achieved directly at far lower cost. The diversion of money and talent from civilian to military technology contributed significantly to the loss of American competitiveness compared to Germany and Japan, while Soviet industry suffered even more. For the period from 1960 to 1988, military expenditures amounted to 10.5 percent of the GNP in the Soviet Union, 6.4 percent in the United States, 3.7 percent in West Germany, and only 0.9 percent in Japan. Annual growth in manufacturing productivity among the four has stood in just the opposite order.[80]

2. *Social Justice*. I have suggested that the justice of a policy can be judged by its impact on the least advantaged members of society. Arms expenditures in the U.S. grew rapidly in the eighties, while funds for housing, social services, and education were drastically cut. The poor, the elderly, and the young—and especially urban black families—were hardest hit. Third World countries spent billions on arms even as malnutrition and poverty became more accute. Military interventions overseas have usually defended narrowly national interests and hindered progress toward more equitable terms of trade and a more just sharing of the world's resources. During the eighties, defense took 71 percent of government R & D funds in the United States, 50 percent in Britain, and 31 percent in the European Community, diverting scientists and engineers from research more directly related to human needs.[81]

3. *Environmental Protection*. U.S. nuclear weapons plants have been immune from national environmental regulations; the cleanup of radioactive contamination from wastes and leaking storage tanks in twelve states will cost $35 billion in the next five years and up to $200 billion before it is completed.[82] Other toxic wastes from military plants are among the worst in the nation. Military operations are damaging to environments (for example, defoliants were widely used in Vietnam, while the Gulf War was highly destructive to fragile water and land ecologies).

Military preparations are thus costly in terms of economic development, justice, and the environment, but they could still be justified if they assured *world peace*. More often, however, they have been inimical to peace. The sale of arms to the Third World has aggravated regional tensions and has strengthened authoritarian regimes. Since 1945, 125 armed conflicts in the Third World have killed 22 million people (of whom three fourths have been civilians) and created 13 million refugees.[83] I will argue below that nuclear deterrence indeed prevented direct military conflict between the United States and the Soviet Union, but that it hindered efforts to halt the arms race. Reliance on military force has led us to neglect the potential of the U.N. for conflict resolution and peacekeeping. Moreover, the proliferation of nuclear weapons to additional nations increases the risk that they will be used. Nuclear war would be the ultimate violation of all human and environmental values.

3. Consequences of Nuclear War

As the first atomic bomb was exploding in the New Mexico desert, the director of the project, Robert Oppenheimer, was reminded of a passage from the

Hindu epic, the *Gita:* "If the radiance of a thousand suns were to burst into the sky, that would be like the splendor of the Mighty One." But as the mushroom cloud rose, he recalled another line, "I am become Death, the Shatterer of worlds."[84] The bomb dropped on Hiroshima killed 80,000 people by burns, blast, and radiation, and many others were maimed or died later. Today the world has 51,000 nuclear warheads, some of them 2,000 times as powerful as the Hiroshima bomb, adding up to the equivalent of five tons of TNT for every man, woman, and child on earth.[85] One large nuclear warhead has more explosive power than all the weapons used by all nations in World War II. A single 25-megaton warhead dropped on New York City would level buildings for thirteen miles in all directions.

A *major nuclear attack* on the United States would result in 70 to 160 million deaths (35 to 77 percent of the population) in the first thirty days, according to a 1979 report. Millions more would die later from lack of medical attention, since surviving doctors and hospitals would be unable to cope with the huge number of casualties; additional millions would die over a period of forty years from radiation-induced cancer. Widespread disease and famine could be anticipated.[86] Such a full-scale attack appears less likely in the 1990s, but smaller nuclear exchanges could escalate, and future political alignments are unpredictable. Even a dozen nuclear weapons could devastate a nation's population centers.

More recent studies have shown that after a major nuclear attack the smoke and dust from fires in cities and forests would cut out 95 percent of the light and create a *nuclear winter* over most of the world. A 1984 report by the National Academy of Sciences acknowledged the great uncertainties entering any estimates, but it concluded that temperatures would drop by 18° to 45° F for 6 to 20 weeks, and ultraviolet radiation would rise by 50 percent because of ozone reduction.[87] Other studies in the United States and Europe conclude that subfreezing temperatures, loss of sunlight, ozone depletion, and radioactive fallout lasting many months throughout the globe would destroy crops and create worldwide famine. Populations in the Southern hemisphere would be decimated, and there would be few survivors in the North. The total extinction of the human race would be unlikely, but it cannot be ruled out, according to these scientists. To make nuclear winter impossible, global nuclear weapons would have to be reduced below 500 warheads—1 percent of the present total, but quite enough to provide minimum deterrence.[88]

It is sobering to realize that we have been willing to accept even a small risk of *the extinction of humanity* and that we might face such a risk again in the future. Jonathan Schell says that while the death of an individual can have meaning within the context of ongoing history, the death of humanity would be the end of human birth, life, and history.[89] It would destroy our genetic and cultural legacy, all that we have inherited from the past and could pass on to the future. To the Christian this would be the ultimate blasphemy, the destruction of that which God has evoked into being over millions of years of evolutionary history. If risk is defined as the probability of an event multiplied by the magnitude of its consequences, then even a very small probability of the extinction of humanity, multiplied by the almost infinitely tragic consequences, amounts to an enormous risk. No possible national interest could justify gambling with the future of the whole human race, along with that of other life forms that might be harmed or destroyed.

We have protected ourselves from facing nuclear war by a kind of *psychic numbing and denial.* Against such awesome forces we have felt powerless, paralyzed, and fatalistic.[90] We continue to be at the mercy of leaders who could misjudge each other and initiate a series of actions and reactions they had not intended. Only a vision of our interconnectedness with all people and all forms of life can empower us to try to make such a catastrophe impossible by getting rid of these weapons.[91]

2. ETHICAL AND THEOLOGICAL ISSUES

It is often said that *deterrence* can be justified because its goal is the *prevention* of war rather than the pursuit of war. The purpose and intention in having nuclear weapons is not to use them but to avoid their use. Several philosophers and theologians have replied that if an action is wrong, it is wrong to threaten the action, even if one hopes not to have to carry out the threat. A threat is empty, a mere bluff, if one does not intend to carry it out under any circumstances. To increase the deterrent effect, the threat is stepped up, and the consequences of failed deterrence become even more horrendous.[92] We are responsible for the risks we take, whatever our intentions and hopes. To be sure, deterrence seemed to work between the superpowers for forty years, and the balance of terror encouraged them to avoid direct military confrontation. But it perpetuated the arms race and it remains a risky policy. Accidents, miscalculations of intentions, even computer malfunctions during crises could lead to irreversible decisions. Moreover, the proliferation of nuclear weapons to additional nations in areas of regional conflict vastly increases the prospects of their use, though on a smaller scale.[93]

Nuclear deterrence remains part of the policy of Western nations, and it continues to raise major ethical issues. Much of the past debate about deterrence has drawn on the *"just war" principles* elaborated by Aquinas and others in the Middle Ages. The principle of proportionality states that the damage and costs of military action must be proportionate to the good expected from it. The principle of discrimination prohibits directly intended attacks on noncombatants; the deliberate killing of innocent persons is judged wrong, regardless of the goals sought.[94] In 1983 the U.S. Catholic bishops published a detailed evaluation of deterrent policy using these criteria. The principle of proportionality led them to reject all "limited nuclear war" strategies and all "first use" in response to a conventional attack in Europe, because such uses would be likely to escalate to all-out nuclear war, with consequences disproportionate to the original objectives. The principle of discrimination led them to reject all deliberate targeting of cities or population centers. They concluded that the use of nuclear weapons could be justified only under very restricted circumstances as a retaliatory strike against strictly military targets. Even this minimal deterrent role, they said, could be accepted only as a temporary measure while the nation vigorously pursues arms control as a step toward disarmament. The bishops reaffirmed these conclusions in 1988.[95]

In 1986 a report adopted by the U.S. Methodist bishops went further in *condemning nuclear deterrence.* It rejected all deployment and use of nuclear weapons. The report pointed out that even a strike against military targets would result in massive civilian casualties from blast, radioactive fallout, and environmental damage. It also discussed the growing economic costs of the arms race and the diversion of funds from urgent needs in health, education, and housing at home and abroad. It supported a comprehensive ban on nuclear testing, a freeze on nuclear weapons production and deployment, and independent initiatives as the beginning of verifiable mutual nuclear disarmament.[96] By contrast, a 1988 study by the Church of England accepted deterrence as a necessary compromise in an imperfect world.[97]

A *pacifist minority* within the churches has continued to reject all use of military force. In the first four centuries the church was predominantly pacifist and most members refused to serve in the army. But attitudes changed after Constantine proclaimed Christianity the official religion of the empire (A.D. 312), and by the Middle Ages the ideal of nonviolence was upheld mainly in the monasteries. Since the Reformation, the Mennonites, Brethren, and Quakers, along with some Catholics and members of mainline Protestant denominations, have stressed the radical demands of the gospel and the call to follow the teaching and example of Jesus without compromise.[98]

Gandhi and Martin Luther King showed the power of *nonviolent resistance* to break the cycle of violence and oppression. Civilian resistance in Norway had considerable effect in hindering the Nazi occupation. More recently the idea of preparing people for effective nonviolent civilian-based defense has been explored.[99] Most Catholic and Protestant leaders today defend conscientious objection as a valid expression of individual conscience and witness, though they do not advocate it as a national policy. But the pacifist and "just war" traditions seem to be converging in opposing most if not all uses of nuclear weapons.

The threat of nuclear war has also prompted *theological reflection* going beyond ethical analysis. Some fundamentalists identify nuclear holocaust with the apocalyptic imagery of the Bible. They say that as we near Armageddon (the final battle between good and evil), we can count on God to ensure the victory of the good (our side, of course). They urge us to accept what happens as part of God's plan.[100] However, most biblical scholars hold that the apocalyptic literature was influenced by Persian dualism and was written in a late period when the land was occupied by foreign powers so that no constructive national action was possible. They point out that the earlier prophets, writing under conditions more like our own, spoke of God's judgment on the sins of their own nation and called for national repentance, renewal, and action to fulfill God's purposes of justice and peace. Today, the assumption that nuclear war is inevitable only increases the likelihood that it will happen.[101]

Gordon Kaufman's *Theology for a Nuclear Age* argues that traditional ideas of divine omnipotence and the providential control of history are dangerous because they lead us to escape responsibility for the enormous destructive powers in our hands.[102] Sallie McFague holds that the monarchical model of God reflects the dominance of male attitudes in Western culture. She says that in a nuclear age we must make greater use of models of God drawn from the experience of women.[103] Several British authors have noted the danger of invoking divine omnipotence in relation to nuclear war. They assert that the future is not predetermined but depends on our choices and how we respond to God. Our actions can advance or impede the realization of God's will.[104]

In the previous volume I presented *the process view* that God's influence is persuasive rather than coercive. God works through the creative processes of nature and history rather than by intervening dramatically from outside. God allows evil and suffering in the world—including the Nazi holocaust and presumably also a nuclear holocaust. The gift of human freedom extends even to the possibility of self-destruction. This view would encourage our responsibility to cooperate with God in averting such a tragedy. The end of humanity would be an immense setback for God's purposes on planet earth, though it would not be the end of God's continuing activity here and elsewhere in the cosmos, nor would it end God's own eternal life.

Religious exclusivism adds to the perils of a nuclear world. The medieval crusades were launched by Christian Europe against the Muslim "infidels." Subsequent imperialism and colonialism were given a religious justification because they "brought Christianity to the heathen world." Religious conflicts today play a major part in some of the most serious regional tensions, such as those between India and Pakistan or between Israel and its Muslim neighbors. Jewish, Christian, Muslim, and Hindu fundamentalism and claims to be "the only true religion" aggravate such national conflicts. In the previous volume I advocated a road between absolutism and relativism that would encourage dialogue and respect between religious traditions, while recognizing the distinctive experience, history, and contribution of each. The world's religious communities must collaborate in furthering a common vision of peace and justice.[105]

3. ARMS CONTROL TREATIES

Arms control treaties between 1960 and 1990 had a relatively small effect on the arms race, but they offer some lessons for the future. The 1968 *Nonproliferation Treaty* required states without nuclear weapons to refrain from acquiring them and to submit their nuclear energy programs to inspection by the U.N.'s atomic energy agency. States with nuclear weapons simultaneously agreed to pursue negotiations for arms control and disarmament. The non-nuclear nations who signed the treaty have subsequently criticized the nuclear states for doing so little to fulfill their part of the agreement.

In the *SALT treaty* of 1972 the Soviets and the Americans agreed to hold the number of missile launchers (excluding bombers) at 1972 levels. However, the limits on launchers did nothing to slow the technology race for improved weapons. The U.S. was ahead in multiwarhead missiles, which circumvented the limits on launchers; but the lead was short-lived, and both sides soon had many more warheads. *SALT II* specifying numerical ceilings on all types of launchers was signed in 1979, but support for it in the Department of Defense and Congress could be obtained only by promises of increased funding for MX missiles, Trident submarines, and other new systems. Both these treaties had the effect of legitimating and encouraging the continued race for new technologies.[106]

During the *START negotiations* in the 1980s the United States proposed a reduction in the number of strategic weapons but insisted that the largest cuts must be in land-based missiles, which were the Soviets' main strength, rather than in submarine-launched missiles, where the Americans led. Both sides were more interested in preserving their advantages than in achieving major reductions. Meanwhile, intermediate-range missiles were being introduced in Western Europe to counter similar Soviet weapons. The Soviets had long opposed on-site inspections, but by 1987 they were willing to accept them. With Gorbachev in power and Reagan under pressure to negotiate more seriously, the *Intermediate Nuclear Forces* treaty to remove these Euromissiles was signed in 1987.[107]

By 1990 the Berlin Wall had fallen, Germany had been reunited, and Soviet forces were withdrawing from Eastern Europe. Frequent on-site inspections, including intrusive and unscheduled ones, were accepted by both superpowers. The 1991 *START agreement* did reduce long-range weapons by 50 percent, but it still allowed vast arsenals to be retained. By early 1992, President Bush had announced further American missile reductions and proposed an agreement with the former Soviet States to eliminate all multiple warheads from ICBMs. Orders for new U.S. systems were scaled down but not terminated (Stealth bombers, Midgetman missiles, and Star Wars technology). As the central authority of the Soviet Union was breaking up, the status of nuclear weapons in the separate republics was a new source of uncertainty.

The U.N. conference reviewing the *Nonproliferation Treaty* (NPT) in 1990 reached agreement on ways of strengthening the international nuclear inspection system. The conference called for unscheduled inspections of nuclear facilities, sanctions against violators, and tighter rules governing the purchase of equipment that could be used to produce nuclear weapons. But no official declaration was adopted because the U.S. would not accept the provision that the NPT would be extended when it expires in 1995 only if by then there is significant progress toward a Comprehensive Test Ban forbidding the testing of all nuclear weapons.[108] The inadequacy of previous inspection procedures used by the International Atomic Energy Agency were evident in 1991 when U.N. inspectors found that Iraq was closer to nuclear capability than anyone had anticipated.

The nations participating in the U.N. Test Ban Review Conference in 1991 voted 75 to 2 to continue efforts to achieve a *Comprehensive Test Ban*. The only two negative votes were those of the United States and Britain. These two nations maintain that testing is needed to ensure deterrence; they will accept a halt to testing only as part of a larger disarmament process. All other nations hold that such a ban is essential to stop the continuing arms race and to make possible a more effective nonproliferation agreement.[109] Multilateral action through the U.N. to halt proliferation and new weapons systems offers greater promise for the future than the earlier pattern of bilateral treaties limiting the number of weapons deployed.

4. GLOBAL SECURITY

Until 1990, the U.N. Security Council was virtually paralyzed by the veto power of the Big Five, and the polarization that put the Soviet Union and the United States on opposite sides of almost every security issue. While the structure of the United Nations contains problems, the biggest obstacle it has faced is that the major powers have ignored it, choosing instead to rely on alliances, balances of power, and direct or indirect military intervention. But the end of the Cold War has opened up new possibilities. The 1990 U.N. resolutions to apply sanctions to Iraq were an encouraging sign, though the United States took military action before the sanctions had time to be effective.

The first steps toward genuine *nuclear disarmament* are the Comprehensive Test Ban and the strengthening of the Nonproliferation Treaty. U.N. inspectors must be allowed access to all nuclear facilities, not just those declared by the government that owns them. Next, a ban on the production and deployment of new weapons systems and a more radical reduction of nuclear warheads and ballistic missiles should be undertaken, accompanied by extensive on-site inspections to verify compliance. One hundred survivable missiles (on submarines or mobile land launchers) would provide minimal deterrence during the transition period.[110] As the demand for nuclear explosives falls, the production of plutonium and highly enriched uranium could be phased out, along with breeders, reprocessing, and other proliferation-prone technologies (see chapter 5). In 1982 and again in 1989 the Palme Commission advocated the internationalization of the critical stages of the nuclear fuel cycle, including regional fuel banks and spent fuel management.[111] Because a nation gaining nuclear weapons in a denuclearized world would acquire great power from the threat to use them, a rigorous inspection and enforcement regime would be essential, and transfer of nuclear components and materials would have to be controlled.

Offensive conventional weapons could be reduced and phased out while nonthreatening *defensive* weapons were retained and improved. Tanks would have to be cut back, but electronically guided antitank weapons could be improved. Long-distance bombers and missiles would have to go, but short-range defensive missiles such as the Patriot could be produced. NATO and officers of former Warsaw Pact nations have talked about a force restructuring that would eliminate the most threatening weapons but retain the non-provocative ones. Such a shift to defensive weapons can best be achieved mutually by a global international treaty. But it can also be undertaken by a nation acting on its own. Both Switzerland and Sweden have substantial armed forces, but they are armed with primarily defensive weapons. Both have maintained a neutral stance and have been active in diplomacy and peacemaking.[112]

The verification of all of these measures would require a considerable expansion of *the United Nation's inspection personnel*. It would also be aided by the establishment of a U.N. agency with its own satellites for global monitoring. Such an agency could verify compliance with treaties and monitor military forces and peacekeeping operations. The technical agencies of the United Nations have established a good record of professionalism and impartiality. Adequate funding of an expanded inspection capability and regular access to production facilities would greatly facilitate these disarmament procedures and the early detection of any violations.[113] Antisatellite weapons should be banned since they would threaten surveillance abilities. Restrictions on the international sale of offensive weapons would greatly reduce regional tensions but would require strict monitoring.

Other measures can contribute to *the strengthening of the United Nations*. Its resources for anticipating conflicts and attempting to resolve them peacefully should be expanded. Much greater use could be made of the World Court and regional institutions in the settlement of disputes. Economic sanctions against treaty violators are difficult to enforce, but if rigorously pursued and accompanied by blockades they can eventually have considerable impact. The U.N. could also be authorized to seize a nation's foreign assets and cut off its loans. In the late 1980s the U.N. peacekeeping forces facilitated the Soviet withdrawal from Afghanistan and the transition to independence in Namibia, and more recently they were sent to Cambodia and Yugoslavia. A permanent U.N. peacekeeping force would require adequate funding, special training, and the stockpiling of equipment for rapid deployment. It should be recruited individually rather than as contingents from national armies.[114] The U.N. needs independent sources of funds such as taxes on seabed mining, international trade, and military expenditures.[115]

True security requires more than mutual military security. It must include worldwide *economic development* and *environmental protection*. Only a demilitarized world with reordered priorities can vigorously address the staggering problems of poverty, hunger, and environmental damage. One percent of the world's present military budget would provide family planning services globally. A quarter of the military budget, put into sustainable agriculture, energy, and reforestation, would alter the future for the Third World—and thereby for the whole world. Lower defense spending and new priorities would encourage progress toward a more just international economic order. The various agencies of the U.N. specializing in food, health, trade, and the environment have all done excellent work, but they could play a much larger role in promoting such sustainable development.

Progress toward global security will also require a new understanding of *national sovereignty*. Most institutions of national governance could be retained, but the ability to make war would gradually have to be given up. Each nation would have to accept the limitation of its own military power, but in return it would not be threatened by the military power of others. Nations would have to agree to the settlement of conflicts by nonmilitary means, as they do now within their own borders. National cultural identity could be retained, and the distinctiveness of each culture preserved, as people extended their identification and acquired a more global perspective and loyalty. None of these changes is achievable overnight, but we can set a new direction by taking at least the first steps toward a less militarized world. The most far-reaching changes will require changes in public attitudes and the mobilization of a worldwide movement for peace and justice (see chapter 9).

IV. CONCLUSIONS

Environmental destruction, genetic modification, and nuclear weapons are products of very different kinds of technology. But they share several characteristics that call for similar responses.

1. Unprecedented Power. Each represents a power over nature and humanity that is new in history. Genetic engineering is only in its infancy, and the dangers it presents are less immediate and dramatic than environmental or nuclear destruction. But in all three cases our technological decisions have far-reaching impacts around the globe and on future generations.

2. Government Policies. Because these technologies entail risks to public health and safety, government regulation is appropriate. Moreover, since human and environmental values are so strongly affected, the development of these technologies should not be left to market forces or the decisions of corporations alone. The petrochemical, biotechnological, and defense industries all have considerable economic and political power, but their interests must be balanced against other interests. The directions of development of these technologies are legitimate matters for public debate, public policy, and research and development funding decisions. The next chapter is devoted to the control of technology in the framework of democratic political institutions.

3. International Action. The impacts in all three cases cross national boundaries and are potentially global in scope. In responding to such problems as global warming, the environmental release of genetically modified organisms, or the proliferation of nuclear weapons, any nation hesitates to act alone if it thinks it might harm its competitive position. Only concerted action by the community of nations, primarily through the United Nations, can deal adequately with these threats.

4. A Broad Definition of National Security. Hunger, poverty, environmental destruction, economic depression, and social unrest are greater threats to the security of most nations than external military aggression. Enormous military budgets and the employment of many scientists and engineers in military work hinder efforts to address these problems. The only security today is mutual security and sustainable development.

5. Humility and Respect. All technologies increase human mastery over nature and increase the power of some people over other people. Power tends to corrupt its users. The technologies discussed in this chapter are particularly prone to distortion in the quest for dominion and absolute control of nature and in the pursuit of the narrow self-interest of organizations and nations. In these areas there is a special need for humility and caution, based on respect for the natural world and other people, and an awareness of both external limits and our own limitations. We have awesome powers, but we remain dependent on fragile ecosystems. We know a lot, but our actions still have unexpected repercussions, which can be disastrous. We can find technical fixes, but often at the price of important human and environmental values. As our technologies become more powerful the vision to use them wisely is more crucial.

CHAPTER 8

Controlling Technology

How can technology be controlled in a democracy? I have said that political participation is one of the main ways in which we can have a voice in decisions that affect our lives. But decisions about technology involve complex technical questions that most citizens feel incompetent to understand. Do citizens and legislators then have to rely on experts from industry and government agencies, even though such experts often have a vested interest in promoting particular technologies? What methods of assessing technology allow human and environmental values to be taken into account in policy decisions? Finally, how can government regulations, political coalitions, and scientists and engineers in their professional work contribute to the redirection of technology?

I. GOVERNING TECHNOLOGY

We look first at the ways in which technological policy is addressed within the legislative, administrative, and judicial structures of government. Then we look more specifically at the role of technical experts in democratic governments, and turn last to the ways in which citizens can participate in policy decisions about technology.

1. THE DIVERSE ROLES OF GOVERNMENT

Governments have three broad functions in relation to technology.[1] First, they provide *funds for research and development*. In 1989, 47 percent of all research and development (R & D) in the United States was government funded.[2] Public funds are channeled through the National Science Foundation, the National Institutes of Health, and various government departments (agriculture, energy, defense, and so forth). The actual research is carried out in industry (under contracts), in universities (under grants), and in government laboratories or agricultural experimental stations. Because the public supports this work with its taxes and is affected by the results, it has the right to establish the R & D priorities that will influence the direction of development of technology toward socially desirable goals.

Second, governments *cooperate with industry* to encourage the growth and diffusion of technology. Many economic and trade policies are designed to promote technologies that are likely to contribute to productivity, competitiveness, and economic growth. Though the free market rhetoric of industrialists calls for the independence of industry and government, in practice they have sought and received subsidies, tax benefits, trade protection, and even financial bailouts. The partnership of government and industry is more overt in Japan and Germany, but it is present in all industrial countries. In Europe, some formerly nationalized industries have been privatized, but many are still publicly owned. Governments also exert some influence on industry by the procurement of equipment (such as vehicles, communication equipment, and of course weapons), but their main supporting role is through tax and trade policies.

A third function of government is *the regulation of technology* in the interests of health, safety, and environmental protection. Some of the impacts of technology are visible and dramatic, but many are delayed and uncertain. The assessment of such risks and the design of strategies to reduce them require the use of scientific and technical information. But in decisions about regulatory standards, the public also has a right to be heard because public health, safety, employment, and economic development are affected.

The relative emphasis among these three functions has changed considerably in recent decades.[3] *During the 1950s and 1960s,* an optimistic view of science and technology prevailed in Western nations. The implicit social contract assumed that the public would support basic and applied research, while scientists and engineers would be free to allocate funds (through review panels) to the most promising proposals for research. Society would ultimately benefit from the technology that would result from such autonomy. When regulation was needed, or when social and technical questions intersected (in nuclear plant safety or in airport site selection, for instance), the public was told to trust the experts.

During the 1970s this reliance on experts was widely questioned. Environmentalists claimed that human and environmental values had been neglected in technocratic decision-making. In the United States, citizens demanded not just greater protection from risks but a greater voice in decisions about risks. The major environmental laws of the late 1960s and early 1970s included provisions for public participation on governmental advisory boards and in regulatory hearings. Citizens were given standing to challenge agency actions in the courts. The new laws mandated that drafts of proposed agency actions be circulated for comment by industry, public interest groups, and other agencies, before their final formulation. Public access to information in government agencies was expanded, and greater openness and accountability in decision-making were required. Some areas of research were questioned on social, economic, and environmental grounds. A few large expensive projects were terminated, including the supersonic transport and the breeder reactor.

In the 1980s, administration policies under Reagan strongly favored R & D that would contribute to military and commercial goals. Public funding for research in energy and agriculture was cut, and the portion allocated to defense rose from 50 percent during the 1970s to 74 percent in 1987.[4] Industrial competitiveness, especially in electronics, biotechnology, and other high-tech fields, was given priority over social and environmental issues. Industrial sponsorship of university research grew considerably, leading to the problems of secrecy and patenting discussed earlier. Opportunities for public participation in decisions were reduced in favor of greater reliance on experts from industry and government and greater dependence on market criteria in policy decisions. Deregulation was designed to relieve industry of "excessive costs of compliance"; proposed new regulations were to be subjected to cost-benefit analysis. Greater scientific proof of risk was demanded before regulatory action, and residual uncertainties were used as a reason to avoid action (on acid rain, for example). But it became more difficult to secure such scientific evidence because funding for the regulatory agencies was drastically cut.

In the American system, the president appoints the top officials in each of the government agencies, but Congress adopts the budgets and formulates the laws under which the agencies operate. The legislation dealing with particular technologies is initially formulated by *congressional committees* (on science, energy, health, defense, agriculture, transportation, environment, and so forth). Each committee is responsible for oversight of one or more agencies. But a "cozy triangle" tends to develop between a committee, an agency, and the industry whose interests are most at stake. We noted earlier the lobbying of the armed services committees by defense contractors and the flow of top leadership between the Department of Defense and the defense industry. Similar triangles supported the nuclear power and oil industries, auto and pesticide manufacturers, and many other industries.[5]

A variety of *special interest groups* is active in political lobbying. One way in which these groups try to influence elections and legislative votes is to contribute to the huge campaign funds that candidates use to finance television advertisements.[6] (This problem is less severe in most European nations because election campaigns are briefer and the media give more time to coverage of issues rather than personalities and images.) *Private interest groups* represent the interests (primarily economic) of particular industries, businesses, or occupations. Trade associations, labor unions, and farmers' organizations, as well as particular corporations, lobby for legislation favorable to the institutions for which they speak.

Public interest groups have smaller financial resources and claim to represent the interests of the wider public, such as civil liberties, consumer safety, or environmental protection. Critics maintain that some of these groups have rather small memberships, mostly middle class and well educated. Even if these nonprofit citizens' organizations do not fully represent the public, they do operate in independence from the main centers of economic power, and they often defend environmental and human values neglected by government agencies and private interest groups.[7]

Until 1970, congressional committees had to rely on experts from industry or agencies for most of the *technical information* on which to base their decisions. But during the seventies a wider range of expertise was available to them, including the Office of Technology Assessment and a greater diversity of witnesses at committee hearings coming from universities, environmental organizations, and other public interest groups. Such independent expertise helped Congress to resist the administration's deregulation program in the eighties. However, with the presidency and Congress controlled by different parties during most of the eighties and early nineties, and with Congress itself often divided by regional loyalties, few new regulatory laws were passed and the enforcement of earlier laws was very uneven.

The courts provide additional checks and balances on the power of government agencies. Judicial review of agency actions in the U.S. was encouraged by the environmental legislation of the 1970s. In technical disputes, courts have usually ruled on the fulfillment of procedural requirements rather than on the substantive decisions made by agencies. In other cases, product liability laws have enabled consumers exposed to risks to sue the corporations responsible for them. In the Ford Pinto case it was found that the company knew in advance that the fuel tanks were prone to explosion in rear-end collisions (in which some people would be burned alive). Redesigning the tanks would have cost $11 per car, but the company calculated that even if it lost some court cases it would be cheaper to pay damages than to correct the defect.[8] The case points to a weakness of liability claims: they are reactive, dealing with harms after they occur rather than anticipating them. But the threat of substantial class action suits does encourage firms to give greater consideration to health and safety.

In most European countries, *government ministries* are less subject to legislative oversight and judicial challenge than agencies in the U.S. The prime minister and the cabinet ministers are drawn from a parliamentary majority or coalition, and thus administrative and legislative powers are more closely integrated. Many European nations have had a long tradition of public deference to bureaucracies and experts ("Nanny knows best," as the British used to say), but this is beginning to change, especially in Germany, the Netherlands, and Scandinavia.

Most European ministries have *advisory committees* drawn from a wide spectrum of affected interests. Meeting behind closed doors, these committees of officials, technical experts, and labor and industry representatives try to reach political compromises. This avoids the public polarization and adversarial confrontation common in the United States, but the committees often depend on data supplied by industry. Consumer and environmental interests are usually not well represented. There are fewer opportunities for the actions of ministries to be challenged in court.[9] France has the most centralized administration of any Western democracy, especially in its management of nuclear power. Greater centralization in decision making is conducive to administrative efficiency but at the expense of citizen participation.[10]

2. TECHNICAL EXPERTS AND POLICY DECISIONS

What role should technical experts have in the governance of technology? Some scientists and engineers are employed *within governmental institutions*. They may take part in the work of an agency or ministry (in defense, agriculture, or health, for example). They may be engaged in the assessment of occupational risks, the regulation of toxic substances, or the management of natural resources. A few are full-time science advisers to political leaders or legislative committees. These "insiders" can have considerable influence when they are trusted and share the assumptions of government leaders. But as part of a team they find it difficult to oppose current policies or to challenge underlying assumptions. They are likely to be co-opted into providing scientific support for policies adopted on primarily political grounds. Despite these limitations, they serve an important function in making scientific information readily available to political decision makers.[11]

Other technical experts from industry and the academic world interact with the political process for limited periods as *witnesses and advisers*. They testify at legislative or regulatory hearings or in the courts, and they serve on review committees, advisory panels, and national commissions. Some work locally or nationally with public interest groups dedicated to environmental preservation and public health. Some scientists speak out through the public media on policy issues related to their field of expertise by writing magazine articles or letters to the editor or by taking part in television or radio interviews or programs. Several journals have been founded to deal specifically with technological policy questions.[12]

Reliance on experts within government agencies has been questioned because of the frequency of *disagreement among experts* in recent controversies. Government experts may claim to be objective and nonpolitical, but most policy options favor some interests more than others. Disagreements can be attributed to four features of these controversies.[13]

1. Scientific Uncertainties.

Many recent controversies have arisen from new technologies or environmental impacts with which there had been little previous experience. The greenhouse effect, for example, is the product of a very complex set of interactions among atmospheric gases, weather conditions, oceans, and forests. Studies of separate components of nuclear reactors reveal little about interactive or very improbable failure modes—to say nothing of human errors. Low-probability events and low-level exposure to chemicals or radiation are particularly difficult to study. If urban smog causes a small increase in fatalities from chronic respiratory disease, or if radiation or chemicals cause cancer fatalities, the specific victims cannot be identified among those who have the same symptoms from other causes. The effects of radiation and cancer-causing chemicals may be delayed 20 to 40 years after exposure.

Such scientific uncertainties can often be reduced by *further research*. In some situations it is wise to postpone decisions until at least the most significant uncertainties have been resolved, especially if the regulation of the risk entails high economic or social costs. But postponing decision may lead to much higher costs at a later time. Moreover, new technologies soon acquire a strong institutional momentum; when more jobs and financial interests are at stake, they become more difficult and disruptive to regulate.

The allocation of *the burden of proof* under conditions of uncertainty is not itself a scientific question. Environmentalists tend to assume that a new chemical or activity is harmful unless there is strong evidence that it is safe, and they make cautious assumptions in their calculations. Proponents assume that a new development is innocent until proven guilty beyond reasonable doubt, and they seek to minimize regulations that might slow economic growth. In some cases, legislation assigns the burden of proof. New pesticides and certain types of chemicals must be tested before marketing in the United States, and evidence for safety must be provided by the manufacturer. But for a chemical or pesticide already in wide use the burden of proof lies with the Environmental Protection Agency (EPA).

2. Disciplinary Assumptions

In the previous volume I suggested that every discipline has its paradigms—sets of intellectual assumptions that influence the selection and interpretation of data. Some of the disagreements among scientists in the interpretation of data can be traced to the differing cognitive and professional frameworks within which they operate. Thus early debates on lead pollution were dominated by experts in industrial medicine who worked with healthy adult males. Only later were the effects on children, women, and ill people studied as carefully.[14] An American review board, holding open meetings, recommended banning the pesticide Aldrin; many of its members were biologists and health experts who were impressed by evidence from animal studies. A British review board had access to the same data but recommended against banning Aldrin; its meetings were closed, and most of its members were toxicologists, many with close ties to the pesticide industry, and they wanted clear evidence from human populations before taking action.[15]

Professional training may influence the way issues are formulated and problems are conceptualized and bounded. The time scale employed may emphasize short-term or long-term effects. Simplifying models selectively represent the aspects of the world with which one's discipline can deal. Implicit assumptions may lead to divergent inferences from the data. An environmental engineer and an ecologist, for example, may bring differing assumptions to their estimates of the environmental impact of a new industrial plant. In some disputes the central issues for one side may be peripheral or irrelevant for the other, so that the opponents talk past each other. People who have devoted most of their lives to a particular technology are predisposed to promote it.[16] The presence of such professional biases is one of the reasons for insisting that technology assessment teams and review panels should be genuinely interdisciplinary.

3. Institutional Affiliation

Most people either agree with the goals of their employer when they take a job, or they adjust their goals to those of the organization for which they work. Loyalties to one's employer, one's career, or one's profession may sometimes conflict with the public interest, but more often they can be reconciled—though perhaps at the price of subtle rationalizations or biases. As one science policy analyst puts it, "Where you stand depends on where you sit."[17] During the congressional hearings on the antiballistic missile (ABM), for example, 79 percent of the scientists who favored the ABM had received research funds from the Defense Department, whereas only 18 percent of the anti-ABM scientists had.[18]

Every institution has *a narrow range of objectives* and tends to encourage the perpetuation of its own power. Information is selectively filtered—deliberately or unconsciously—as it is transmitted within an organization. People like to tell their superiors what they want to hear. Institutional biases are particularly likely to enter risk assessments that involve major scientific uncertainties or judgments about human errors and actions. Occasionally witnesses or advisers have direct financial interests in decisions on which they are testifying. More commonly they may benefit personally in indirect ways, including future employment opportunities, grants, or contracts. Few people are impartial when their own advantage is at stake. Witnesses and panelists can be required to declare their financial and professional interests in a policy under debate, but more subtle biases can be revealed only by drawing experts from a variety of institutional contexts.

4. Value Judgments

Technological policy choices affect many economic, social, and environmental values. The weighing of incommensurable costs, benefits and risks, and issues of justice in their distribution are never simply scientific questions. Frequently one group of people will get most of the benefits, and another will bear the indirect costs and risks. As we will see shortly, many factors enter judgments of the acceptability of a risk. One study of science advisers concludes, "Policy-relevant studies and advice can never be value-free, even when carried out by scientists and engineers."[19]

Scientists have no greater wisdom than other people concerning *conflicting value judgments;* these should be resolved primarily through democratic political processes. Scientists offering testimony inevitably express their own value commitments but should try to make them explicit. Similarly, the assumptions used in the interpretation of data should be made clear and the presence of uncertainties acknowledged. The credibility of scientific witnesses is suspect when they claim a greater certainty than the often-ambiguous evidence warrants. Scientific integrity requires both careful documentation and appropriate tentativeness in stating conclusions.

Advisory panels of experts drawn from diverse professional and institutional affiliations can sometimes clarify policy controversies. In the 1970s the National Academy of Sciences was criticized for appointing panels on which most of the members had close ties with the industry or agency under study.[20] Since then the NAS has tried to achieve greater diversity and has instructed broadly based review committees to scrutinize early drafts of reports, looking for biases and self-serving conclusions. Some types of government hearing allow experts with differing affiliations an opportunity to cross-examine each other in order to establish points of agreement and disagreement. A science court has been proposed in which a panel of scientist-judges would hear testimony and issue a report on the established facts, free of political biases.[21] However, this proposal assumes that facts and values can be completely separated, which I do not believe is possible in controversial policy disputes.

Provision for input by experts from *diverse disciplines and institutions* can help in distinguishing facts from values, though they are never completely separable. Focusing on technical problems sometimes obscures the underlying political issues or reinforces previously existing viewpoints. Experts can seldom offer a clear-cut resolution of policy disputes. "Science cannot provide the sort of legitimacy which actors in the policy drama so crave."[22] Even when most of the scientific uncertainties have been resolved, the final judgment on these issues must be made through political processes.

3. CITIZENS AND POLITICAL PARTICIPATION

Elected representatives are the main channel through whom citizens in a democracy can influence policy decisions. Significant policy issues are present in most election campaigns, though they are often overshadowed by facile slogans and personality images presented in television advertisements and news "sound bites." Working within a party, a citizen can help select local and national candidates and encourage real debate on policy issues. Concerned citizens can write to their elected representatives; an outpouring of mail is taken seriously by legislators.

Public debate by *an informed electorate* is an essential element in democracy. Scientific illiteracy is a barrier to political participation in a technological society. The whole educational system has a responsibility to foster greater public understanding of science, and the media can bring to adults a greater awareness of modern science and technology. Television programs such as Nova and news coverage of events involving technology and the environment reach millions of people. To be sure, the media tend to dwell on dramatic crises and controversial or alarmist views, but a variety and balance of views can be presented. The media have helped to expose governmental mismanagement and to publicize what turned out to be real dangers ignored by industry or government agencies. Publicity about the leakage of toxic chemicals at Love Canal, for instance, was a major factor in the passage of the 1980 Superfund bill.[23] The Freedom of Information Act has allowed greater access to government files in order to make agency officials more accountable to the public.

Citizens feel rather helpless as individuals, but they can be effective working together through *environmental and consumer organizations* on both local and national issues. Seven million citizens in the U.S. belong to environmental organizations, most of which combine educational and political programs.[24] Some are action oriented and mobilize demonstrations and protests at local sites to block projects that threaten to harm people or the environment. Others work through the political system, seeking to influence legislators or agency officials by lobbying or by providing testimony at hearings. As indicated above, they can bring in experts with perspectives differing from those in industry or government agencies. Sometimes they can present creative ideas early in the formulation of plans rather than simply protesting actions after they have been initiated.

Experience with *public hearings* in the United States has been mixed. Agencies holding hearings have often tightly controlled the range of participants and the questions that may be discussed. In some cases the hearings were held so late in the planning process that most of the important decisions had already been made; the purpose of the hearings was only to enlist public support. When experts set the boundaries of discussion and use technical jargon to define the issues it is difficult for laypersons to affect decisions. But other hearings have been held early enough that public participants could really discuss alternative options.[25] In Canada, the Berger Inquiry on pipeline routing held hearings in native villages in the MacKenzie Valley to evaluate the social and cultural as well as environmental impacts of the pipeline.[26]

A two-year *public inquiry* on plans to build the first pressurized-water reactor in Britain heard extensive testimony from government officials and public interest groups. The final report issued in 1987 by the hearing officer, a prominent barrister, expressed confidence in the safety licensing inspectorate, even though the licensing data was not available during the inquiry as had been promised. The report accepted government figures on economic costs and concluded that the new reactors would be cost-effective.[27] Two years later, when the government tried to sell its nuclear plants to private investors, independent accountants found that if government subsidies and future costs of decommissioning and waste disposal are included, nuclear power is twice as expensive as coal-generated electricity. It appears that the public inquiry had not examined government claims with sufficient care, and it did not evaluate alternative means of producing or conserving energy.

High-level commissions usually operate with greater independence from government agencies. Their membership is diverse, including prominent public leaders as well as academic and industrial experts. The U.S. presidential commissions on Three Mile Island, the Challenger disaster, and ethical issues in biomedicine each took testimony from a wide range of witnesses at public hearings and issued comprehensive and balanced reports. In Britain, royal commissions have a more diverse membership than public inquiries and are better equipped to examine broad policy issues. They receive written and oral testimony, but their meetings are not open to the public.[28]

The *referendum* has been used in some American states and European nations to allow citizens to vote directly on technological issues. In California, citizens voted in 1989 to shut down a nuclear plant with a history of problems. In Sweden, after an extensive program of study groups and public discussion, the electorate voted not to build any more nuclear plants.[29] Critics say that the referendum weakens representative government and presents yes-or-no alternatives with no opportunity for negotiation and compromise. Moreover, a referendum can be influenced by the side with the most money to spend on a publicity campaign, though this can be partly offset by deliberate efforts at public education and discussion.

The main objection to public involvement in government decisions is that it can lead to *delay and obstruction*. Consider the siting of new plants or facilities. Prior to 1970, local officials usually cooperated with industry in encouraging new enterprises that would augment local taxes and jobs. But since 1970, with mounting fears of toxic substances and other risks, plans for many projects have been virtually paralyzed by protracted protest and litigation. In the siting of improved hazardous waste treatment and disposal facilities, every community says, "Not in My Backyard." A locality can effectively block a project from which a wider public would benefit. Critics of public participation say that regulation can be carried out efficiently only by agency officials in consultation with their own experts, under policies established by elected representatives. If citizens want to object, let them write their representatives or elect new ones.

Closed, centralized administration is indeed more efficient in the short run, but it can neglect factors potentially significant in the long run. Concerned citizens may see problems and defend a range of values missed by officials. For example, intervenors at hearings on the trans-Alaska pipeline maintained that running hot oil pipes underground through permafrost was risky; plans were altered to put the pipe above ground.[30] In other cases citizens expanded the framework of analysis by pointing out social impacts or ethical issues that technical experts had ignored. Regional or national authorities in land-use planning must have the power to override local opposition, but public acceptance requires that the selection process and the sharing of burdens be perceived as fair. One problem with large-scale facilities is that benefits are widely distributed while risks are concentrated in a small area. Smaller facilities produce a more equitable distribution of risks and benefits. When possible, those who benefit should compensate those at risk for the extra costs and services needed for safety.

Public participation in provisions for *monitoring and enforcing safety standards* can also encourage community acceptance. Public access to plant data and local involvement in plans for detection and mitigation can be part of the siting contract, subject to court action if violated. In much of Europe and in the province of Ontario, hazardous waste management is a public utility like sewage treatment plants; Houston, Texas, has a quasi-governmental waste authority. California's plan for equitable distribution of hazardous waste facilities is based on each county's share of waste generation.[31] Siting decisions require adequate provision for early public input and fair selection and compensation procedures but also sufficient authority to override local opposition after these procedures have been carried through.

In sum, *citizen participation* in a technological society is a difficult but not impossible task. Even a relatively small number of informed and active citizens can contribute to greater public awareness and can enhance the accountability of legislators and officials. Public debate may delay decisions, but it is the lifeblood of democracy.

II. ASSESSING TECHNOLOGY

In formulating policies for technology and in setting standards for the protection of health and the environment, government agencies have extensively used two formal techniques: cost-benefit analysis and risk assessment. After looking at the strengths and limitations of these quantitative analytic methods, I will suggest that technology assessment provides a broader framework within which the human and environmental values discussed in this volume can more readily be considered.

1. COST-BENEFIT ANALYSIS

Cost-benefit analysis (CBA) requires a monetary estimation of the costs and benefits of a proposed project or regulation. In general the goal is to maximize the balance of aggregated benefits over total costs. CBA has been used extensively in planning large public facilities such as dams and airports. For a dam, for example, the economic costs of construction and operation can readily be calculated, but it is more difficult to estimate the value of its diverse benefits: irrigation water, flood control, hydroelectric power, lake recreation, and regional development.[32]

The marketplace already assigns *an economic value* to some of the benefits and costs of such projects. But how are monetary equivalents to be assigned for those that do not normally enter the market? One method is to make estimates based on the prices of the closest market substitute. In some cases, however, there simply are no comparable items on the market, especially for such common-property resources as clean air and recreational water, which are not bought or sold by individuals. Another method is to infer willingness to pay from indirect evidence concerning people's behavior or to conduct surveys in which respondents are asked what they would be willing to pay for a particular benefit or to accept as compensation for a particular loss. For instance, they can be asked what they would pay for fishing or swimming opportunities in a public lake.[33]

Cost-benefit analysis provides useful information to decision makers, but its limitations must be kept in mind:[34]

1. Distributive Justice. CBA deals with aggregate costs and benefits. It does not ask who bears the costs and who reaps the benefits. The economist's concern for efficiency rather than equity has been defended on various grounds. Sometimes it is said that inequities will tend to average out; a person who is a gainer in one project may be a loser in the next one. However, this neglects the structures of economic and political power through which some people are consistent gainers and others are habitual losers. Again, it is said that if there is a net benefit, the gainers could compensate the losers and everyone would still benefit. But such hypothetical reasoning is not convincing if there are no mechanisms for the actual compensation of the losers. It is often asserted that if individual projects are judged by criteria of total costs and benefits, any inequities that ensue can be mitigated by adopting income transfer policies, such as progressive taxes and welfare payments. I have urged, by contrast, that issues of justice should be part of project evaluation itself. In a society in which wealth is unevenly distributed, willingness to pay is a dubious criterion of social benefit. The preferences of a small minority may be heavily weighted because of the economic resources at their disposal.

2. Discounting the Future. Over how long a time period, and at what discount rate, should future costs and benefits be calculated? We have noted (chapter 3) some of the reasons for discounting the future. People prefer present benefits to future ones. A dollar now is worth more to me than the promise of a dollar ten years hence because I could invest the first dollar during the interim. But the use of current interest rates to discount future environmental and health costs is dubious. Questions of sustainability, resource depletion, and harm to future generations are effectively excluded; any effects more than twenty years off are treated as insignificant for decisions today. The Rawlsian concept of justice and some versions of utilitarianism treat present and future people equally. This would imply a zero discount rate except where there are actual mechanisms by which present investment will produce augmented future benefits. There are no such mechanisms for compensating our distant descendants for long-lived risks.[35]

3. Environmental Values. A number of economists have maintained that the market process in general and cost-benefit analysis in particular systematically undervalue irreplaceable natural assets. In the past, technology could be counted on to reduce the cost of natural resources or to provide substitutes. But nonrenewable resources are becoming scarcer, and often there are no substitutes. Wilderness and unspoiled scenic areas will be more and more scarce, and technology will only make them scarcer. While the decision to preserve such areas is reversible (since the option to develop remains open), the decision to develop may be irreversible; wilderness once destroyed can seldom be recovered. The extinction of an endangered species is irreversible; the genetic information once lost is irretrievable. CBA makes no allowance for the importance of keeping future options open or for preserving unique or rare natural environments. Nor can ecological balance and diversity be converted into market equivalents.[36] While environmental regulations such as pollution emission standards do "internalize" some market externalities, other environmental impacts enter neither market costs nor CBA calculations.

4. Human Values. Cost-benefit analysis is oriented to material values. The requirement of quantification leads to the neglect of intangible human values. Attempts to quantify aesthetic values are of limited applicability. One can ask people what they would pay to visit an area of natural beauty or to prevent its destruction, but this may not always be a valid way of judging its importance to them.[37] In response to the Roskill Commission's CBA of alternative sites for a third London airport, there was a strong public reaction, prompted in part by the small weight given to natural beauty in comparison to the travel time of the affluent. The loss of Norman churches and other historical sites did not even enter the calculations.[38]

Cost-benefit analysis is carried out in the framework of utilitarian assumptions (see chapter 2). It deals with consequences and with quantitative totals. It cannot directly include *duties and rights*. It is anthropocentric and excludes duties to animals, the welfare of nonhuman beings, and the stewardship of creation, except as they result in human economic benefits. Moreover, individual rights play no role, except as they are embodied in legal constraints that limit the options analyzed. The analyst may be firmly committed to justice or participation, but these commitments will have to be expressed at other points in the decision-making process and not among the costs and benefits calculated.

5. Institutional Biases. The previous four problems are inherent limitations in CBA as a method; the fifth and sixth are problems arising from the institutional setting in which it usually is carried out. In practice, CBA is almost always a means by which an agency justifies and promotes its own programs. The formulation of problems and the selection of alternatives, which are frequently the most important decisions, occur before the analysis is made. In the analysis itself, an agency typically overstates benefits and understates costs (because of subsidized capital, preferential tax treatment, and neglected environmental effects). Secondary benefits are assiduously sought, while secondary costs are conveniently overlooked. One study speaks of "the self-serving assumptions" of agency CBAs.[39]

6. The Role of the Expert. While the assigning of monetary values appears to be a technical question, it often reflects the biases of analysts or their judgments of what the public wants. The Department of Transportation's CBA of the supersonic plane overestimated the demand for high-speed travel and the dollar value of the time saved. Differing weights would be assigned by various social groups to the incommensurable benefits of such projects. Value conflicts that should be resolved politically are concealed in what look like rational, neutral, and objective calculations. This may appeal to administrators, but it hinders public debate of the policy issues and lessens the accountability of bureaucratic officials. Numbers carry an unwarranted authority when used to legitimate decisions that are basically political in character. In theory, CBA yields a document with explicit assumptions, open to public discussion, forming one among other inputs to democratic decision making. In practice it is often an instrument of technocratic planning, accessible only to specialists, and difficult for citizens or legislators to challenge. Because it usually is expressed in the language of professionals, public participation is inhibited.

Some of the criticisms above are attacks on inadequate or biased cost-benefit analyses; they can be answered by requiring *better analyses* and modifications in their *institutional context*. Some of the neglected costs can be included if indirect impacts on the environment and various population groups are considered. The problem of institutional bias can be mitigated if CBAs are carried out in a variety of institutional settings. Some authors have urged that analyses be done by disinterested experts on behalf of the public, perhaps at regional centers.[40] Others question whether any organization is disinterested, and they recommend that alternative plans and evaluations be prepared by several organizations representing diverse interests, including government agencies, independent consulting firms, university teams, and public interest groups.[41]

Other criticisms above refer to *inherent limitations* of cost-benefit analysis as a method. These limitations are not serious when a project has narrow and clearly defined objectives, when the main impacts are physical and readily quantifiable, and when there are a small number of options for achieving widely accepted objectives. Even when benefits and costs are more diverse, CBA can be a useful input into decision processes, provided people are aware of its limitations. In general, it introduces a bias toward development because many of the indirect costs are difficult to quantify. In the past, it often has been used late in the planning process to justify agency decisions already made on other grounds. Usually only options within a relatively narrow set of assumptions have been considered. But when it is used cautiously, as one input into an open and pluralistic decision-making process, it can contribute to rational discussion, the accountability of agency officials, and accessibility to review by legislatures, the judiciary, and the public.

My own conclusion is that cost-benefit analysis can be a useful component of project planning, provided (1) there is pluralistic participation, public discussion, and recognition of assumptions and limitations, and (2) it is viewed as only one input into wider decision-making processes that can take unquantifiable consequences and intangible values into account.

2. RISK ASSESSMENT

In industrial nations, overall risks to human life and health have decreased over the last two centuries, as indicated by longer life expectancy. But recent technologies have created new and pervasive risks that have caused widespread public anxiety. Hazardous chemicals are used in many agricultural and industrial processes; they appear in workplaces, food, air, and water. Chemical, nuclear, and other industrial accidents are rare but their consequences can be devastating. How can such risks be assessed and how can appropriate safety standards be established?[42]

During the 1970s and 1980s, risk assessment came to be widely used in setting regulatory standards in industrial nations. According to most of its proponents, risk assessment should be separated into two components.[43] *Risk analysis* is the objective, scientific estimate of risks by technical experts. In general it requires estimation of the probability of an event (exposure to a chemical, occurrence of a plant accident, and so forth) and estimation of the magnitude of the consequences (number of people affected and level of harm). *Risk evaluation,* on the other hand, is a more subjective, value-laden, and pragmatic process in which the results of risk analysis are used by public officials to establish policies after taking into account public perceptions and attitudes (risk acceptability), legal and political constraints, and value trade-offs. The second stage goes beyond science because there is no scientific answer to the question: How safe is safe enough?

Risk analysis for chemicals is carried out primarily by dose-response experiments on animals, mainly at moderate doses, since the effects of low doses would require thousands or millions of animals to be statistically significant. A typical study requires five hundred mice, two years, and $250,000.[44] The results must then be extrapolated to low doses in humans on the basis of assumptions about thresholds and biological similarities. Data on human exposure to chemicals is scarce, and small changes in the incidence of cancer are difficult to establish or to interpret, since individuals are exposed to many pollutants and a latency period of many years may elapse between exposure and illness. Risk analysis of accidents in industrial systems requires laboratory experiments or historical data on the malfunction of system components and estimates of the probability and consequences of human errors.

Risk analysis can be combined with *cost-benefit analysis* if the economic costs of the potential harm are compared with the economic cost of compliance with regulations that would reduce the risk. Property damage, medical costs, and wages lost from illness or injury can be directly estimated. Calculation of the economic value of a human life is more problematic. Projected future earnings can be estimated—perhaps $400,000 for a typical American worker. But this is a dubious method, since it implies that the life of the elderly is valueless. Court awards for serious injury or death can be used as a basis of calculation, but they vary widely.[45]

I would maintain that distinctive characteristics of *human life* should make us hesitant to treat it as if it were a commodity on the market. Life cannot be transferred, and its loss to a person is irreversible and irreplaceable. To speak of human life as sacred does not mean that it can be assigned an infinite value in practice or that it cannot enter trade-offs with other costs and benefits. However, it does suggest that a cost in human lives should not be aggregated with economic costs but kept as a separate kind of cost concerning which accountable decision makers, rather than technical analysts, should make the inescapable value judgments.[46]

The *public acceptability of risks* is influenced by many factors that are not considered in risk analysis. Voluntary risks (in sports, automobiles, and smoking, for instance) are widely accepted at levels roughly one thousand times higher than involuntary ones. We have a greater responsibility to protect people from hazards imposed by others than from hazards voluntarily chosen. Risks to identifiable individuals (such as a trapped miner or a critically injured child) are taken more seriously than statistical risks in which the victims cannot be identified in advance (such as carcinogens), though this distinction is difficult to justify except as a symbolic affirmation of the value of persons. An invisible and unfamiliar threat, such as radiation, also creates greater anxiety than visible and familiar dangers. Fairness in the distribution of risks and benefits also strongly affects public judgments about risk acceptability.[47]

The public also fears *large-scale catastrophes,* even when their probability of occurrence is very low. In formal analysis, a risk is defined as the probability of an occurrence multiplied by the magnitude of its consequences; very low probability can compensate for high magnitude to give a low overall risk. But I believe we should take greater precautions than risk analysis suggests in order to avoid large-scale disasters, because their social disruption would be greater than a succession of smaller accidents producing the same total damage. Again, great uncertainty is usually present in estimating low probabilities, and the consequences of error or unexpected sequences of events in large-scale systems can be enormous. In tightly coupled complex systems, such as nuclear reactors, unlikely combinations of unforeseen circumstances and human errors can be disastrous.[48] Other risks are exceptionally serious because their consequences are irreversible or can be reversed only with great difficulty (such as endangered species, global warming, germ-line genetic changes, and nuclear war). Options that are flexible and easier to correct offer more opportunity to learn from experience.

Clearly many *value judgments* enter risk evaluation and policy formulation. But some critics have pointed to the presence of value judgments even in the supposedly value-free process of scientific risk analysis.[49] Analytical results are based on many assumptions about problem definition, boundaries of analysis, and the degree of certainty and level of confidence expected in accepting statistical evidence. As with cost-benefit analysis, the analysts' assumptions are likely to be influenced by policy preferences and disciplinary training. According to one study of risk assessment: "Beliefs about the facts of the matter shape our values; those values in turn shape the facts we search for and how we interpret what we find."[50] Another author found that analysts from industry consistently used assumptions that downplayed the magnitude of risks.[51] It is therefore advisable that risk analysis, like cost-benefit analysis, be carried out in a variety of institutional settings or by panels with experts drawn from diverse institutions.

The separation of *risk analysis* from *risk evaluation* can be defended, despite the inseparability of facts and values. Risk analysts should not be asked to recommend policies or to balance value trade-offs. The formalization of risks allows the analysis to be submitted for peer review; its assumptions and inferences can be scrutinized by others. It can help to focus research that might reduce the most problematic uncertainties. Risk analyses can help decision makers in setting priorities when a wide spectrum of values are brought together in formulating policies to manage risks.[52] Greater openness in decision making and greater citizen participation on risk advisory committees enhances the accountability of government agencies. After a history of unanticipated accidents, citizens today are less willing to trust the experts in decisions made behind closed doors. They want to know what is going on, and they wish their voices to be heard.[53]

3. TECHNOLOGY ASSESSMENT METHODS

Technology assessment (TA) deals with a broader set of human and environmental impacts than either cost-benefit analysis or risk assessment. TA is the study of the direct and indirect consequences of a new technology (or significant changes in an existing technology) and analysis of policy options that might affect its development. These are the distinctive characteristics of TA:[54]

1. Early Anticipation. Technology assessment is an attempt to anticipate consequences beforehand rather than waiting for them to become evident. It is a future-oriented inquiry, an "early warning system." The crucial decisions in the social management of a new technology should be made at an early point in its deployment, before heavy financial investments and employment patterns have built up pressures for its perpetuation. The benefits of a technology usually are immediate and obvious; the indirect costs and risks often are delayed, remote, and cumulative. Careful study and foresight can identify some of these effects before the new technology has acquired a momentum that is difficult to control.

2. Diverse Impacts. A wide range of impacts are considered, beneficial as well as adverse, social and political as well as environmental and economic. Particular scrutiny is given to unintended effects, second-order consequences, indirect costs, and long-term repercussions. Because these impacts are diverse, assessment should be carried out by an interdisciplinary team including sociologists, economists, and other social scientists along with biological and physical scientists and engineers. The analysis should deal with both quantifiable and unquantifiable variables.

3. Diverse Stakeholders. In the past, representatives of government and industry have presented the benefits of a new technology, whereas people who may have to face the indirect costs often have had no effective voice. TA requires the identification of the main parties who may be affected and analysis of the consequences for each party. The impacts on different social and economic groups may vary widely. Provision can be made for public participation at certain stages of the assessment process to ensure that all major interests and viewpoints are represented.

4. Alternative Policies. Assessments not only trace current trends but also analyze the effects of alternative policies. As a policy-making tool, TA is designed to present decision makers with information about the probable consequences of various options for legislative or agency decision. If the studies are to be useful to decision makers they must be understandable to the nonspecialist, and some thought must be given to possible modes of implementation and regulation.

The *Office of Technology Assessment* (OTA) was established by the U.S. Congress in 1972. Congress had depended on federal agencies for most of its technical information and wished to have its own assessment institution. Most requests for assessments originate in congressional committees, and they must be approved by the OTA Board (six members from the Senate and six from the House, evenly divided between the two political parties). Members of the Advisory Council are drawn from diverse academic fields, labor and industrial organizations, and public interest groups. OTA has a core professional staff that ensures continuity, but much of the analysis is carried out under contract by interdisciplinary teams at universities, research institutes, or consulting firms. Before a final report is prepared, a preliminary version is sent to perhaps a hundred reviewers, representing a broad spectrum of stakeholders, for comments on its adequacy in presenting alternative options.[55]

The Office of Technology Assessment has built up a reputation for *comprehensive and balanced assessments.* In previous chapters I have cited some of these OTA reports (on energy, computers, genetic engineering, and the Strategic Defense Initiative, among others). OTA does not in general make policy recommendations but rather lays out a range of policy options and the likely consequences of each. It informs Congress about the main stakeholders and the reactions that can be anticipated from them. It does not ignore disagreements but tries to explore the reasons for them. The direct involvement of the parties affected introduces a variety of perspectives and allows a greater diversity of impacts to be identified and studied. Such public involvement also makes clear the political character of policy choices, in contrast to technocratic modes of analysis in which value judgments are often disguised as technical decisions.[56]

Some critics maintain that OTA has been too preoccupied with *short-term issues,* neglecting long-run issues and the "early warning" functions originally envisaged.[57] However, to be useful to Congress, OTA reports must be relevant to pending legislation and immediate policy decisions. The reports have often been quoted in committee deliberations and occasionally in congressional debates. Moreover, OTA has given some attention to problems with a long-term horizon, such as global warming, endangered species, and the allocation of research and development funds.

Other critics maintain that the Office of Technology Assessment implicitly accepts *the dominant values of industrial society,* such as technical efficiency and economic growth. It seldom raises fundamental questions about consumerism or appropriate technology. In other words, OTA tends to look at alternative means of achieving generally accepted ends, without challenging those ends.[58] I would agree that broader social assessment is needed, but OTA cannot move too far from prevailing values without losing its usefulness to Congress. In the

1970s, the National Science Foundation sponsored some more comprehensive assessments, but funding for them was cut off in the eighties. A few private institutes, foundations, universities, and public interest groups have given critical assessments in which wide-ranging social and ethical questions were raised about particular technologies. I would support the expansion of technology assessment in a variety of institutional settings where it can benefit from wider reflection on the goals of industrial society.

In the European *parliamentary system,* parties in power have worked closely with government ministries and have seldom sought other sources of information. Many of the ministries have promoted technological innovation as a source of productivity and national competitiveness. Parties out of power have proposed technology assessment offices with little success. However, the proliferation of environmental and social problems attributable to technology in the eighties has encouraged the establishment of several agencies similar to OTA, though on a smaller scale. In Britain, the Parliamentary Office of Science and Technology opened in 1989, though without public funds. Parliamentary select committees continue to hold inquiries into controversial issues such as radioactive waste disposal; these committees serve some of the functions of TA. In the former West Germany, members of parliament and outside experts have formed study commissions to examine long-term technological and environmental questions, and more recently a technology assessment institute was established. France, Denmark, and the Netherlands also have parliamentary technology offices.[59]

In summary, I have maintained that the basic decisions about technological policies are *value laden and political.* In a democracy, these decisions should be made by elected representatives and carried out by the government agencies accountable to them. Formal analytic techniques such as cost-benefit analysis and risk assessment can contribute significantly to such policy decisions (especially in setting health and safety standards), but the trade-offs among competing and incommensurable values should be made by accountable officials and not by technical experts. Technology assessment provides a broad analysis of diverse impacts on major stakeholders, which can be particularly helpful to elected representatives. If assessment teams are multidisciplinary, look at a wide range of options, and encourage stakeholder participation, the environmental and human values explored in this volume can be brought to the attention of decision makers.

III. REDIRECTING TECHNOLOGY

In this section some conflicting values in the redirection of technology are considered. Various methods for setting standards and achieving compliance are compared. Then the apparent conflict between justice, employment, and environmental protection in the regulation of technology is explored, and strategies and coalitions to combine these values are proposed. The final topic is the social responsibility of scientists and engineers and their opportunities to redirect technology in their professional work.

1. REGULATORY STRATEGIES

Three ways of setting standards for health and safety have been used.[60]

1. "Zero Risk" or "Health Only" Standards

The Clean Air Act (1970) required EPA to establish national ambient air standards to protect public health "allowing an adequate margin of safety," without consideration of economic costs. The Delaney Amendment states that a food additive is to be banned if it is found to induce cancer in humans or animals; the benefits of an additive, no matter how large, are not to be weighed against a confirmed cancer risk, no matter how small. This approach makes sense if it is assumed that (1) there are thresholds for the concentration of each pollutant, below which there are no hazards to health, and (2) technology can be developed to keep pollution levels below these thresholds at a reasonable cost.

However, if there is no safe threshold, and if the cost of removing a pollutant increases rapidly as 100 percent removal is approached, very large costs may be incurred to achieve very small additional benefits in the pursuit of zero pollution. Moreover, detection techniques have improved so that very minute traces of many chemicals can be detected. In practice, therefore, economic trade-offs usually do affect the setting of standards, or special exceptions are made (as in the case of saccharin, a very weak carcinogen, which was exempted from control at a time when few sugar substitutes were available). Some balancing of costs and benefits seems preferable to the search for zero risks regardless of cost.

2. Technology-based Standards

The Clean Water Act of 1972 required the best practicable technology by 1977 and the best available technology by 1983. Effluent standards for a given type of industry were to be uniform despite variations in the assimilative capacity of the rivers into which plants discharge, so some effluents were overtreated and others undertreated. Huge federal subsidies were provided for the construction of municipal treatment plants, a practice that was popular politically since it brought funds to the home districts of members of Congress.

But the same water quality improvement could have been achieved at 30 to 35 percent lower costs if effluent limits were not uniform but varied according to local water quality and control costs.[61] Moreover, technology-based standards give little incentive to improve control technology, since such improvements would only result in tighter standards. Companies hesitate to look into cheaper alternatives whose approval is uncertain. In addition, this method specifies waste treatment technology and does not encourage attempts to cut down on the generation of wastes in the first place. Once again, consideration of costs and benefits seems preferable.

3. Balancing Costs and Benefits

U.S. laws regulating toxic substance production and pesticide licensing require that risks and benefits must be "in reasonable relationship," leaving EPA to set permissible levels of risk. Formal cost-benefit analysis is not mandated. The Occupational Safety and Health Administration is required to set standards to protect the health of workers "to the extent feasible." In 1981 the Supreme Court ruled that in setting cotton dust standards OSHA does not have to balance risks to workers against the employer's cost, as long as the corrective technology is available at a reasonable cost, since in writing the law Congress had put health above other considerations. But controls that would bankrupt the industry would be unreasonable, according to the court.[62]

Such balancing of costs and benefits requires an immense amount of data and places a huge burden on underfunded agencies. Even though the burden of proof for testing new chemicals falls on industry, EPA needs a much larger budget to test chemicals already in use and to monitor air and water quality. Moreover, public officials have to consider unquantifiable effects such as environmental impacts and justice in the distribution of costs and benefits. Economists estimate that the benefits of current air pollution standards clearly exceed the costs of compliance, but water pollution abatement is more dubious and its benefits fall more unevenly. The urban poor, for example, benefit greatly from clean air standards, but some water improvement projects— especially those aimed at water recreation—benefit mainly upper and middle income families.[63]

Once regulatory standards have been set, the next question is how to achieve them. Three differing regulatory strategies have been used.

First, *limits on air emissions and water effluents* from particular sources can be set and enforced by penalties. The total reduction in pollutants needed to meet air and water quality standards in a given region is calculated and emission limits are then assigned to each source. Unless the penalties are substantial and monitoring is effective, industry is likely to think delay or noncompliance are cheaper options than compliance. This strategy offers no incentive to reduce emissions below the assigned limits. Such direct "command and control" regulations have been the main form of pollution strategy in both the United States and Europe because their results are fairly predictable.

A second strategy, *emission or effluent taxes or charges,* has not been attempted in the United States (except in the case of chlorofluorocarbons) but is advocated by many economists. The tax would be proportional to the quantity of a pollutant emitted. Polluters could decide their own type and level of control, but the tax rate would be steep enough that average emissions would be brought down to the desired levels. For example, the tax per pound of sulfur dioxide emitted might be very high in regions of poor air quality; industries would have incentives for diverse responses in fuel use, technological research, plant location, and so on. We have seen that in order to slow global warming several European nations have introduced small taxes on the carbon in fuels, and other nations are considering them. Denmark has a small pesticide tax—both to discourage use and to support research and education on alternatives. Water pollution charges are levied in several European countries, primarily as a way of raising revenue for water treatment; the charges are too low to provide much incentive to reduce pollution. The European Community is seeking common standards so that no nation will have a competitive advantage in the common market.[64]

Emission and effluent taxes offer a number of advantages. They rely on *economic incentives* that allow greater freedom of choice. Total costs are lower; more controls will be installed on plants with low control costs than on those with higher costs. In contrast to subsidies, taxes can provide revenue, which could be used for pollution control research. Above all, there would be incentives for industrial research on a variety of technologies and for other pollution-reducing responses.[65]

Such taxes do have some *disadvantages*, however. The setting of tax rates is difficult and controversial. The Netherlands tried such a system and found that fee setting required extensive information, much of which was difficult to get. Estimates must be made of probable average costs and benefits, which require data on technology and health effects. But it would be relatively easy to adjust the tax rate in light of further data, since the taxes would not be tied to specific technologies. The main obstacles to emission taxes are political. The present systems in both the United States and Europe have been widely accepted, and major changes would involve further uncertainty and delay.

A third strategy is the use of *marketable permits*. Tradeable discharge permits for a given amount of pollution would be either auctioned to the highest bidders or given out free and subsequently bought and sold. In the midseventies in the U.S. an "offset" program was introduced in which a plant could reduce air emissions below its allowed limit and sell the emission credits to another company that wished to build a new plant in the same air district. Clean air legislation adopted in the United States in 1990 allows plants that reduce their sulfur dioxide emissions below mandated levels to sell the credits to utilities planning to build new plants.[66] Like pollution taxes, marketable permits set social goals but leave some room for market forces and individual choices in meeting those goals. Such economic incentives lead to greater efficiency and greater flexibility in seeking technological improvements.

In a number of areas, however, economic incentives alone will not be effective. Endangered species and wilderness preservation affect future generations; only regulations and penalties can prevent their destruction. Again, some chemicals are so toxic that absolute bans are appropriate, especially when there are substitutes for them. Finally, the effort to prevent the generation of pollutants in the first place, rather than to control them after they have been produced, will require a combination of regulations and economic incentives designed to change industrial processes and recycle materials. A variety of strategies are needed for effective conservation and reduction of resource use, as we will see in the next chapter.

2. JUSTICE, EMPLOYMENT, AND THE ENVIRONMENT

In regulating technology, the values of social justice, meaningful work, and environmental protection often seem to conflict. For factory workers, smokestacks represent employment, but for environmentalists they are symbols of acid rain falling on forests and lakes. The environmental movement has been accused of being elitist.[67] Defenders of wilderness areas are mainly educated, white, and middle class; there are few blacks in the Sierra Club. In the past, many environmentalists have been uninterested in urban problems and have been insensitive to fears of plant closings. But in fact justice, jobs, and the environment are not as incompatible as is often assumed.

First, *strong support for the environment is found in all socioeconomic groups.* To be sure, in environmental organizations the leadership and even the membership are drawn disproportionately from the educated middle class, but this is true for most social movements, including the civil rights movement. One author wrote in 1984, after summarizing several public opinion polls, "In short, while environmentalists certainly constitute a socioeconomic elite, the elitist label misses the mark because their policies receive wide support from all socioeconomic categories."[68] By the 1990s public environmental support was greater and more broadly based than at any time since the early seventies.[69] Environmentalism is clearly not a class issue.

Second, *low-income families bear a disproportionate share of the burden of environmental degradation.* The urban poor are almost always exposed to higher levels of air pollution, water pollution, noise, and lead poisoning than more affluent citizens. In one study of neighborhoods in three cities, high levels of carbon monoxide (mainly from autos) and sulfur oxides (mainly from coal-burning plants) were highly correlated with low average neighborhood income.[70] Another study found that 60 percent of all blacks and Hispanics in the United States live in communities with toxic waste sites.[71] Low-skill jobs often involve extensive exposure to a variety of pollutants. The poor have little economic or political power to defend themselves from such risks. Environmentalists have worked with urban groups on air pollution, traffic congestion, toxic waste problems, and the planning of parks and recreation areas accessible to inner city residents. For example, a conference sponsored by the Sierra Club and the National Urban League (a major voice for city blacks) was devoted to the improvement of urban environments through land-use plans, community self-help, recycling and solar energy projects, and locally controlled, environmentally sound programs to create new jobs.[72] Pollution is indeed a social justice issue.

Third, *environmental regulations have not contributed to unemployment.* In a time of high unemployment, the threat of layoffs is a powerful political weapon. Companies have often said that they would have to close a plant if a particular regulation were passed, but in most cases they did not actually do so.[73] EPA estimates that in a workforce of 100 million, only 30,000 workers were laid off between 1971 and 1981 in plant closings influenced by environmental regulations, and in many of these closures, plant obsolescence or declining sales were more important factors. In the same period, a far greater number of new jobs was created by environmental activities. EPA estimated employment in water pollution control (equipment production, operation, and regulation) at 222,000 and air pollution control at 120,000. And of course many jobs would be jeopardized by air and water deterioration (in fishing, tourism, and agriculture, for instance).[74] Some jobs have been lost when plants were moved to other countries, but the moves were motivated more often by cheap labor than by differences in environmental standards.

Overall, the total annual expenditures necessitated by all U.S. environmental regulations in 1990 were estimated at $85 billion. The benefits to human health, agricultural output, recreational opportunities, and the preservation of ecosystems are very difficult to calculate but are thought to exceed this figure. One recent study concludes, "Environmental regulation adds slightly to the inflation rate, has a negligible effect on the unemployment rate, and somewhat reduces the rate of productivity growth."[75]

Although environmental regulations have resulted in a small net gain in jobs, and most threats of plant closure have not been carried out, some layoffs have caused great hardship to individuals and local communities. As in the case of layoffs caused by automation, job retraining and adjustment assistance would reduce the human impact of plant closings related to emission standards. Several European countries (notably Sweden and Germany) have employee protection laws requiring several months' advance notice of shutdowns, and they provide funds for extensive retraining programs.

Fourth, *labor unions and environmentalists have many common interests*. They have cooperated on laws for occupational health and safety, including the regulation of asbestos, benzene, and other chemicals in the workplace. Both groups have advocated right-to-know laws requiring employers to inform workers about the toxic chemicals to which they are exposed. The unions of machinists, autoworkers, and steelworkers have supported clean air and water legislation. Labor and environmentalists both seek greater accountability on the part of corporations and government bureaucracies. They want broader public access to information and decision processes. Environmental legislation in the seventies that opened government agency decisions to public scrutiny and legal challenge have been used by unions as well as environmental and consumer groups.[76]

However, in some situations the interests of unions and environmentalists inescapably conflict. Construction unions have supported plans for large industrial plants and federal projects that citizens' groups have opposed because of their environmental impacts. The protection of the spotted owl, an endangered species, requires curtailing logging in some old timber stands on public lands on the West Coast, threatening the livelihood of local loggers.

However, the decline in timber-related jobs is primarily the product of many years of overcutting on private lands and the introduction of automated equipment; moreover, the spotted owl is only one of the plant and animal forms that need protection in virgin forest areas. In such cases there is no alternative to searching for compromise through political processes and providing compensation for those hardest hit by new policies. When the Redwood National Park in California was expanded, the unions worked out an agreement under which Congress provided funds for four years to cover full salary and health and pension benefits to the 2,500 timber workers laid off.[77] However, little was done for retraining, relocation, or bringing in new jobs, which would have helped more in the long run.

Fifth, in more general terms, *the exploitation of nature and the exploitation of workers are typically products of the same economic and political forces.* In Appalachia, the landscape and the people have suffered for the same reasons: the economic profits of coal companies and the political power that they wield in state legislatures. Strip-mined hillsides, polluted streams, declining land values, and inadequate mine safety precautions are all products of the pursuit of financial gain and the exercise of political influence. The market neglects external costs, whether imposed on the environment or on human health. A society pursuing affluence destroys human community along with natural resources. The manipulative mentality leads to the control of both things and people in the interest of efficiency. Technologies are often geared to the interests of organizations rather than the welfare of persons or the ecosystem.

A *political strategy* dedicated to both justice and the environment will thus require cooperation between environmentalists, labor unions, community organizations, and urban and civil rights groups.[78] These organizations, together with the peace movement, the women's movement, and church groups, could form effective coalitions to support proposed legislation and candidates favorable to both human and environmental values.

3. THE SOCIAL RESPONSIBILITY OF SCIENTISTS AND ENGINEERS

We have been looking at the ways in which both citizens and experts can redirect technology through political processes. Scientists and engineers can also influence the design and use of technology directly *through their professional work.* In earlier chapters I described their role in several types of research including agricultural research, nuclear reactor design, computer systems, genetic engineering, and military research. Let us ask in more general terms how scientists and engineers can enhance human and environmental values through their work.[79]

In *choosing a career,* each person must make decisions, first about education and then about employment. Such choices are constrained by one's abilities, interests, and finances, as well as by fortuitous circumstances and available opportunities. But they can include some reflection on the social consequences of one's future work. The applications of science cannot be predicted, but often one has some idea of potential uses; moreover, the lines between pure and applied science are often blurred today. In choosing employers, one knows the kinds of products or services they provide. A person might use some of the same skills working for a defense contractor or for a solar energy company.

Many scientists and a few engineers work in *universities*. Here they have greater freedom than in industry, more choice of research topics, and once they have tenure they can speak out on controversial issues without fear of losing their jobs. Academia does have its own rewards and punishments, its internal politics and departmental power structures, and its pressures to "publish or perish." Many fields of university research depend on government grants, and these are determined by public priorities and by the assumptions and interests of peer review panels. Other grants come from corporations, raising the issues of secrecy and distortion of priorities that I pointed out in discussing biotechnology. But choices among research topics are still present in the academic world. And in their teaching—both in technical courses and in interdisciplinary programs in Science, Technology, and Society—professors can help students to see science and technology in their wider social settings.

Some scientists and most engineers work in *industry*, where they have diverse and sometimes conflicting obligations to employers, colleagues, and the public. They are part of a team to which they are expected to be loyal. But they also have loyalties to their families and careers, to the wider scientific and professional community, and to consumers and the public. Company decisions may give considerable weight to public safety and environmental protection because of government regulations, technical standards in the industry, fear of litigation, concern for the public relations image, or genuine concern on the part of management. In their work, engineers can try to use a broad range of design criteria, including efficiency, resource use, and environmental impacts. Sometimes they can consider the durability of a product or ways in which it might be adapted to the needs of the Third World.[80]

When engineers believe that a product on which they have been working will expose the public to *a significant risk*, they can call it to the attention of their superiors or even appeal to higher management if the risk appears serious. But the decision to "go public" in reporting such a risk is more serious because their careers and jobs are so vulnerable. Engineers usually have less opportunity for independent judgment than doctors, lawyers, or professors.

The *codes of conduct* of the professional societies offer little help in such cases. Their wording is usually vague and general, and the only sanction is expulsion from the society, which has rarely occurred. Some of the codes deal with relationships to clients; for example, members of the society should preserve confidentiality and should not misrepresent a product or their own qualifications. The code of the Association for Computer Machinery says, "An ACM member shall express his professional opinion to his employer or clients regarding any adverse consequences to the public which might result from work proposed for him" (note the masculine pronoun). But what if the employer does nothing to remedy the risk? The code of the Institute of Electrical and Electronics Engineers goes further by including the duty "to protect the safety, health, and welfare of the public and speak out against abuses in these areas affecting the public interest."[81] A few professional societies have offered legal services or mediation procedures for members caught in such disputes, but most have been hesitant to get involved, partly because their members include employers as well as employees.

Three examples illustrate the importance of *"going public" concerning a potential risk.* In 1972 several American industrial scientists knew of Italian studies showing that Vinyl Chloride causes cancer in rats, but they did not notify the occupational safety agency or the public. Of course, uncertainties often enter the interpretation of data, as we have seen, and it is easy to rationalize doing nothing. But in this case the evidence was strong, and it should at least have been disclosed so that other scientists could conduct further studies before thousands of workers were exposed to dangerous concentrations of the chemical.[82] In another case, three engineers working on plans for the Bay Area Rapid Transit system pointed to serious problems in its computer control system. They were fired for speaking out in public. Dangerous failures after it had been installed vindicated their judgment, and the control system had to be abandoned. The Society of Professional Engineers came to their defense and they were eventually reinstated.[83]

A third example of *whistle-blowing* involved the Challenger space shuttle disaster. Morton Thiokol produced the rockets used by the space agency (NASA). In several earlier flights, O-ring seals between rocket sections had been severely eroded by escaping gases, and the design team had sent several memos to their superiors warning that a disaster was imminent unless the seals were redesigned. The night before the Challenger launch in 1986, a temperature of 18° F was forecast (35° below any previous launch); tests had shown that the seals lose resiliency at low temperatures. In a telephone conference with NASA, all fourteen engineers and the company mangement recommended postponing the launch, but NASA was behind schedule in its launch program and asked them to reconsider. The engineers still opposed launch, but the four senior executives overruled them and made a "management decision" to approve it. When one of the engineers, Roger Boisjoly, later revealed these facts to the presidential commission investigating the disaster, he was demoted to another job.[84]

Some protection of whistle-blowers from such retaliation can be provided by *independent review procedures.* Employers must of course also be protected from exaggerated claims by disgruntled or unreasonably alarmist employees (though these are rare because employees are so vunerable to subtle forms of retaliation for dissent, even when they are not fired). The Whistleblower Protection Act of 1989 strengthens procedures for reviewing cases where federal employees have been penalized for disclosing dangers to the public or violations of the law. Several environmental and occupational safety laws offer some recourse for employees in private industry who report violations of these laws. But stronger legislation is needed to protect industrial employees who act in the public interest.[85]

These occasional dramatic cases of dangers to the public should not blind us to the positive opportunities for scientists and engineers to further human and environmental values in their choice of jobs, in their daily work on the job, and in their influence on the decisions made by the organizations for which they work.

IV. CONCLUSIONS

The democratic governance of technology is not easily achieved because policy decisions require information from experts and because the structures of economic and political power favor particular technologies. The main channel of public participation is through elected representatives and the government agencies accountable to them, but citizens can also interact directly with these agencies. I argued that formal analytic techniques, such as cost-benefit analysis and risk assessment, can contribute to the formulation of policies and regulatory standards, but technology assessment allows consideration of a wider range of values and stakeholders. The final decisions must be made by legislators and officials responsive to the electorate. The values of justice, meaningful work, and environmental protection often seem to conflict, so we are challenged to develop policies and political coalitions that combine them.

What, then, can you do as a citizen in a technological society?

1. If you already have expertise relevant to a current policy decision, you can inform other citizens through the media, public forums, and articles. You might be able to use your expertise on a government advisory board, as an independent witness at regulatory or legislative hearings or in court, or by working with a public interest group. As a scientist or engineer, you also have opportunities to exercise social responsibility on the job, despite the constraints of your employment situation.
2. If you work in education or the media, you can help students or the public to gain some understanding of the relation of science and technology to policy issues.
3. If you have no relevant expertise, you can try to become better informed about one or two issues. Ask your librarian for publications from academic, government, industry, and public interest sources. Look especially for well-documented, policy-oriented studies. Subscribe to a couple of magazines dealing with the issues on which you wish to concentrate.
4. Encourage your church, club, labor union, or civic group to sponsor a public forum or discussion series and include speakers with varied institutional affiliations. Look at local and regional issues that are related to national and global problems, such as food, health, environmental protection, land use, or hazardous substances.

5. Work through a political party to select and elect candidates for local and national positions and to present platforms that reflect your political agenda. Write your elected representatives when key legislation is being debated.

6. Join and support a national organization that deals with the human and environmental impacts of technology. In addition to publishing and lobbying, many such groups testify at hearings and review government reports and drafts of regulations.

7. Finally, look at the human and environmental impacts of your own lifestyle and resource use. Could you adopt in your own home, business, or community a decentralized technology that can be locally controlled, preferably one that uses sustainable resources? Such issues in resource use are explored in the final chapter.

CHAPTER 9

New Directions

Three groups of values were presented in part 1. Among the individual values, *food and health* are today most critical globally, though meaningful work and personal fulfillment are always important. Among the social values, *economic development* is a high priority, especially for the two-thirds of the world locked in poverty, but growth must be selective and sustainable. I have also said that economic growth must be accompanied by *justice* and *participation* if its benefits are to be widely shared. Of the environmental values, I have given greater emphasis to *resource sustainability* than to environmental protection or respect for all forms of life, though I have urged action to prevent pollution and to preserve wildlife and endangered species.

Public concern about environmental pollution has grown rapidly since 1970, largely because of its connection with health. But concern about *resource sustainability* is more recent and less widespread. Only in the late eighties were people beginning to be aware of the extent to which both high consumption in the First World and desperate poverty in the Third World have resulted in the destruction of soils, forests, and other natural resources. In the poorer nations, the 1980s were virtually a lost decade for development. Agricultural production barely kept up with population growth overall, and food per capita actually declined in the poorest nations. The gap in living standards between North and South grew wider.

Current practices of *resource use* around the world are neither just nor sustainable. Soil erosion and massive deforestation are rapidly depleting our biological capital and will impose on future generations the damage we have caused. If all nations were using resources at the rate of industrial nations today, the ecosystem would soon collapse. Yet industrial nations seek even higher consumption levels. The framework of our political and economic institutions is short-term and national rather than long-term and global.

In this chapter we ask first what types of production technology would contribute to a more just, participatory, and sustainable world. Next we look at industrial nations and ask what social policies and individual life-styles would encourage more just and sustainable patterns of consumption. In the final section, we consider the changes in values and institutions that would be needed to effect such policies, and we look at the distinctive contribution the biblical tradition might make to the motivation for change.

I. TECHNOLOGY AND HUMAN VALUES

What priorities in technology policy would express the values we have been discussing? Are smaller-scale appropriate technologies a viable alternative to the large-scale technologies prevalent in industrial nations? Finally, what mixture of centralized and decentralized technologies would allow greater participation without jeopardizing economic development?

1. POLICY PRIORITIES

If we were to take seriously the values defended in part 1, and the state of the world portrayed in subsequent chapters, what priorities would we seek in public policies concerning technology?

1. *Technology for Basic Human Needs.* Adequate food, shelter, and health are the most universal needs and, when not fulfilled, the most urgent. The Rawls principle tells us to seek the welfare of the least privileged, while the biblical commitment to social justice demands action concerning the causes and not merely the symptoms of hunger, disease, and poverty. Technologies of agriculture, public health, and low-cost housing are crucial to developing nations as well as to low-income groups in industrial nations. Humanity has the technical capacity to eliminate absolute poverty and hunger if substantially increased international aid were directed to sustainable development designed to meet basic human needs.

2. *Reordered Research Priorities.* We have seen that both public and private research has been dictated more by bureaucratic and industrial interests than by human needs or ecological wisdom. Half of the world's scientists and engineers are in defense-related research, and many of the remainder are working on projects that will provide luxuries for the privileged.[1] Of the world's total research expenditures, only 6 percent are in developing countries.[2] If energy, food, and population are urgent global problems, higher priority should be assigned to such research areas as low-cost solar energy, high-protein crops, and family planning in all its dimensions.

3. *Resource Use and Environmental Impact.* A more ecologically sensitive technology would seek human life in harmony with nature, not domination of nature. Efficient use of energy and resources and low levels of pollution should always be considered in technological policy and design. Products that are recyclable cut down on pollution and resource depletion and in most cases also conserve energy. Product durability contributes to all these conservation objectives. Waste is reduced further when several processes can be integrated, as in the cogeneration of heat and electricity. I have said that in industrial nations selective growth should be concentrated in the technologies related to services, such as education, health care, and communications, rather than in

the more resource-intensive and heavily polluting manufacturing and consumer goods industries.

4. Economic Efficiency. Economic criteria have usually determined technological decisions, and they are obviously very important. The consumer benefits if a cheaper process makes a product available at a lower price. A company benefits if it can keep its costs at least as low as its competitors'; it has to earn a profit to stay in business. Regulations to protect the environment and the safety of workers add to costs, but if widely adopted they affect competitors equally; the added costs can be passed on to consumers as part of the true cost of the product when environmental and health costs are internalized. The economic efficiency of a particular process will vary greatly according to local conditions, including the cost and availability of capital, labor, technical skills, natural resources, and transportation facilities.

5. Job Satisfaction. Work provides society with needed goods and services, and it should provide the worker with enough income to support a family. But meaningful work can also add immensely to a person's life. In chapter 6 the alienating character of many industrial jobs was described. Most workers derive little personal reward from production; they work to obtain money to spend, seeking satisfaction primarily in consumption. Among the factors in job satisfaction are humane labor conditions, participation in job-related decisions, and opportunity for some creativity, self-expression, or pride in one's work. Such factors are primarily the result of institutional structures and management practices, but they may also enter into technological design and policy.

6. Democratic Control. Several political channels for wider participation in policy decisions were outlined in the previous chapter, including communication from an informed public to elected representatives, citizen action through the courts, and technology assessment that includes long-range environmental and social impacts. State and local governments also present many opportunities for the democratic redirection of technology toward broader human and environmental goals.

In some cases *large-scale capital-intensive technologies* present such significant economies of scale that economic efficiency outweighs other considerations, as I will suggest below. But large scale often hinders the achievement of other goals. Large-scale technologies often have massive environmental impacts, though these can be reduced by suitable regulations. Large installations are more vulnerable to sabotage, terrorism, or catastrophic accidents. Large-scale systems require centralized hierarchical organization, which militates against creativity and participation by workers. Large organizations result in a concentration of economic power that translates into political power, making democratic control more difficult. Each of these goals seems easier to achieve with smaller-scale technologies.

2. APPROPRIATE TECHNOLOGY

Let us consider first the appropriate technology movement in developing nations, where progress in rural development has been so disappointing. By

1970 industrialization was occurring in urban areas of the Third World, but poverty, malnutrition, and unemployment prevailed for the rural majority. The transfer of capital-intensive technologies from industrial nations increased the power of urban elites, and it perpetuated dependence on foreign experts and transnational corporations (TNCs). Development strategies began to give greater emphasis to basic human needs and rural development, but implementation was slow and ineffective. In this situation, E. F. Schumacher and others advocated *intermediate technology* (IT) with the following characteristics:[3]

1. Intermediate Scale. Between the sickle and the tractor, or between handcrafts and mass-production factories, is a range of machines of intermediate size and cost. They are more efficient than traditional methods, but cheap enough to be widely affordable. If the equipment for one worker costs $1 with traditional tools and $1,000 with high technology, an intermediate technology might cost $10 or $100 per worker. It could be obtained by scaling down large industrial processes or by scaling up and modifying traditional techniques—or, better yet, by applying scientific knowledge and practical experience to a middle range of community-based production processes.

2. Labor Intensity. Many modern technologies substitute capital for labor (for example, an automated factory or a highly mechanized farm). But in the Third World labor is plentiful and capital is scarce. In that context, technology should be aimed at increasing the number of jobs and the generation of rural income. Employment and increased income would in turn create expanded local markets for locally produced industrial products, including equipment needed in agriculture.

3. Relative Simplicity. The person using highly complex and sophisticated technology must depend on experts to produce and repair equipment. By contrast, simpler systems are more understandable, less vulnerable to breakdown, and have less serious consequences when breakdowns do occur. Equipment can be locally adapted and repaired, with opportunities for diverse skills. Simpler technical systems usually require simpler organizational support. Simplicity is a relative term, and in some cases reliable design depends on sophisticated principles.

4. Local Control. Intermediate-scale units provide opportunities for participation in work-related decisions. Production for local markets is more likely to be directed to basic human needs. The use of local materials encourages greater autonomy and self-reliance. A technology should be adaptable to its particular social and cultural environment. Institutions facilitating such local control include producers' cooperatives, credit unions, village-level organizations, and small farms and businesses.

Among examples of *successful IT projects* is the small power tiller developed by the International Rice Research Institute in the Philippines. It was a great improvement on ox-drawn plows, but it cost less than half the cheapest imported tiller, and it was soon being manufactured in six Asian countries.[4] In Thailand, microelectronic load regulators have improved the efficiency of small hydroelectric generators. In Kenya, improved open-pan equipment for sugar processing is smaller, less capital-intensive, and more labor-intensive than the vacuum-pan process, and the overall costs per pound of sugar are lower. Compared to the larger process, the smaller one requires (per unit of production) 27 percent of the capital, 5.9 times the labor, and the overall costs are 34 percent lower. In that context, the additional labor costs are more than compensated for by the saving in interest on capital. For the same output, far more jobs are created with a much smaller investment; the capital cost per employee is only 5 percent of that in the larger plants.[5]

By the 1980s, intermediate technology was widely accepted by *Third World governments,* but its implementation has been very slow. A survey of development plans from 1975 to 1985 in forty-one developing nations found that more than half had specific goals for IT dissemination, but in many cases levels of funding were low and programs ineffective.[6] But several nations, with varied political systems, have had strong IT programs for rural industrial development. China wants to "walk on two legs" with both small labor-intensive industries and large capital-intensive ones. Taiwan and South Korea have encouraged small farms and rural industries. Tanzania, Kenya, and Zambia have tried to strengthen village-based community production. In India the government gives preference to small enterprises in the purchase of 241 products, including transistor radios and small electric motors. Regional technical institutes in India have pioneered in IT research and education.[7]

The main obstacles to the spread of intermediate technology have been political and institutional rather than technical or economic. In the past, international agencies and transnational corporations favored large-scale capital-intensive technologies that were familiar and easily administered. Very little R and D funding has been available for improving the efficiency of IT. Third World scientists and engineers trained in the North usually return to urban areas and are out of touch with rural needs. Tariff, subsidy, and price policies are geared to urban populations that have disproportionate political power. Credit policies favor large established farms or firms. By the late 1980s international agencies and national governments were giving greater attention to small industries and rural development, but many institutional biases in technology selection continued to favor large-scale systems and more affluent markets.[8]

This underscores the importance of political action to change *national policies* as well as technical work to improve IT hardware. We would be succumbing to a new form of technological determinism if we thought that IT itself could solve the problems of the Third World. Here again the contextualists' insights concerning the two-way interactions of technology and society are helpful.

The *appropriate technology* (AT) movement in the North had similar goals, but it arose in very different circumstances. The issues in the North were the problems of overdevelopment, not underdevelopment; personal fulfillment, not unmet basic needs; and meaningful work, not the creation of jobs. One root of AT in the North was environmentalism. Smaller-scale technologies were seen as less harmful to the environment than larger ones. Sustainable agriculture, renewable energy sources, and resource conservation were prominent themes in AT literature. A second root was disillusionment with industrial society, alienating work, and impersonal bureaucracies. Some people advocated alternative forms of industrial organization with greater participation and workplace democracy. Others sought more satisfying forms of personal life in alternative life-styles in rural areas.[9]

Appropriate technology in Europe and America in the 1970s was thus varied. The revolt against bureaucracy led some people to stress individual creativity and autonomy. Others were more interested in community development and grass-roots organizations (rural or urban). Still others advocated decentralized technologies within a radically decentralized society.[10] Much of the communication among AT proponents occurred through newsletters, networks, and popular do-it-yourself manuals such as the *Whole Earth Catalogue* and its successors. A federally funded National Center for AT was established in the United States in 1977, but its funding was severely cut in the eighties by a more conservative administration. The London-based Intermediate Technology Development Group has worked on equipment for Britain as well as for the Third World. "Green" political parties in many European nations in the eighties have supported AT.

Behind these diverse movements, South and North, are some *common principles* that are more fundamental than the particular technologies advocated in varying circumstances. The first is that there is always a choice among technologies and that technologies are not neutral in their social and environmental consequences. Environmental concerns have been more prominent in recent writing in the South, including the "sustainable development" literature discussed earlier. The social and environmental impacts of large-scale technologies have been more apparent worldwide. The second general principle is that a technology should fit particular local conditions, which would lead us to expect considerable diversity among the specific characteristics appropriate in different regions and nations. A corollary is that local participation in decisions should be encouraged. Once again, the relation between organizational structure and engineering design is seen to be crucial. The challenge is to integrate the technical, economic, social, and environmental dimensions of technology choice in particular cases.[11]

Appropriateness, I suggest, must always be judged by a particular set of *values* and under particular *conditions*. One must always ask: Appropriate for what? Appropriate where? We can take seriously the insights of the appropriate technology movement, without making any list of specific characteristics absolute or universal. "Labor-intensive" and "simple" technologies may be appropriate in rural development in the South, but not necessarily everywhere. Some AT devices being promoted in the North may be too expensive for the South or even for low-income families in the North. AT in the North has been a predominantly middle-class movement, and its proponents often add AT components to a relatively affluent life-style that is supported by an industrial society.

In many instances, application of the above principles of appropriateness would lead us to conclude that "small is beautiful," but this may not always be the case. Let us consider in more detail the question of scale.

3. SCALE, EFFICIENCY, AND PARTICIPATION

The concept of *scale* can be applied to individual machines, industrial plants, and organizations (businesses, corporations, or government agencies). Scale can be measured by physical size or by output (number of units produced or total sales). These various levels and measures of scale tend to be correlated. A nuclear reactor is a huge machine in a large structure, which only a large utility or government agency can afford. However, there are exceptions to such correlation that we will have to consider.

Economies of scale resulting in lower cost per unit output are found in many industrial processes. Material inputs can be purchased in larger quantities at bulk rates. Production is speeded up by subdividing labor tasks and installing specialized machines and production lines. Advertising and distribution systems create mass markets for mass-produced products. In some cases it appears technically difficult if not impossible to design efficient small-scale processes (steel mills, oil refineries, and auto production, for example). Materials extraction and processing often offer substantial savings for large operations (mining, cement, glass, paper, and chemicals, among others). Most of these economies of scale require substantial investment capital, managerial and technical skills, and a good transportation system.[12]

Diseconomies of scale have received more attention in recent years and frequently seem to outweigh the economies, even by purely economic criteria. Organizational costs increase faster than size because additional management and staff are needed to coordinate complex operations. Large systems are frequently less reliable and more vulnerable to breakdown, as we saw in the previous chapter. When errors or failures occur, they can be catastrophic (Chernobyl or the Exxon Valdez, for instance). Environmental pollution is more concentrated. Siting of large facilities arouses intense local opposition because a small region bears the risks while the benefits are widely distributed. Larger plants usually have longer construction times, and they are less flexible in adapting to changing markets. As the costs of transportation rise, smaller plants have the advantage of being nearer to raw materials and markets.[13]

Because these diseconomies become more pronounced at large scales, an *intermediate scale* is often most efficient from a purely economic viewpoint. In chapter 4 we saw that middle-sized farms are more efficient (in yield per acre and per dollar invested) than either small or large farms. A study of electrical generators finds that the optimal size calculated from construction and capacity costs alone is 350 to 500 megawatts, but that even smaller units—far below the current industry average—are optimal if reliability, siting flexibility, and shorter transmission lines are taken into account.[14] In chapter 6 some examples were cited in which decentralized smaller offices or plants were linked by computers and communications to central offices, combining features of large and small organizations. Computers have helped many smaller enterprises to remain competitive (for example, the small-town printer can use a computer for composition).

Thus the *prevalence of large-scale technologies* cannot be attributed to economies of scale alone. Organizations seek growth to increase their total profits and their power as much as to increase their efficiency, and this influences their choice of technology and the type of research they conduct. In addition, tax structures, credit provisions, subsidies, and tariffs have usually favored large scale. Earlier chapters cited biases toward large producers in agricultural price supports, nuclear and fossil fuel energy subsidies, and military contracts with computer companies. In the Third World, government policies and international agencies have favored the politically powerful urban population over rural populations; food prices are kept low and investment is directed more often to large plants than to agriculture and rural industry.

Lastly, purely economic criteria omit the *environmental and human consequences* of scale-related choices (except when indirect costs are internalized because of government regulations). Economic efficiency ignores the distribution of costs and benefits among various income groups. Many mass-produced products do serve basic human needs and are available to low-income families. But in other cases (especially in the Third World) capital-intensive production caters primarily to the more affluent urban market or the export market. In the First World, large systems tend to be detrimental to job satisfaction, the motivation of workers, and the quality of work life. Small work teams and greater autonomy on the shop floor have sometimes been established within large industrial plants (see chapter 6), but in general job satisfaction is greater in smaller enterprises. In many cases there are trade-offs between economic efficiency and other values, but often these diverse goals can be furthered at the same time by imaginative technological design. Each part of a technological system must be examined separately in relation to its particular context. The goal is not to *maximize* efficiency but to *optimize* production according to plural criteria. Production must be sufficient to meet basic human needs, but in ways that encourage wider participation and more just distribution of benefits.

My own conclusion is that *a mix of small and large technologies* is desirable. There are many viable "niches" in an industrial society; diversity here permits experimentation and flexibility in adapting to changing conditions, as it does in biological ecosystems. For some products, mass production is substantially more efficient, even after some of the environmental and social costs are internalized. Often products that are decentralized in use (such as photovoltaic panels) must be mass-produced to be cheap enough for wide use. Private autos are decentralized in ownership, use, and servicing, but their production must be centralized. However, the greater fuel efficiency and lower pollution of mass transit are desirable, despite the fact that its ownership is more centralized. Urban transportation, utilities, and sewer systems are inescapably large scale. Long-distance transportation and communication must be centrally planned and administered. But a mix of large and small systems today requires deliberate emphasis on opportunities for small enterprises in order to overcome the built-in biases toward the large, inherited from past policies.

Let us ask finally about the relation between *scale* and *participation*. I have defended three forms of participation in the decisions that affect our lives:

1. Political Participation. National policies are crucial in technological development. Local grass-roots movements can affect local decisions but can also influence national policies through democratic processes. Environmental regulation has to be centralized because pollution crosses local jurisdictional lines. However, even when policies are legislated nationally, some of the details of implementation can be left to regional and local governments, with latitude for diversity in the ways in which national goals are met. Similarly, resource allocation requires long-range comprehensive planning. Greater local self-reliance is desirable, but natural resources are unevenly distributed and no region can be self-sufficient. The infrastructure for transportation and trade can be established only at the national level. Moreover, the regulation of national and international corporations can be achieved only through the powers of the state.

We have seen that large corporations with large technologies represent a concentration of economic and political power that is difficult to control democratically. Decentralized organizations with smaller technologies offer greater opportunities for political participation. Decisions can be made in a plurality of centers; users as well as producers can have a direct voice, and adaptation to local conditions and cultural differences can be more easily achieved, encouraging greater diversity. But when the public interest or economies of scale require centralized management, democratic legislation may still be effective if vigorously pursued, as I argued in the previous chapter.

2. Participation in the Marketplace. An ideal free market is a model of decentralized participation. Consumers make individual choices according to their preferences, and producers make individual decisions in response to market conditions. Through prices the market feeds back information on both supply and demand, and no central planning authority is needed. But as we have repeatedly noted the market is far from ideal. It is distorted by monopolies; it neglects externalities, future consequences, and distributional inequities; and it often fails to provide the basic necessities for those with low purchasing power. But the alternative of state ownership has produced large centralized bureaucracies and inefficient industries. When prices are fixed there is no feedback of market information and no competition to improve efficiency. The recent popular uprisings in Eastern Europe and Russia were demands for greater productivity as well as for greater political and economic participation. Most advocates of appropriate technology favor private or community ownership, preferably on a small scale: small businesses, farms, or cooperative enterprises.

Government policies set the conditions within which markets operate. I have supported a balance between government intervention and reliance on market forces. Political processes should be used to set broad goals, regulations, and economic policies, including taxes, subsidies, and research support. Within that framework, market forces can govern the decisions of individual consumers and producers in response to changing prices. Where large scale is desirable there may be some merit in the Swedish model of plants owned by governmental or quasi-governmental agencies operating in competition with private industries, with prices determined by the market. However, even Sweden is planning to sell some of its government-owned plants to private investors, and few nations today are moving toward increased government ownership.

3. Participation in the Workplace. Big systems can be managed only by hierarchical organizations, while small ones can be more egalitarian. But we have seen that unions, work teams, and shop floor reforms permit some participation even in large industrial plants. Attempts at worker ownership and greater industrial democracy have had only limited success in industrial nations. In the Third World, access to any work at all is a higher priority than meaningful work or participation in workplace decisions. Efficiency may dictate a scale and a mode of production that workers might not prefer. Nevertheless, workplace participation should be sought not only for the sake of job satisfaction but because it can influence safety, health, and production goals. All three forms of participation are important for technologies of all scales, and none is ruled out by the need for economic efficiency.

II. SUSTAINABLE CONSUMPTION

Sustainability requires attention to patterns of consumption as well as patterns of production. Let us consider two approaches to slowing the growth of consumption in industrial nations. First, institutional policies and technological systems can be designed to reduce waste and to increase the efficiency of resource use. Such conservation and recycling techniques can be introduced without major changes in individual values and behavior. Second, if individuals in affluent societies adopted simpler and more frugal life-styles the demands on the world's resources would be reduced. A frugal life-style does not mean a more primitive one; it can in fact be more healthy and personally fulfilling than prevalent activities and purchases. If widely practiced, such life-style changes would effect a substantial reduction in the consumption of resources, but this would require major shifts from the values current in industrial societies.

1. A CONSERVER SOCIETY

Previous chapters have described many changes that would *conserve resources:* sustainable agriculture that uses lower inputs of petrochemicals and fuel; energy efficiency and conservation measures (including cogeneration, mass transit, and smaller autos); the substitution of communication for transportation; and the shift from manufacturing to information technology and a service economy. Resource conservation can also be furthered by *effective waste management.* The following five ways of managing wastes are listed in order of their potential for conserving resources.

1. *Landfills* are the least desirable alternative. Many sites are nearing capacity, and most are not well designed to prevent the leaching of harmful substances into surface waters or aquifers. Half of the volume of household wastes comes from some form of packaging (paper, glass, aluminum, or plastic). The United States discards seventy billion beverage cans annually.[15] This mountain of garbage is costly to dispose of safely, and it is a huge waste of energy and materials.[16]

2. *Incineration* reduces the volume of wastes by 90 to 95 percent, and some energy is recovered if the heat is used for industrial processes or for the generation of electricity (or both). But no materials are recovered (except for residual ash that can be used in concrete or asphalt for roads). The energy recovered in burning is usually only a fraction of the energy that went into producing the materials burned, and of course carbon dioxide is emitted. Recycling materials saves twice as much energy as burning and replacing them, on average. Moreover, capital costs for incinerators are high. For the same investment of $8 billion, one-fourth of U.S. municipal solid waste could be burned or three-fourths could be recycled and composted.[17] Finally, toxic air emissions (such as the release of dioxin when plastics are burned) can be controlled by using high temperatures and smokestack devices, but careful operation and monitoring is required to be sure standards are not violated. It is even more difficult to be sure that toxics (especially heavy metals) in the ash will not leach into aquifers.[18]

3. *Recycling* saves natural resources and the energy that goes into extracting and processing them. Making a beverage can from recycled aluminum uses only 5 percent of the energy needed to make a can from aluminum ore. Steel from scrap saves more than half the energy and 80 percent of the air and water pollution as compared to making steel from iron ore. If the U.S. recycled its Sunday papers it would save 500,000 trees every week, plus the energy required to transport them and make pulp from them. Glass can be recycled, though this takes more energy than reusing returnable bottles. Some plastics can be recycled, but the different types of plastic must be treated separately. Japan recycles more than half its garbage, Western Europe 30 percent, and the United States only 10 percent.[19]

Recycling of metals, paper, glass, and plastics can be combined with *composting* of organic materials. Leaves, grass clippings, and organic kitchen wastes are composted by individuals or in community facilities, and the residue is used as fertilizer. Seattle is aiming for a 60 percent recovery of solid wastes by intensive recycling and composting; up to 90 percent recovery is theoretically possible.[20] But such results require extensive citizen education and cooperation in sorting garbage for curbside pickup. As landfill costs rise and markets for recovered materials are developed, recycling will be increasingly attractive financially. Some of the wastes that cannot be recycled or composted could be burned, but the amount would be relatively small, the energy content low, and the capital costs high.

4. *Reusing* a product is sometimes an improvement on recycling the materials in it. Cleaning and refilling glass bottles uses less energy and materials than recycling them. Reusable container programs create more new jobs than are lost in the disposable container industry. Standardized sizes would allow the same stock of containers to be used for diverse products. Denmark has banned one-way containers. Germany and nine American states now require refundable deposits on beverage bottles to encourage reuse, and other nations and states are considering them.[21] Within an industrial plant, materials can sometimes be reused on-site. An "industrial ecosystem" is designed so that the waste from one process is an input to another process, either in the same plant or in another industry (preferably without having to melt or dissolve the material as required in recycling programs).[22]

5. *Design for low resource use* is the most far-reaching option. In chapter 7, I discussed industrial changes aimed at preventing the generation of hazardous wastes in the first place (rather than removing them from effluents after they have been generated). Similar principles can be used in industry to reduce the input of materials and energy (for example, by substituting optic fibers for copper wires in communications). Processes can be deliberately designed to use recycled materials. Products can also be designed for durability, repairability, and recyclability (for instance, with durable components that can be easily separated for repair and eventually for recycling).[23] Basic styles in autos and clothes should be retained, avoiding the planned obsolescence of frequent style changes. The throwaway society catering to short-run convenience and short-run profits must give way to a society more concerned about long-run sustainability and the indirect costs of our actions.

Consumers can reduce both resource depletion and waste generation by selective purchases. They can choose products that are durable or recyclable and containers that can be reused or recycled. They can use paper instead of styrofoam cups and plates. They can buy food from farmers' markets or in bulk or with minimal wrapping rather than with heavy packaging. They can select low-phosphate detergents, energy-saving small fluorescent bulbs, and latex instead of oil-based paints (to reduce disposal hazards). Recommendations in guides for "green buying" are based on the chemical ingredients of various products and the wastes they generate, and also on conservation goals.[24] The advertising of supposedly "eco-friendly" products is sometimes misleading, but in Germany a Blue Angel seal appears on a product only after it has been studied and certified by a jury of independent scientists. Similar green labeling systems are under consideration by several other nations. Consumer boycotts have been effective against a number of environmental and health hazards (such as aerosol spray cans and tuna caught in nets that kill dolphins). Consumers seem to be willing to make some sacrifice of convenience when they are sufficiently concerned about the environmental impacts of their purchasing habits, and they are beginning to look also at resource depletion.

When making a major purchase, consumers in a conserver society would also consider the annualized *life-cycle cost:* initial cost, operating cost, repair cost, and disposal cost, spread over the probable life span. Consumer education and the testing of appliances by government agencies would encourage consumers to estimate life-cycle costs (including the cost of gas or electricity to operate them) rather than to judge by purchase price alone.[25] We could also recover our grandparents' commitment to maintaining, mending, and repairing the things in our homes rather than throwing them away and replacing them.

Many current *tax and price policies* work against resource conservation. Tax breaks for the extraction of raw materials were intended to stimulate rapid industrial growth and greater national self-sufficiency. Mineral, oil, and timber depletion allowances, lower capital gains taxes, and "current expensing" of development costs are subsidies encouraging use of these resources. According to one estimate, investment in extraction industries is 50 percent higher than it would be with normal taxation.[26] Public subsidies in the United States for the use of fossil and nuclear fuels total $44 billion a year.[27] Freight rates for used iron, paper, and glass are higher than those for virgin materials, thereby discriminating against recycling. As a minimum, the various forms of subsidy and pricing that accelerate resource depletion should be abolished.

Tax structures could in fact be deliberately designed to *encourage resource conservation.* Tax breaks could be given to recycling industries to help them develop new technologies. A resource extraction tax would reverse present policies and in general would reduce pollution and energy use as well as resource depletion. Such a tax would be a move toward full-cost pricing to internalize long-term social and environmental costs. Even the gradual introduction of such taxes would be a major market intervention, but economic incentives would allocate resources once the taxes were in place. Herman Daly and others have proposed a system of depletion quotas or severance taxes on nonrenewable raw materials. For example, licenses to mine a given quantity of ore could be auctioned off, or a tax could be levied on each ton mined.[28] Such policies would encourage resource-saving and recycling technologies. They would also raise the price of resource-intensive goods and discourage consumption of them.

In more general terms, a conserver society would put a smaller fraction of its income into short-term consumption and a larger fraction into *research, education,* and *investment* in efficient, productive technologies—all of which bring future benefits. The United States lags behind its competitors in both private and public support for civilian research, and its educational standards are below those of other nations, especially in mathematics and the sciences. Once a creditor nation, the U.S. is now the world's largest debtor. Instead of saving for the future, both the nation and many individual families have gone heavily into debt, improving the present by mortgaging the future.

2. INDIVIDUAL LIFE-STYLES

A further reduction in resource use could be effected by changes in individual life-styles (that is, patterns of daily life and work and ways of spending time and money). The prevailing pattern in industrial nations can be characterized as consumerism, but some people are seeking more frugal and more personally fulfilling life-styles.

There are powerful institutional pressures toward *the escalation of consumption*. Corporations promote increased consumption for the sake of their own growth and profits. Mass production requires mass markets. We are told to buy on credit and pay later. $125 billion was spent on advertising in the U.S. in 1989.[29] By the age of twenty, the average person has already seen 350,000 TV commercials.[30] The mass media hold before us the images of a high-consumption society. Personal identity and self-worth are defined by possessions, and happiness is identified with the purchase of products. The insatiable wants of a consumer society create an ever-expanding demand for resources, which the richest countries have the greatest power to secure. Consumerism has all the characteristics of an addiction, including the denial of its influence.

The erosion of community in industrial nations is both an effect and a cause of such consumerism. Millions of people have moved from rural communities to large urban centers. American families move every five years, on average, often at the request of an employer or to find employment elsewhere. While mobility gives greater freedom and opens up new opportunities, it also undermines the security of continuing traditions, stable relationships, and community ties. Paul Wachtel argues that the loss of these traditional sources of security has led people to seek security more individualistically through possessions and the symbols of material success.[31] William Leiss maintains that industrial culture encourages people to try to satisfy all their diverse psychological needs through consumption.[32] Yet in surveys within nations, and in comparisons between nations, it has repeatedly been found that happiness has very little correlation with income or wealth, whereas it has a strong correlation with marriage and family relationships and meaningful work.[33] Divorce, alcoholism, drug use, and suicide rates have been rising in Western nations, especially in the United States.

Contemporary consumerism contrasts strongly with the *simpler life-style* that some people have advocated throughout American history. David Shi has described several early versions: Puritan frugality, Quaker egalitarianism, Jefferson's vision of an agrarian nation, and the transcendental harmony with nature advocated by Emerson and Thoreau. Other voices continued to uphold these ideals, but they were drowned out in a culture preoccupied with industrialization and material progress. Shi says that two new factors contributed to a revival of interest in simpler life-styles since the late 1960s. First, the human and environmental costs of compulsive materialism were personally experienced by many people. The youth counterculture, the human potential movement, and the midlife crises of some business executives expressed the search for less stressful and more humane life patterns. Second, a new frugality was motivated by a recognition of finite resources and the limits to growth, and by a concern for sustainability. Shi maintains that simpler life-styles today can be both personally gratifying and socially constructive.[34]

Duane Elgin also sees *a double motive for simplicity.* On the one hand it is a response to the alienation and spiritual impoverishment attending the pursuit of material wealth. On the other, it is a response to the poor at home, the needs of the Third World, and the future generations with whom we share the world's resources. But Elgin's outlook is life affirming rather than ascetic or life denying. He seeks neither poverty nor excess, but a balance of material and spiritual goals in a rewarding and nourishing pattern of life. He favors decentralized appropriate technologies, local decision making, and the cultivation of human relationships and the inner life. These changes, he says, must be sought voluntarily and intentionally: "Poverty is involuntary, whereas simplicity is consciously chosen. Poverty is repressive, simplicity is liberating. Poverty generates a sense of helplessness and despair; simplicity fosters personal empowerment."[35]

Here are three examples of simpler life-styles:

1. *Diets for Health and for Justice.* Ninety-eight million American adults are overweight.[36] We have the diseases of overeating (such as heart disease), while other countries have the diseases of malnutrition. For the sake of health we need to reduce our food consumption, particularly of fats and sugars. We should eat more natural foods in place of the overprocessed, overpackaged, and artificially flavored commercial products that fill the supermarkets.[37] For the sake of justice we should reduce our use of such nonfood crops as coffee, tea, tobacco, and grains for alcohol, which tie up land in developing countries that could be used for food crops. Seven pounds of grain are needed to produce one pound of beef, so a reduction in meat consumption or a vegetarian diet would encourage more efficient use of land.[38] Here, clearly, individual action and national policy can support each other.

2. *Community Self-Help.* A variety of participatory projects have been initiated to meet basic needs, especially among low-income families. In "urban homesteading," families can renovate old houses and apply their own labor toward purchase of the houses. Community gardening can provide better nutrition and lower food bills, as well as exercise and recreation. Community health care programs stress preventive medicine, nutrition, family planning services, and individual responsibility for health. Through physician's assistants, paramedical workers, and public health nurses, access to health care can be more widespread. Some of these community efforts have failed from poor organization, narrowly defined interests, or opposition from established economic or political institutions. But many have succeeded and have brought social as well as economic benefits to local groups.[39]

3. *Alternative Institutions.* A variety of decentralized institutions in which people have greater control of their own lives have sprung up in recent years, including producer cooperatives, worker-owned businesses, printing and craft collectives, and farmers' markets. On the consumption side, grocery and food cooperatives are common. Neighborhood clinics and alternative schools are usually autonomous and self-managed. These local groups are in touch with each other through networks and informal publications without any overall organization. They typically combine personal participation, shared decision making, and wider political action. Such small-scale institutions provide viable patterns of economic activity under local control, with some degree of independence from the larger economic structures.

A more frugal life-style, in short, can affirm several of the values presented in this volume. It contributes to *sustainability* by reducing resource use and the environmental damage that goes with resource extraction, processing, and waste disposal. It supports *justice* if it is an act of solidarity with the world's poor, a commitment to a more equitable distribution of resources. I have traced some of the connections between overconsumption in rich nations and underdevelopment elsewhere, and the injustices of past exploitation and the present exercise of economic and military power. *Participation* is expressed in taking greater control of one's life and refusing to be manipulated by the "hidden persuaders" of a consumer society. We are not powerless; each person can make choices in his or her own life. Alternative life-styles are a means of consciousness raising, providing examples of what is possible. Both the exercise of purchasing power within existing institutions and the creation of alternative institutions can help redirect production toward the satisfaction of genuine human needs for all rather than artificially generated wants and luxuries for those who can afford them.

Personal fulfillment is the final consideration. New life-styles arise not only from a rejection of prevailing patterns but also from a positive vision of human fulfillment and a redefinition of the good life. A person's life can be built around the intrinsic satisfaction of activities and relationships rather than the extrinsic rewards of possessions. As Erich Fromm says, happiness depends more on who one is, what one does, and how one relates to other people than on what one owns.[40] One can seek significance in work itself, beyond earning income for consumption. Priority can be given to persons rather than things. Such a life-style is less competitive and more cooperative, less hectic and more serene, and provides more opportunity for personal growth. It acknowledges the spiritual emptiness within a materialistic culture. Here is a response not simply to an environmental and resource crisis but also to the modern crisis of meaning. For many people in industrial nations, a simpler life would free time and resources for the human relationships that are most significant. For most of the Third World, and for those trapped in poverty in the First World, levels of consumption must of course rise substantially if basic needs are to be met, but this could occur without the excesses of consumerism.

In order to effect wider changes, concern for individual life-styles and community life must be coupled with *political action* at the local and national levels. Only through public policies can resources saved here benefit those in need abroad. A more just international order will not arise without concerted worldwide commitment. Only political processes can be effective in influencing corporations or government agencies. We can act individually and in small communities and at the same time try to affect legislation and social policies. By both word and deed, each person can influence public attitudes and help to create new political constituencies.

III. CHANGING VALUES

The reorientation of technology toward justice, participation, and sustainability will require a major change from past attitudes and values. I note first some evidence that for a significant minority in Western industrial societies such a change has already begun. Next, I suggest that the biblical tradition could make a distinctive contribution, provided the churches are willing to reexamine some of their doctrines and practices. Education, political action, catalytic crises, and the vision of new alternatives offer hope of averting the environmental and social disasters toward which current policies seem to be leading.

1. A NEW SOCIAL PARADIGM

In the previous volume I discussed Thomas Kuhn's idea of a *paradigm:* a cluster of conceptual, metaphysical, and methodological presuppositions embodied in a tradition of scientific research. Kuhn describes a paradigm shift as a "radical transformation of the scientific imagination" in which the old data is interpreted in a new way.[41] Social scientists have described a *dominant social paradigm* as "the collection of norms, beliefs, values, habits and survival rules that provide a frame of reference for members of a society."[42] It is a mental image of social reality that guides behavior and expectations.

Writing in 1976, Willis Harman identified the following components of *the industrial era paradigm:* expectation of unlimited material progress and ever-growing consumption; faith in science and technology to solve all problems; goals of efficiency, growth, and productivity; mastery of nature; and competition and individualism. He held that this paradigm has led to environmental degradation, resource depletion, loss of meaningful work roles, inequitable distribution, and ineffective control of technology.[43]

Harman outlined a *postindustrial paradigm* that he said was beginning to emerge: material sufficiency in the satisfaction of basic needs; frugality in resource use and transition to renewable resources; ecological ethics and stewardship of nature; dedication to human development, self-realization, and growth in awareness and creativity; and cooperation and community solidarity in place of competition and individualism. Harman and his colleagues at the Stanford Research Institute estimated in 1980 that one in five adults in the

United States (mainly younger, middle-class persons) supported most of these new values.[44]

In a questionnaire survey of three nations in 1982, Lester Milbrath found wide support for what he called the *New Environmental Paradigm*. It included a high valuation of nature, acceptance of resource limitations and simpler life-styles, and commitment to greater participation in political and work deci-sions. Respondents in the U.S., Britain, and West Germany all gave higher priority to environmental protection than to economic growth, by roughly 3-to-1 ratios.[45] In another series of surveys in the U.S., people were asked whether they agreed or disagreed with the following statement: "Protecting the environment is *so* important that requirements and standards cannot be too high, and continuing environmental improvements must be made *regardless* of cost." Those agreeing rose steadily from 45 percent in 1981 to 80 percent in 1989.[46]

Ronald Inglehart's *Culture Shift in Advanced Industrial Society* (1990) gives data from surveys in twelve nations since 1970. He found a continuing shift from *materialist goals* (such as high economic growth, keeping prices down, and maintaining order) toward what he calls *postmaterialist goals* (such as free-dom of speech, a less impersonal society, and a greater voice in decisions on the job and in politics). He found a growing concern for environmental pro-tection, the quality of life, and personal self-realization. The shift occurred in the 1970s and continued throughout the 1980s, despite periods of economic recession and the greater conservatism of Western political leaders. In 1970 materialists outnumbered postmaterialists by 4 to 1, but by 1988 this ratio had fallen to 4 to 3. Postmaterialism was strongest among younger and more edu-cated respondents. The shift was greatest in northern Europe (especially West Germany, Denmark, and the Netherlands), somewhat smaller in Britain and the United States, and much smaller in southern Europe, but it was present in all these countries.[47]

Inglehart attributes these value shifts to the *unprecedented prosperity* of the fifties and sixties when many of these respondents were growing up. He says that people who were socialized during a time of economic and physical secu-rity adopted broader goals, which they retained even during later periods of recession and inflation. Stephen Cotgrove finds a similar postmaterialist trend in Britain, but he attributes it to new ideals and values in schools, homes, and political parties, rather than to changes in economic security.[48]

We must recognize, however, that people's *actual behavior* in voting or in daily life may depart considerably from their responses to public opinion polls. Environmental attitudes are now *widely* held, but they may not be *deeply* held. Most people do not vote for candidates primarily because of their envi-ronmental positions. Moreover, unresolved conflicts may be present among a person's values.[49] In particular, prospects for economic prosperity seem to be a dominant criterion in actual voting, whatever people say in theory. Riley Dunlap concludes in a recent study that "decisions in the ballot booth will be influenced more by economic than by environmental conditions."[50] But he thinks this is beginning to change as environmental threats are becoming more noticeable and urgent.

Many of these new values are found in the *Green parties* that appeared in Europe in the 1980s. Drawing initially from groups that had protested against nuclear power plants, missile installations, and industrial pollution, Green candidates were elected first to municipal councils and regional legislatures, and by the late 1980s they had seats in eight national legislatures. In the West German election of 1987, the Green party received 8.3 percent of the national vote, entitling it to forty-four seats in the Bundestag under a system that gives proportional representation to all parties achieving more than 5 percent of the vote.[51] In 1990, in the first election after reunification, the Greens fell below 5 percent in the former West Germany, losing their seats, but this partly reflects greater environmental emphasis in the platforms of the major parties. While there are some variations among countries, the Greens in general support environmental protection, participatory organizational structure, disarmament, and solidarity with the Third World. As a small party they can influence the political agenda and the stance of other parties, and occasionally their votes affect the balance of power between larger parties.

The Greens say that they are *"neither right nor left but in front."* In the past, the political right has favored economic growth and the free market, with little concern for the environment. The traditional left has supported economic growth and state ownership (or political intervention in the market to achieve greater distributive justice); it has had a working-class base and has also tended to ignore the environment. By contrast, the Greens reject unlimited economic growth (in the name of both sustainability and global justice), and they criticize both corporate and state centralization (in the name of grassroots democracy and decentralized technologies). Because they attack the status quo, they are sometimes called "the new left," but they differ greatly from the old left. They draw primarily from the middle class, especially professionals, students, and others outside the industrial economy. Their political effectiveness has been weakened by factional divisions: the reformists seek parliamentary respectability, piecemeal reforms, and strategic compromises to form coalitions, while the radicals are more confrontational and uncompromising, taking an oppositional role both in parliaments and in protest actions in the public arena.[52]

Greens have seldom been elected in Britain and the United States where only one person can be elected in each district. It is much more difficult for a minority party to achieve representation under a winner-take-all system than under proportional representation. However, in the 1989 election of British representatives to the European parliament, Greens received 15 percent of the seats.[53] In the U.S., third parties have had great difficulty in achieving representation, so environmentally oriented groups have usually either worked within existing parties or focused on particular local or national issues. A wider constituency is needed than the Greens have been able to mobilize in any nation.

2. A BIBLICAL PERSPECTIVE

What contribution can the Western religious traditions make to a postindustrial paradigm for a just and sustainable global society? The churches have usually supported the status quo, but they also have contributed to social change. The churches themselves will have to change drastically if they are to facilitate the transition to a sustainable world; however, a new religious consciousness could be a strong force in social transformation. Biblical images still have a latent power to evoke response. The prophets of ancient Israel lived in a world very different from ours, yet they lived, as we do, in times of national crisis and international conflict.

The prophets' commitment to *justice* was rooted in a belief in the fundamental equality of all persons before God. Speaking in the name of a God of justice, Amos denounced the inequalities of his day: "For three transgressions of Israel, and for four, I will not revoke the punishment; because they sell the righteous for silver, and the needy for a pair of shoes—they that trample the head of the poor into the dust of the earth, and turn aside the way of the afflicted. . . . But let justice roll down like waters, and righteousness like an everflowing stream" (Amos 2:6 and 5:24). The biblical God is identified with the dispossessed and portrayed as Redeemer and Liberator. This same commitment to social justice is evident in recent statements from the Roman Catholic Church and the World Council of Churches concerning a more equitable distribution of global resources. The biologist Charles Birch, addressing a WCC assembly, said, "The rich must live more simply that the poor may simply live."[54] Today, as in anciennt Israel, the sharing of resources is a demand of justice, not an act of charity.

The prophetic view of *a created order* that is inclusive in space and time is also relevant today. The whole creation is part of God's purpose. Because all forms of life are within God's plan, we are accountable for the way we treat them. I suggested that stewardship of nature is more typical of the Bible than dominion over nature, though it was often ignored in subsequent Western history. Moreover, the prophets used an extended time scale because they believed that God's purposes extend into the future. We have obligations to posterity and to a God who spans the generations. There is a solidarity in time, a covenant "from generation to generation." The idea of creation is a great unifying framework, encompassing all forms of life and all time from past to future. The theme of sustainability is prominent in recent WCC documents.[55] Environmentalists have often neglected social justice, and advocates of justice have often neglected the environment, but a Christian concern for "eco-justice" represents a distinctive combination of these ideas.[56]

Next, *a broad view of human fulfillment* is expressed in the biblical literature. The good life is identified not with material possessions but with personal existence in community. The prophets upheld the dignity of the individual and the importance of interpersonal relationships. They portrayed harmony with God and neighbor as the goal of life. They recognized the dangers of both poverty and affluence. They saw the harmful consequences of affluence—for the rich as well as for the poor. Jesus, in turn, stressed the importance of feeding the hungry, but he also said that "man does not live by bread alone," and he vividly pictured the dangers of wealth. The earliest Christian community, as described in Acts, "had all things in common." Distribution was made to each "as any had need" (Acts 4:35).

Over the ensuing centuries the monastic orders preserved the ideals of *simplicity* and *community*. The Reformation and then the Puritan movement upheld frugality and simplicity and were critical of "the luxuries of the rich." Mennonites, Hutterites, Amish, and other dedicated groups have practiced these ideals in their own communities separated from the wider society; Quakers have made greater efforts to change social institutions. Other religious groups have participated more fully in the prevailing social order and have challenged it less radically but have still upheld distinctive goals for personal life.

Today, in an overconsumptive society, we need both a rejection of the dominant materialism and a positive witness to the priority of the personal and the quality of the life of the community. New life-styles arise not only from a concern for global justice but also from *a new vision of the good life,* a focus on sources of satisfaction that are not resource consumptive. We have turned technology into a religion, seeking meaning and salvation through new technologies. Here a reformulation of traditional religious faith can speak to the crisis of meaning that underlies the pursuit of affluence. One example is the book *Enough Is Enough* by the British theologian John Taylor, who urges restraint in consumption and a level of material sufficiency that is neither affluence nor poverty.[57] I believe that the biblical vision of human fulfillment can strengthen the search for more frugal life-styles today. The goal is not asceticism, but true fullness of life.

The Bible also offers a distinctive view of *persons in community* which avoids both collectivism and individualism. Industrial society has been dominated by large impersonal organizations, mass advertising, and urban anonymity. In reaction, many people have sought personal meaning in self-actualization as autonomous individuals. Instead we could seek out voluntary associations of intermediate size: neighborhood and civic groups, unions and participatory structures on the job, local political parties and public interest groups. The local church remains a unique opportunity for personal interaction and mutual support. At its best it can be a community of forgiveness, celebration, and common search for both personal and social renewal, starting with the renewal of the church itself.

Finally, the Hebrew prophets brought a double message of judgment and hope. On the one hand, they spoke of God's *judgment* on human greed. They saw military defeat and national catastrophe as forms of divine judgment on

the materialism, idolatry, and injustice of national life. The prophets were realistic about human sinfulness and aware of the misuse of economic and political power concentrated in the hands of a group or a nation. Their first
word was a call to repentance and humility. Today such humility would be an
antidote to the Promethean pride to which industrial nations are prone. The
starting point would have to be repentance for our complicity in structures of
power, greed, and injustice.[58] A greater awareness of our limitations would
help us to recover a sense of the sacred. Obviously the prophets said nothing
about technological centralization, but I suspect that if we accepted their view
of human nature we would be hesitant to rely too heavily on large-scale systems that are vulnerable to human frailty and the abuse of institutional power.

The other side of the prophetic message is *hope*. Beyond judgment and repentance is the prospect of reconciliation and redemption. Reconciliation is
restoration of wholeness, the overcoming of alienation from God, from other
persons, and from nature. Redemption is creative renewal and response to
God's redemptive activity. The ultimate symbol of hope is the vision of a future Kingdom of peace and community. In Micah's words: "They shall beat
their swords into plowshares and their spears into pruning hooks; nation shall
not lift up sword against nation, neither shall they learn war any more; but
they shall sit every man under his vine and under his fig tree, and none shall
make them afraid; for the mouth of the Lord of hosts has spoken" (Mic.
4:3–5). What better image to inspire efforts for global peace and sustainable
agriculture than "swords into plowshares"? The prophetic imagination pictured a future harmony that would include all humankind and all nature. It envisioned a day when sharing by all would end scarcity for any. The biblical
themes of justice, creation, human fulfillment, judgment, and hope can thus
make an important contribution to the search for a just and sustainable society.

In the previous volume I argued that among contemporary reformulations
of Christian thought, *process theology* is particularly relevant to the current
crises of industrial society. Against a prevailing anthropocentrism, it asserts
the interdependence of all beings and the intrinsic value of nonhuman life.
Against reductionism it presents a holistic and multileveled view of reality. It
avoids the dualisms of mind/body, reason/emotion, and man/woman. It
overcomes the sharp separation of humanity from nature and the separation
of God from the world. It affirms divine immanence as well as transcendence,
and it understands God's power as empowerment. Process theology does not
lead us to expect a sudden and dramatic intervention of God in the world, but
it does hold that God is a creative power in cosmic and human history. We are
not determined by the past; God presents us with new opportunities for harmony and unity but does not coerce us to accept them. In the process view,
God has been at work in nature, in history, in the religious community, and in
religious experience (each of these is an expression of the activity of the
Spirit, in traditional terms). I suggested that God is also at work in other religious traditions, so the door is open to genuine dialogue in a pluralistic world.

Such a theology encourages confidence that there are new possibilities, and it motivates responsible action, even when we are discouraged by prevailing trends.

3. SOURCES OF CHANGE

Looking to the future, I see *new opportunities* for change toward a more just and sustainable world. Many people already have greater respect for nature, departing from the dominant attitudes of modern industrialism. Recognition of our global interdependence and our vulnerability to events distant in time and space has been growing. The end of the Cold War offers a new opening for efforts to reduce military spending. I have noted that if a quarter of the trillion dollars spent annually on arms around the world were put into family planning, sustainable agriculture, renewable energy, and reforestation in the Third World, the prospects for the future would be dramatically transformed. The global character of the new environmental threats and the possibility of nuclear disarmament give us new reasons to strengthen the agencies of the United Nations. National and international policies could indeed redirect technology toward basic human needs.

But enormous *obstacles to change* are also present. We should not underestimate the strength of individual or institutional greed or the political power of corporations and bureaucracies with a vested interest in the status quo. Third World nations harbor great resentment of past exploitation and growing global disparities. Ours is a world of political instability and potential for violence. We already have gone so far down unjust and unsustainable paths that we cannot expect to avoid conflict and suffering. Changes in power structures will not be achieved without struggle and upheaval. Despite such obstacles, I see four promising sources of social change today.

1. Education

In many nations, the schools have begun to nurture greater environmental and global awareness, but they could do much more. Curriculum materials and class projects are being developed around ecological and resource themes. Colleges and universities have suffered from departmental fragmentation; we need more emphasis on integrative education and greater support for interdisciplinary teaching and research. Centers for environmental studies and programs in Science, Technology, and Society have brought the insights of the natural sciences, the social sciences, and the humanities to bear on policy decisions like those discussed in this volume. Ethical questions and the role of values in personal and social choices can be considered within courses in many curricular departments.[59]

Formal and informal adult education is also an important means of social change. Citizens' movements and voluntary community groups provide information and raise the consciousness of the public. Churches and labor unions also have a considerable potential for fostering global citizenship. Education in an age of environmental degradation and inequitable resource use must deal with facts and values, information and attitudes, cognitive processes and affective responses. We must be better informed, but we must also learn to see the world from the standpoint of other people, to see ourselves as part of nature, and to gain skills for community action.

The media have reported on dramatic environmental incidents and technological controversies, but they could give us more extensive analyses of the social and political forces leading to such events. Newspapers typically have a large business section but only occasional articles on environmental issues just as important to our future. Global communications, if used more equitably and imaginatively, could be a vehicle of international understanding rather than of cultural domination. Education is a rather slow instrument of change, but its effects are far-reaching because it leads to new forms of perception and interpretation. We are beginning to think of ourselves as part of an interdependent world.

2. Political Action

Policy changes at the local and national level can be brought about by coalitions of diverse interest groups. Single-issue organizations should cooperate in broader programs that include their overlapping interests. As indicated earlier, labor unions and environmental groups have sometimes taken opposing positions, but they have cooperated on toxic chemical and occupational health legislation and efforts to broaden participation in corporate and governmental decision making. Most of the members of the environmental movement, the peace movement, and the women's movement support each other's goals.[60] Civil rights and urban renewal organizations share with environmentalists an interest in the improvement of the urban environment. Many religious groups have formed task forces to examine particular policy options and to cooperate with other groups concerned about justice, peace, and sustainability.

In some nations these varied groups could cooperate to support new political parties. In other nations the electoral system favors working within the existing political parties. In the past, citizens' groups have been more effective in cooperating to oppose proposed legislation than in promoting alternative policies, but with greater experience and mutual understanding they could be a strong political force. In general this will lead to a series of small changes, following the path of reform and "pragmatic incrementalism" within existing political and economic institutions. But as such groups examine the causes as well as the symptoms of technological impacts on nature and society, they will be led to propose changes in the institutions themselves.

Can democratic systems affect individual and national behavior rapidly enough to avoid catastrophic environmental damage, resource scarcity, and international conflict? Most politicians are reluctant to call for any kind of sacrifice or to raise taxes despite huge budget deficits and national debts. Some authors hold that only authoritarian governments or political revolutions can introduce institutional changes on a scale adequate to cope with mounting threats. I believe that democratic systems can respond in time if further changes in public attitudes occur. Popular mobilization leading to legislative changes would preserve a more open society that would outlast coerced obedience. Authoritarian governments often start out acting on behalf of the public good, but they tend to produce a new privileged group protecting its own interests and suppressing dissent.

3. Crisis as Catalyst

Perhaps fuel and water shortages, widespread famines, and civil unrest will have to occur on a more massive scale before enough people wake up. Further urban decay and unemployment may lead people to question the assumptions of industrial society. Taken alone, however, crises may lead to undesirable changes. Governments may rely on technical fixes whose indirect costs only make matters worse. Amid social turbulence and conflict they may adopt more authoritarian measures. But in combination with the vision of positive alternatives, crises can be catalysts for constructive action. In such emergencies as war or natural disaster, people will make sacrifices for the common good—provided the sacrifices are shared, the common good is clear, and a more hopeful future is envisaged. Might they also do so in the face of global crises?

Recognition of the inadequacy of old patterns plus the vision of positive alternatives could produce major changes without widespread violence. If we do not accept these changes voluntarily, they will be forced on us by environmental constraints or revolutionary movements—and at a very high cost in social disruption and human suffering. In their day, the Hebrew prophets were convinced that catastrophe would occur if their nation persisted in its ways. Perhaps we can only prepare foundations for a new order beyond catastrophe or plant healthy social seeds that will grow later. We can provide examples to which people will turn when they are disillusioned with previous patterns and are searching for alternatives. But as long as such catastrophe might be averted we should make every effort to prevent it. For despair is a self-fulfilling prophecy when it leads people to think that action is futile. Committed people can be a powerful inspiration for change.

4. A Vision of Alternatives

New visions can provide the motivation and direction for creative social change. Moral exhortation seldom inspires action among those who are reluctant to change. Visions, on the other hand, present positive alternatives in an imaginative way. They summarize a set of values, using concrete images rather than abstract principles. Visions of alternative futures offer hope instead of despair, a sense of the possible rather than resignation to the inevitable. Most movements of social reform started from utopian imagination, new images of the good life, vivid portrayals of what might be. Dom Helder Camara recalls a Brazilian saying: "When we dream alone it is only a dream. When we dream together, it is no longer a dream but the beginning of reality."[61] Far-reaching social and political changes will be possible as people begin to think of themselves as citizens of the globe and as part of nature. A wider concept of self-interest and self-identity are products of recognizing our interconnection and interdependence with human and nonhuman beings around the world in the community of life.

I believe that the combination of education, political action, catalytic crises, and vision can bring about a more just and sustainable world. Changes that once seemed impossible have occurred surprisingly rapidly when people recognized the inadequacy of past patterns and held a vision of an alternative future. The U.S. civil rights movement of the sixties was a product of the courage of black activists and a commitment to social justice by crucial political leaders with wide public support. The protests of students against the Vietnam War, together with the war's slow progress and mounting casualties, influenced public opinion within a few years. Feminist thought and action changed the career expectations of women within a couple of decades, and employment opportunities have changed dramatically, despite continued discrimination. Again, who would have guessed in the summer of 1989 that within a year Germany would be reunited and Eastern Europe would have moved toward democracy? Each of these events involved dissatisfaction with the past and hope for a new future.

As a symbol of global and ecological awareness, think of the pictures of the earth that were taken by astronauts on the moon. For the first time, the earth could actually be seen as a single unit. There it is, a spinning globe of incredible richness and beauty, a blue and white gem among the barren planets. It has been proposed that we should think of it as *spaceship earth.* The earth is a fragile life-support system. Like a spaceship, it has limited resources that must be conserved and recycled. Its inhabitants are interdependent, sharing a mutual responsibility and a common destiny. This is a striking image that forcefully represents the importance of life support and cooperation for human survival.

But we must extend the spaceship image if it is not to mislead us. A spaceship is a mechanical environment, devoid of life except for human beings. *Planet earth,* however, is enveloped in a marvelous web of life, a natural environment of which humanity is a part and on which it is dependent. We must think not just of life support but of ecological sustainability on a long time scale. So, too, the social order on a spaceship is relatively simple, with only a few people interacting in highly structured ways. Complex relationships between groups and nations are present on planet earth, and we face crucial issues of distributive justice and participation in the allocation of scarce resources. As I see it, we must move toward more globalism, more localism, and less nationalism.

Let us keep before us that image of the spinning globe with its natural environments and its social order. Let us imagine technology used in the service of a more just, participatory, and sustainable society on planet earth.

Notes

Chapter 1. Views of Technology

1. Charles Susskind, *Understanding Technology* (Baltimore: Johns Hopkins University Press, 1973), p. 132.
2. John Zerman and Alice Carnes, eds., *Questioning Technology* (Santa Cruz, CA: New Society Publishers, 1991), p. 217.
3. C. S. Lewis, *The Abolition of Man* (New York: Macmillan, 1965), p. 69.
4. Among the volumes dealing with broad attitudes toward technology are Albert H. Teich, ed., *Technology and the Future,* 5th ed. (New York: St. Martin's Press, 1989), and Carl Mitcham and Robert Mackey, eds., *Philosophy and Technology* (New York: Free Press, 1972).
5. This is close to the definition given by Arnold Pacey in *The Culture of Technology* (Cambridge: MIT Press, 1983), p. 6. Pacey adds "living things" among the "ordered systems" (in order to include agriculture, medicine, and biotechnology), but I suggest that these are already included under the rubric of "practical tasks." Frederick Ferré, *Philosophy of Technology* (Englewood Cliffs, NJ: Prentice-Hall, 1988), defines technology as "the practical implementation of intelligence" and argues that intelligence itself has both practical and theoretical forms.
6. Emanuel Mesthene, *Technological Change: Its Impact on Man and Society* (New York: New American Library, 1970).
7. Melvin Kranzberg, "Technology the Liberator," in *Technology at the Turning Point,* ed. William Pickett (San Francisco: San Francisco Press, 1977). See also Charles Susskind, *Understanding Technology.*
8. Emanuel Mesthene, "Technology as Evil: Fear or Lamentation?" in *Research in Philosophy and Technology,* vol. 7, ed. Paul Durbin (Greenwich, CT: JAI Press, 1984).
9. Daniel Bell, *The Coming of Postindustrial Society* (New York: Basic Books, 1973).
10. Buckminster Fuller, *The Critical Path* (New York: St. Martin's Press, 1981); Herman Kahn et al., *The Next 200 Years* (New York: William Morrow, 1976); Alvin Toffler, *Future Shock* (New York: Bantam, 1971) and *The Third Wave* (New York: William Morrow, 1980).
11. Samuel Florman, *The Existential Pleasures of Engineering* (New York: St. Martin's Press, 1977) and *Blaming Technology: The Irrational Search for Scapegoats* (New York: St. Martin's Press, 1981).
12. Florman, *Blaming Technology,* p. 183.
13. Cf. Alvin Weinberg, "Can Technology Replace Social Engineering," in *Technology and the Future,* ed. Teich.
14. Samuel Florman, "Science for Public Consumption: More Than We Can Chew?" *Technology Review* 86 (April 1983): 12–13.
15. Florman, *Blaming Technology,* p. 193.
16. Harvey Cox, *The Secular City* (New York: Macmillan, 1965), and "The Responsibility of the Christian in a World of Technology," in *Science and Religion,* ed. Ian G. Barbour (New York: Harper & Row, 1968).
17. W. Norris Clarke, S.J., "Technology and Man: A Christian Vision," in *Science and Religion,* ed. Barbour.
18. Pierre Teilhard de Chardin, *The Future of Man,* trans. Norman Denny (New York: Harper & Row, 1964), chaps. 8, 9, and 10. See also "The Place of Technology in a General Biology of Mankind," and "On Looking at a Cyclotron," in *The Activation of Energy* (New York: Harcourt Brace Jovanovich, 1971).
19. George Wise, "Science and Technology," *Osiris,* 2d ser., 1 (1985): 229–46.

20. See for example Lewis Mumford, *The Myth of the Machine*, vol. 1, *Technics and Human Development*, and vol. 2, *The Pentagon of Power* (New York: Harcourt Brace Jovanovich, 1967 and 1969).
21. Studs Terkel, *Working* (New York: Pantheon, 1972); Robert Schrag, *Ten Thousand Working Days* (Cambridge: MIT Press, 1978); William A. Faunce, *Problems of an Industrial Society*, 2d ed. (New York: McGraw-Hill, 1981).
22. Theodore Roszak, *The Making of a Counter Culture* (New York: Doubleday, 1969), and *Where the Wasteland Ends* (New York: Doubleday, 1972); see Ian G. Barbour, "Science, Religion, and the Counterculture," *Zygon* 10 (1975): 380–97.
23. Jacques Ellul, *The Technological Society*, trans. J. Wilkinson (New York: Knopf, 1964); also *The Technological System*, trans. J. Neugroschel (New York: Continuum, 1980), and *The Technological Bluff*, trans. G. Bromiley (Grand Rapids: Eerdmans, 1990).
24. Darrell Fasching, "The Dialectic of Apocalypse and Utopia in the Theological Ethics of Jacques Ellul," in *Research in Philosophy and Technology*, vol. 10, ed. Frederick Ferré (Greenwich, CT: JAI Press, 1990).
25. Langdon Winner, *Autonomous Technology* (Cambridge: MIT Press, 1977) and *The Reactor and the Whale* (Chicago: University of Chicago Press, 1986).
26. Hans Jonas, *The Imperative of Responsibility: In Search of an Ethics for the Technological Age* (Chicago: University of Chicago Press, 1984), p. x.
27. Albert Borgmann, *Technology and the Character of Contemporary Life* (Chicago: University of Chicago Press, 1984); Martin Heidegger, *The Question Concerning Technology*, trans. William Lovitt (New York: Harper & Row, 1977).
28. David Kipnis, *Technology and Power* (Berlin: Springer-Verlag, 1990).
29. Langdon Gilkey, *Religion and the Scientific Future* (New York: Harper & Row, 1970).
30. Paul Tillich, "The Person in a Technological Society," in *Social Ethics*, ed. Gibson Winter (New York: Harper & Row, 1968).
31. Gabriel Marcel, "The Sacred in the Technological Age," *Theology Today* 19 (1962): 27–38.
32. Martin Buber, *I and Thou*, trans. R. G. Smith (New York: Charles Scribner's Sons, 1937).
33. P. Hans Sun, "Notes on How to Begin to Think about Technology in a Theological Way," in *Theology and Technology*, ed. Carl Mitcham and Jim Grote (New York: University Press of America, 1984).
34. Thomas Misa, "How Machines Make History, and How Historians (and Others) Help Them Do So," *Science, Technology & Human Values* 13 (1988): 308–31.
35. Arnold Pacey, *Culture of Technology*.
36. Victor Ferkiss, *Technological Man* and *The Future of Technological Civilization* (New York: George Braziller, 1969 and 1974).
37. Bernard Gendron, *Technology and the Human Condition* (New York: St. Martin's Press, 1977).
38. Norman Faramelli, *Technethics* (New York: Friendship Press, 1971).
39. J. Edward Carothers, Margaret Mead, Daniel McCracken, and Roger Shinn, eds., *To Love or to Perish: The Technological Crisis and the Churches* (New York: Friendship Press, 1972); Paul Abrecht and Roger Shinn, eds., *Faith and Science in an Unjust World* (Geneva: World Council of Churches, 1980).
40. Thomas Derr, "Conversations about Ultimate Matters: Theological Motifs in WCC Studies on the Technological Future," *International Review of Missions* 66 (1977): 123–34.
41. Egbert Schuurman, *Technology and the Future* (Toronto: Wedge Publishing, 1980), also "The Modern Babylon Culture," in *Technology and Responsibility*, ed. Paul Durbin (Dordrecht, Holland: D. Reidel, 1987), and "A Christian Philosophical Perspective on Technology," in *Theology and Technology*, ed. Mitcham and Grote. Schuurman was also a contributor to Stephen Monsma, ed., *Responsible Technology: A Christian Perspective* (Grand Rapids: Eerdmans, 1986).
42. Roger Shinn, *Forced Options: Social Decisions for the 21st Century*, 3d ed. (Cleveland: Pilgrim Press, 1991).
43. H. Richard Niebuhr, *Christ and Culture* (New York: Harper & Brothers, 1951). See also Carl Mitcham, "Technology as a Theological Problem in the Christian Tradition," in *Theology and Technology*, ed. Mitcham and Grote.
44. Ferré, *Philosophy of Technology*, p. 44.
45. Wise, "Science and Technology."

46. Trevor Pinch and Wiebe Bijker, "The Social Construction of Facts and Artifacts: Or How the Sociology of Science and the Sociology of Technology Might Benefit from Each Other," in *The Social Construction of Technological Systems*, ed. Wiebe Bijker, Thomas Hughes, and Trevor Pinch (Cambridge: MIT Press, 1987).

47. John W. Staudenmaier, *Technology's Storytellers* (Cambridge: MIT Press, 1985), p. 165.

48. Cynthia Cockburn, "The Material of Male Power," in *The Social Shaping of Technology*, ed. Donald McKenzie and Judy Wajcman (Milton Keynes, England: Open University Press, 1985).

49. Roslyn Feldberg and Evelyn Nakano Glenn, "Technology and Work Degradation: Effects of Office Automation on Women Clerical Workers," in *Machina Ex Dea: Feminist Perspectives on Technology*, ed. Joan Rothschild (New York: Pergamon Press, 1983); see also articles by Cheris Kramarae, Anne Machung, and others in *Technology and Women's Voices*, ed. Cheris Kramarae (New York and London: Routledge & Kegan Paul, 1988).

50. Cynthia Cockburn, *Machinery of Dominance: Women, Men, and Technical Know-How* (London: Pluto Press, 1985).

Chapter 2. Human Values

1. Edward O. Wilson, *On Human Nature* (Cambridge: Harvard University Press), p. 167.

2. John Rawls, *A Theory of Justice* (Cambridge: Harvard University Press, 1971), p. 11.

3. Kurt Baier and Nicholas Rescher, eds., *Values and the Future* (New York: Free Press, 1969), chaps. 1 and 2; William Frankena, "Value and Valuation," in *Encyclopedia of Philosophy*, ed. Paul Edwards (New York: Macmillan, 1965).

4. Loren Graham, *Between Science and Values* (New York: Columbia University Press, 1981).

5. Stephen Toulmin, "How Can We Reconnect the Sciences with the Foundations of Ethics?" in *The Roots of Ethics*, ed. Daniel Callahan and H. Tristram Engelhardt (New York: Plenum Press, 1981).

6. David Dickson, *The New Politics of Science*, 2d ed. (Chicago: University of Chicago Press, 1988).

7. Jacob Bronowski, *Science and Human Values* (New York: Julian Messner, 1956). See also Bentley Glass, *Science and Ethical Values* (Chapel Hill: University of North Carolina Press, 1965), chap. 3.

8. Robert Merton, *The Sociology of Science* (Chicago: University of Chicago Press, 1973).

9. André Cournand and Michael Meyer, "The Scientist's Code," *Minerva* 14 (1976): 79–96; Robert S. Cohen, "Ethics and Science," in *Science, Technology, and Freedom*, ed. W. H. Truitt and T. W. G. Solomons (Boston: Houghton Mifflin, 1974).

10. Jerome Ravetz, *Scientific Knowledge and Its Social Problems* (New York: Oxford University Press, 1971), chap. 11.

11. H. Richard Niebuhr, *Radical Monotheism and Western Culture* (New York: Harper & Brothers, 1960), chap. 6 and essay 4.

12. Michael Polanyi, *Science, Faith and Society* (Chicago: University of Chicago Press, 1946).

13. William Broad and Nicholas Wade, *Betrayers of the Truth: Fraud and Deceit in the Halls of Science* (New York: Simon and Schuster, 1982).

14. Julian Huxley, *Evolutionary Ethics* (London: Oxford University Press, 1943); C. H. Waddington, *The Ethical Animal* (Chicago: University of Chicago Press, 1967).

15. G. G. Simpson, *The Meaning of Evolution* (New Haven: Yale University Press, 1949), chap. 10; Theodosius Dobzhansky, *The Biological Basis of Human Freedom* (New York: Columbia University Press, 1956).

16. George Williams, "Huxley's Evolution and Ethics in Sociobiological Perspective," *Zygon* 23 (1988): 383–408.

17. A. G. N. Flew, *Evolutionary Ethics* (London: Macmillan, 1967); Anthony Quinton, "Evolutionary Ethics," in *Biology and Personality*, ed. Ian Ramsey (Oxford: Basil Blackwell, 1965).

18. Thomas Nagel, "Morality as an Autonomous Theoretical Subject," in *Morality as a Biological Phenomenon*, ed. Gunther Stent (Berkeley and Los Angeles: University of California Press, 1980).

19. Edward O. Wilson, *On Human Nature*, p. 167.

20. See the discussion of sociobiology in Barbour, *Religion in the Age of Science*, chap. 7. Also Richard Alexander, *The Biology of Moral Systems* (New York: Aldine De Gruyter, 1987).

21. Michael Ruse, "Response to Williams: Selfishness Is Not Enough," *Zygon* 23 (1988): 416.
22. Michael Ruse, *Taking Darwin Seriously* (Oxford: Basil Blackwell, 1986), p. 253.
23. Abraham Maslow, *Toward a Psychology of Being*, 2d ed. (Princeton: Van Nostrand, 1968), and *Motivation and Personality* (New York: Harper & Row, 1970).
24. J. S. Mill, *Utilitarianism* (1863; reprint New York: E. P. Dutton, 1914); J. J. C. Smart and Bernard Williams, *Utilitarianism: For and Against* (New York: Cambridge University Press, 1973); Amartya Sen and Bernard Williams, eds., *Utilitarianism and Beyond* (New York: Cambridge University Press, 1984).
25. William Frankena, *Ethics*, 2d ed. (Englewood Cliffs, NJ: Prentice-Hall, 1971).
26. See Frankena, *Ethics*, chap 2.
27. W. D. Ross, *The Right and the Good* (Oxford: Clarendon Press, 1930).
28. Nicholas Rescher, *Distributive Justice*, (Indianapolis: Bobbs-Merrill, 1966); Hugo Bedau, ed., *Justice and Equality*, (Englewood Cliffs, NJ: Prentice-Hall, 1971).
29. Arthur Okun, *Equality and Efficiency* (Washington, DC: Brookings Institute, 1975).
30. John McHale and Magda McHale, *Basic Human Needs* (Houston: University of Houston Press, 1977).
31. John Rawls, *A Theory of Justice*, p. 83.
32. Robert Nozick, *Anarchy, State, and Utopia* (New York: Basic Books, 1974).
33. Brian Barry, *The Liberal Theory of Justice* (Oxford: Oxford University Press, 1973); Norman Daniels, ed., *Reading Rawls: Critical Studies in a Theory of Justice* (New York: Basic Books, 1974). See also Charles R. Beitz, *Political Theory and International Relations* (Princeton: Princeton University Press, 1979).
34. Joel Feinberg, *Social Philosophy* (Englewood Cliffs, NJ: Prentice-Hall, 1973), chap. 1; P. H. Partridge, "Freedom," *Encyclopedia of Philosophy*, ed. Paul Edwards.
35. H. Richard Niebuhr, *The Responsible Self* (New York: Harper & Row, 1963); Edward L. Long, *A Survey of Christian Ethics* (New York: Oxford University Press, 1967), chaps. 3–10, and *A Survey of Recent Christian Ethics* (New York: Oxford University Press, 1982), chaps. 1–3.
36. Alasdair MacIntyre, *After Virtue*, 2d ed. (Notre Dame: University of Notre Dame Press, 1984); Stanley Hauerwas, *Character and the Christian Life* (San Antonio: Trinity University Press, 1975).
37. George F. Thomas, *Christian Ethics and Moral Philosophy* (New York: Charles Scribner's Sons, 1955), chaps. 20–22; L. Harold DeWolf, *Responsible Freedom: Guidelines for Christian Action* (New York: Harper & Row, 1971); Roger Shinn, *Forced Options*, 3d ed. (Cleveland: Pilgrim Press, 1991).
38. Paul Lehmann, *Ethics in a Christian Context* (New York: Harper & Row, 1963), and *The Transfiguration of Politics* (New York: Harper & Row, 1975).
39. H. Richard Niebuhr, *The Responsible Self*; see also James Gustafson, *Ethics in a Theocentric Perspective*, vol. 2 (Chicago: University of Chicago Press, 1984).
40. Alan Richardson, *The Biblical Doctrine of Work* (London: SCM Press, 1952).
41. Martin Luther, *Works* (Frankfurt am Main: Erlanger, 1862–85) 5: 100.
42. John Paul II, "Laborem Exercens," *Origins* 11 (Sept. 24, 1981): 226–44.
43. For example, Paul Abrecht, ed., *Faith and Science in an Unjust World*, vol. 2 (Geneva: World Council of Churches, 1980), chap. 7; Howard Davis and David Gosling, *Will the Future Work?* (Geneva: World Council of Churches, 1985).
44. Among recent books discussing Christian values in relation to technology are Bruce Birch and Larry Rasmussen, *The Predicament of the Prosperous* (Philadelphia: Westminster Press, 1978); David Young, *The Speed of Love: An Exploration of Faithfulness in a Technological World* (New York: Friendship Press, 1986); and Robert Stivers, *Hunger, Technology & Limits to Growth* (Minneapolis: Augsburg Publishing House, 1984).
45. Reinhold Niebuhr, *An Interpretation of Christian Ethics* (New York: Harper & Brothers, 1935), chap. 6; George Thomas, *Christian Ethics and Moral Philosophy*, chap. 11.
46. *Estimation of Human and Financial Resources Devoted to R & D at the World and Regional Level*, UNESCO Publication CSR-S-7, 1979.
47. Reinhold Niebuhr, *The Children of Light and the Children of Darkness* (New York: Charles Scribner's Sons, 1944), p. xi.

48. Max Weber, *The Protestant Ethic and the Spirit of Capitalism,* trans. Talcott Parsons (1930; republished New York: Charles Scribner's Sons, 1958); R. H. Tawney, *Religion and the Rise of Capitalism* (New York: Harcourt Brace & Co., 1926).
49. Donella H. Meadows, et al., *The Limits to Growth* (New York: Universe, 1972).
50. United Nations Development Program, *Human Development Report 1990* (Oxford: Oxford University Press, 1990).
51. Alan Durning, "Ending Poverty," in *State of the World 1990,* ed. Lester Brown (New York: W. W. Norton, 1990), p. 151.
52. Herman E. Daly and John B. Cobb, Jr., *For the Common Good: Redirecting the Economy Toward the Community, the Environment, and a Sustainable Future* (Boston: Beacon Press, 1989). See also Yusuf Ahmad et al., eds., *Environmental Accounting for Sustainable Development* (Washington, DC: World Bank, 1989); David Pearce et al., *Blueprint for a Green Economy* (London: Earthscan Publications, 1989).
53. See Barbour, *Religion in an Age of Science,* pp. 204–9.

Chapter 3. Environmental Values

1. Aldo Leopold, *A Sand County Almanac* (New York: Oxford University Press, 1949), p. 25.
2. John Passmore, *Man's Responsibility for Nature* (New York: Charles Scribner's Sons, 1974), p. 187.
3. Lynn White, Jr., "The Historic Roots of our Ecologic Crisis," *Science* 155 (1967): 1206.
4. E. J. Dijksterhuis, *The Mechanization of the World Picture* (Oxford: Clarendon Press, 1961).
5. See William Leiss, *The Domination of Nature* (New York: George Braziller, 1972).
6. Marjorie Nicolson, *Mountain Gloom and Mountain Glory* (Ithaca: Cornell University Press, 1959); Hans Huth, *Nature and the American* (Berkeley and Los Angeles: University of California Press, 1957), chap. 3.
7. Henry David Thoreau, *Walden* (1854; reprint, New York: W. W. Norton, 1951); Roderick Nash, *Wilderness and the American Mind,* 3d ed. (New Haven: Yale University Press, 1982), chap. 5.
8. Carolyn Merchant, *The Death of Nature: Women, Ecology, and the Scientific Revolution* (New York: Harper & Row, 1979); Susan Griffin, *Woman and Nature* (New York: Harper & Row, 1978); Karen Warren, "Feminism and Ecology: Making Connections," *Environmental Ethics* 9 (1987): 3–20; Judith Plant, ed., *Healing the Wounds: The Promise of Ecofeminism* (Santa Cruz, CA: New Society, 1989).
9. Rachel Carson, *Silent Spring* (Boston: Houghton Mifflin, 1962); Carroll Pursell, ed., *From Conservation to Ecology: The Development of Environmental Concern* (New York: T. Y. Crowell, 1973); Samuel Hays, *Beauty, Health, and Permanence: Environmental Politics in the United States, 1955–1985* (Cambridge: Cambridge University Press, 1987).
10. G. Tyler Miller, *Living in the Environment,* 4th ed. (Belmont, CA: Wadsworth, 1985); Charles H. Southwick, *Global Ecology* (Sunderland, MA: Sinauer Associates, 1985); Eugene Odum, "The Emergence of Ecology as a New Integrative Discipline," *Science* 195 (1977): 1289–93.
11. Donella H. Meadows et al., *The Limits to Growth* (New York: Universe, 1972); Edward Goldsmith et al., *Blueprint for Survival* (Boston: Houghton Mifflin, 1972).
12. Aldo Leopold, *Sand County Almanac,* pp. 224–25.
13. Ibid., Foreword.
14. Holmes Rolston III, "Is There an Ecological Ethic?" *Ethics* 85 (1975): 93–109. Cf. Holmes Rolston III, *Environmental Ethics: Duties to and Values in the Natural World* (Philadelphia: Temple University Press, 1988), p. 232.
15. J. Baird Callicott, "Animal Liberation: A Triangular Affair," *Environmental Ethics* 2 (1980): 319.
16. J. Baird Callicott, "Hume's Is/Ought Dichotomy and the Relation of Ecology to Leopold's Land Ethic," *Environmental Ethics* 4 (1982): 174. Also "Animal Liberation."
17. Bill Devall and George Sessions, *Deep Ecology* (Salt Lake City: Peregrine Smith Books, 1985), p. 67.
18. Robin Attfield, *The Ethics of Environmental Concern* (New York: Columbia University Press, 1983).

19. J. Baird Callicott, "Conceptual Foundations of the Land Ethic," in *Companion to A Sand County Almanac*, ed. Callicott (Madison: University of Wisconsin Press, 1987).

20. Don Marietta, "Ethical Holism and Individuals," *Environmental Ethics* 10 (1988): 251-58.

21. Arne Naess, "The Shallow and the Deep, Long-Range Ecology Movement," *Inquiry* 16 (1973): 95-100; "A Defence of the Deep Ecology Movement," *Environmental Ethics* 6 (1984): 265-70.

22. See Barbour, *Religion in an Age of Science*, chap. 7.

23. See Eric Katz, "Organism, Community, and the 'Substitution Problem,' " *Environmental Ethics* 7 (1985): 241-56.

24. Passmore, *Man's Responsibility for Nature*.

25. William Frankena, "Ethics and the Environment," in *Ethics and Problems of the 21st Century*, ed. K. M. Sayre and K. E. Goodpaster (Notre Dame: University of Notre Dame Press, 1979).

26. Roderick Nash, *Wilderness and the American Mind*, chaps. 3, 4, 5, and 8; Holmes Rolston, *Environmental Ethics*, chap. 1; Max Oelschlaeger, *The Idea of Wilderness: From Prehistory to the Age of Ecology* (New Haven: Yale University Press, 1991).

27. Mark Sagoff, "On Preserving the Natural Environment," *Yale Law Journal* 84 (1974): 205-67, and *The Economy of the Earth: Philosophy, Law, and the Environment* (Cambridge: Cambridge University Press, 1988).

28. Sigurd Olson, "The Spiritual Significance of Wilderness," in *Voices for the Wilderness*, ed. William Schwarz (New York: Ballantine, 1969); Linda Graber, *Wilderness as Sacred Space* (Washington, DC: Association of American Geographers, 1976).

29. Kent Gill, "Preservation and Recreation," *Sierra Club Bulletin* 60 (Feb. 1975): 15.

30. William Godfrey-Smith, "The Value of Wilderness," *Environmental Ethics* 1 (1979): 309-19; David Ehrenfeld, "The Conservation of Non-Resources," *American Scientist* 64 (1976): 648-56. References on endangered species are included in chapter 7 below.

31. See John Black, *The Dominon of Man* (Edinburgh: Edinburgh University Press, 1970), chap. 8.

32. J. J. C. Smart and Bernard Williams, *Utilitarianism: For and Against* (New York: Cambridge University Press, 1973), pp. 62-67.

33. Martin Golding, "Obligations to Future Generations," *Monist* 56 (1972): 85; R. J. Sikora and Brian Barry, eds., *Obligations to Future Generations* (Philadelphia: Temple University Press, 1978); Ernest Partridge, ed., *Responsibilities to Future Generations* (Buffalo: Prometheus Books, 1981).

34. Peter Wenz, "Ethics, Energy Policy and Future Generations," *Environmental Ethics* 5 (1983): 195-210; Derek Parfit, "Energy Policy, and the Further Future: The Social Discount Rate," in *Energy and the Future*, ed. Douglas MacLean and Peter Brown (Totowa, NJ: Rowman and Littlefield, 1983).

35. John Rawls, *A Theory of Justice* (Cambridge: Harvard University Press, 1971); Ronald Green, "Intergenerational Distributive Justice and Environmental Responsibility," *Bioscience* 27 (1977): 260-65; D. Clayton Hubin, "Justice and Future Generations," *Philosophy and Public Affairs* 6 (1976): 70-83.

36. R. Routley and V. Routley, "Nuclear Energy and Obligations to the Future," *Inquiry* 21 (1978): 133-79.

37. Mary B. Williams, "Discounting versus Maximum Sustainable Yield," in *Obligations to Future Generations*, ed. Sikora and Barry.

38. Brian Barry, "Intergenerational Justice in Energy Policy," in *Energy and the Future*, ed. MacLean and Brown. See also Attfield, *Ethics of Environmental Concern*, chap. 10.

39. Talbot Page, *Conservation and Economic Efficiency* (Baltimore: Johns Hopkins University Press, 1977), chap. 9.

40. Brian Barry, "Intergenerational Justice," in *Energy and the Future*, ed. MacLean and Brown.

41. Smart and Williams, *Utilitarianism*, p. 67. On animal rights in general, see Tom Regan and Peter Singer, eds., *Animals Rights and Human Obligations* (Englewood Cliffs, NJ: Prentice-Hall, 1976); Tom Regan, *The Case for Animal Rights* (Berkeley and Los Angeles: University of California Press, 1983).

42. Peter Singer, *Animal Liberation* (New York: New York Review, 1975).

43. Office of Technology Assessment, U.S. Congress, *Alternatives to Animal Use in Research, Testing, and Education* (Washington, DC: GPO, 1986); Committee on the Use of Laboratory Animals

in Biomedical and Behavioral Research, *Use of Laboratory Animals in Biochemical and Behavioral Research* (Washington, DC: National Academy Press, 1988).

44. Kenneth Goodpaster, "On Being Morally Considerable," *Journal of Philosophy* 75 (1978): 308-25.

45. Donald Scherer, "Anthropocentrism, Atomism, and Environmental Ethics," *Environmental Ethics* 4 (1982): 115-24.

46. Attfield, *Ethics of Environmental Concern.*

47. References on process thought are given in chap. 8 of Barbour, *Religion in an Age of Science.*

48. John B. Cobb, Jr., *Is It Too Late?* (Beverly Hills: Bruce, 1972); John B. Cobb, Jr., and David Griffin, *Process Theology* (Philadelphia: Westminster Press, 1976), chap. 9; John B. Cobb, Jr., "Process Theology and Environmental Issues," *Journal of Religion* 60 (1980): 440-58.

49. Charles Birch and John B. Cobb, Jr., *The Liberation of Life* (Cambridge: Cambridge University Press, 1981), chap. 5; Daniel Dombrowski, *Hartshorne and the Metaphysics of Animal Rights* (Albany: State University of New York Press, 1988).

50. Susan Armstrong-Buck, "Whitehead's Metaphysical System as a Foundation for Environmental Ethics," *Environmental Ethics* 8 (1986): 241-69.

51. Jay McDaniel, "Christian Spirituality as Openness to Fellow Creatures," *Environmental Ethics* 8 (1986): 33-46, and *Of God and Pelicans: A Theology of Reverence for Life* (Louisville: Westminster/John Knox Press, 1989).

52. Huston Smith, "Tao Now: An Ecological Testament," in *Earth Might Be Fair,* ed. Ian G. Barbour (Englewood Cliffs, NJ: Prentice-Hall, 1972); Chung-ying Cheng, "On the Environmental Ethics of the Tao and the Ch'i," *Environmental Ethics* 8 (1986): 351-70; Po-keung Ip, "Taoism and the Foundations of Environmental Ethics," in *Religion and Environmental Crisis,* ed. Eugene Hargrove (Athens: University of Georgia Press, 1986); Tu Wei-Ming, "The Continuity of Being," in *Nature in Asian Traditions of Thought,* ed. J. Baird Callicott and Roger Ames (Albany: State University of New York Press, 1989).

53. Joseph Needham, *Science and Civilization in China,* 6 vols. (Cambridge: Cambridge University Press, 1954-1984); Colin Ronan and Joseph Needham, *The Shorter Science and Civilization in China,* 3 vols. (Cambridge: Cambridge University Press, 1978-1986).

54. Yi-Fu Tuan, "Our Treatment of the Environment in Ideal and Actuality," *American Scientist* 58 (1970): 246-49.

55. Leo Orleans and Richard Suttmeier, "The Mao Ethic and Environmental Quality," *Science* 170 (1970): 1173-76.

56. H. Byron Earhart, "The Idea of Nature in Japanese Religon and its Possible Significance for Environmental Concerns," *Contemporary Religions of Japan* 11 (1970): 1-26; Masao Watanabe, "The Conception of Nature in Japanese Culture," *Science* 183 (1974): 279-82.

57. Norie Huddle and Michael Reich, *Island of Dreams: Environmental Crisis in Japan,* rev. ed. (Cambridge, MA: Schenkman Books, 1987); David Vogel, "Environmental Policy in Europe and Japan," in *Environmental Policy in the 1990s,* ed. Norman J. Vig and Michael E. Kraft (Washington, DC: Congressional Quarterly Press, 1990); Ellis Kraus and Bradford Simcock, "Citizen's Movements: The Growth and Impact of Environmental Protest in Japan," in *Political Opposition and Local Politics in Japan,* ed. Kurt Steiner et al. (Princeton: Princeton University Press, 1980).

58. James Whitehill, "Ecological Consciousness and Values: Japanese Perspectives," in *Ecological Consciousness,* ed. Robert Schultz and J. Donald Hughes (Washington, DC: University Press of America, 1981).

59. Lynn White, Jr., "The Historic Roots of Our Ecologic Crisis," *Science* 155 (1967): 1203-07, reprinted with a reply to his critics in *Western Man and Environmental Ethics,* ed. Ian G. Barbour (Reading, MA: Addison-Wesley, 1973). See also David and Eileen Spring, eds., *Ecology and Religion in History* (New York: Harper & Row, 1974).

60. Clarence Glacken, *Traces on the Rhodian Shore* (Berkeley and Los Angeles: University of California Press, 1967); Passmore, *Man's Responsibility for Nature,* chap. 1.

61. Lewis W. Moncrief, "The Cultural Basis of our Environmental Crisis," *Science* 170 (1970): 508-12.

62. Bernhard W. Anderson, "Creation in the Bible," in *Cry of the Environment,* ed. Philip Joranson and Ken Butigan (Sante Fe: Bear, 1984); William Dryness, "Stewardship of the Earth in the

Old Testament," in *Tending the Garden: Essays on the Gospel and the Earth,* ed. Wesley Granberg-Michaelson (Grand Rapids: Eerdmans, 1987).

63. Glacken, *Traces on the Rhodian Shore;* Attfield, *Ethics of Environmental Concern;* H. Paul Santmire, *The Travail of Nature: The Ambiguous Ecological Promise of Christian Theology* (Philadelphia: Fortress Press, 1985).

64. See Barbour, *Religion in an Age of Science,* chap. 9.

65. H. Richard Niebuhr, *The Purpose of the Church and Its Ministry* (New York: Harper & Row, 1956), p. 38.

66. H. Paul Santmire, *Brother Earth* (New York: Thomas Nelson, 1970); Wesley Granberg-Michaelson, *A Worldly Spirituality: The Call to Take Care of the Earth* (San Francisco: Harper & Row, 1984); Dieter Hessel, ed., *For Creation's Sake: Preaching, Ecology, and Justice* (Philadelphia: Geneva Press, 1985); Loren Wilkensen et al., *Earthkeeping: Christian Stewardship of Natural Resources* (Grand Rapids: Eerdmans, 1980); Douglas John Hall, *Imaging God: Dominion as Stewardship* (New York: Friendship Press, 1986).

67. Paulos Mar Gregarios, *The Human Presence: An Orthodox View of Nature* (Geneva: World Council of Churches, 1978); Sean McDonagh, *To Care for the Earth: A Call to a New Theology* (London: Geoffrey Chapman, 1986); John Habgood, "A Sacramental Approach to Enviromental Issues," in *Liberating Life: Contemporary Approaches to Ecological Theology,* eds. Charles Birch, William Eakin, and Jay B. McDaniel (Maryknoll, NY: Orbis Books, 1990).

68. Matthew Fox, *Original Blessing: A Primer in Creation Spirituality* (Sante Fe: Bear, 1983) and *Creation Spirituality: Liberating Gifts for the Peoples of the Earth* (San Francisco: Harper SanFrancisco, 1991).

69. See Barbour, *Religion in an Age of Science,* chap. 8.

70. Paul Abrecht and Roger Shinn, eds., *Faith and Science in an Unjust World,* 2 vols. (Geneva: World Council of Churches, 1980).

71. World Council of Churches, *Gathered for Life* (Geneva: WCC, 1983). A more recent WCC consultation is reported in Birch, Eakin, and McDaniel, eds., *Liberating Life.*

72. National Conference of Catholic Bishops, "Renewing The Earth," *Origins* 21 (1991): 425–32.

73. National Conference of Catholic Bishops, *The Challenge of Peace: God's Promise and Our Response* (Washington, DC: United States Catholic Conference, 1983), pp. 4–5.

Chapter 4. Agriculture

1. Lester Brown, "The New World Order," in *State of the World 1991,* ed. Lester Brown et al. (New York: W. W. Norton, 1991), pp. 11–15.

2. "1991 World Population Data Sheet" (Washington, DC: Population Reference Bureau, 1991).

3. Linda Starke, *Signs of Hope: Working Towards Our Common Future* (Oxford: Oxford University Press, 1990), chap. 8.

4. Oxfam America, "The Causes of Hunger," *Seeds* 11 (Aug. 1988): 18. See also *Hunger 1990: A Report on the State of World Hunger* (Washington, DC: Bread for the World Institute, 1990), p. 8.

5. Erik Eckholm, *Losing Ground* (New York: W. W. Norton, 1976); Sandra Postel, "Halting Land Degradation," in *State of the World 1989,* ed. Lester Brown et al. (New York: W.W. Norton, 1989).

6. Literature in the debate on the effects of the Green Revolution on land ownership in the Third World is cited in section III below.

7. Data from Frances Moore Lappé and Joseph Collins, *Food First* (Boston: Houghton Mifflin, 1977). More recent literature in the debate on export versus food crops is discussed in section III below.

8. Ervin Lazlo, ed., *Goals for Mankind* (New York: E. P. Dutton, 1977), p. 275.

9. Lester Brown and John Young, "Feeding the World in the Nineties," in *State of the World 1990,* ed. Lester Brown et al. (New York: W.W. Norton, 1990), p. 60.

10. Harvey Arden, "Iowa, America's Middle Earth," *National Geographic* 159 (May 1981): 618. See also Pierre Crosson, ed., *The Cropland Crisis: Myth or Reality?* (Baltimore: Johns Hopkins University Press, 1982).

11. Lester Brown and Edward Wolf, *Soil Erosion: The Quiet Crisis in the World Economy* (Washington, DC: Worldwatch Institute, 1984).
12. Sandra Postel, "Halting Land Degradation," in *State of the World 1989;* Lester Brown, "The New World Order," in *State of the World 1991.*
13. Sandra Postel, "Saving Water for Agriculture," in *State of the World 1990.*
14. Lester Brown, "Reexamining the World Food Prospect," in *State of the World 1989.*
15. Wayne Jordan, ed., *Water and Water Policy in World Food Supplies* (College Station: Texas A & M Press, 1987).
16. See Sandra Postel, "Saving Water for Agriculture," in *State of the World 1990.*
17. Kenneth Frederick, "Water Supplies," in *Current Issues in Natural Resource Policy,* ed. Paul Portney (Washington, DC: Resources for the Future, 1982). See also Kenneth Frederick, "Irrigation and the Future of American Agriculture," in *The Future of American Agriculture as a Strategic Resource* (Washington, DC: Conservation Foundation, 1980).
18. National Research Council, *Regulating Pesticides in Food: The Delaney Paradox* (Washington, DC: National Academy Press, 1987), p. 57.
19. World Commission on Environment and Development, *Our Common Future* (Oxford: Oxford University Press, 1987), p. 126.
20. Molly Coye, "The Health Effects of Agricultural Production," in *New Directions for Agriculture and Agricultural Research,* ed. Kenneth Dahlberg (Totowa, NJ: Rowman & Allanheld, 1986).
21. Robert Metcalf, "Benefit/Risk Considerations in the Use of Pesticides," *Agriculture and Human Values* 4 no. 4, (1987): 15–25.
22. Raymond Poincelot, *Toward a More Sustainable Agriculture* (Westport, CT: AVI Publishing, 1986), chap. 2. See also David Pimental and Carl Hall, eds., *Food and Energy Resources* (Orlando: Academic Press, 1984).
23. Richard Kirkendall, "A History of the Family Farm," in *Is There a Moral Obligation to Save the Family Farm?,* ed. Gary Comstock (Ames: Iowa State University Press, 1987).
24. William Heffernan, "Review and Evaluation of Social Externalities," in *New Directions,* ed. Dahlberg, p. 217.
25. Alessandro Bonanno, "Changes, Crisis, and Restructuring in Western Europe: The New Dimensions of Agriculture," *Agriculture and Human Values* 6, no. 1 (1989): 2–10.
26. *Issues and Challenges for OECD Agriculture in the 1980s* (Paris: Organisation for Economic Cooperation and Development, 1984).
27. Thomas Daniels, "A Rationale for the Support of the Medium-Sized Family Farm," *Agriculture and Human Values* 6, no. 4 (1989): 47–53. See also H. Wayne Moyer and Timothy Josling, *Agricultural Policy Reform: Politics and Process in the European Community and the U.S.* (Ames: Iowa State University Press, 1990); E. Wesley Peterson and Clare Lyons, "The Perpetual Agricultural Policy Crisis in the European Community," *Agriculture and Human Values* 6, no. 1 (1989): 11–21.
28. Jim McNeil, "Strategies for Sustainable Economic Development," *Scientific American* 261 (Sept. 1989):159.
29. Office of Technology Assessment, *Technology, Public Policy, and the Changing Structure of American Agriculture* (Washington, DC: OTA, 1986).
30. Daniels, "Medium-Sized Family Farm."
31. Bonanno, "Changes, Crisis, and Restructuring."
32. Ruth Gasson, "Family Farming in Britain," in *Family Farming in Europe and America,* ed. Boguslaw Galeski and Eugene Wilkening (Boulder: Westview Press, 1987).
33. Heffernan, "Review and Evaluation," in *New Directions,* ed. Dahlberg.
34. Alain de Janvry and E. Phillip Le Veen, "Historical Forces that Have Shaped World Agriculture: A Structural Perspective," in *New Directions,* ed. Dahlberg. See also Ingolf Vogeler, *The Myth of the Family Farm: Agribusiness Dominance of U.S. Agriculture* (Boulder: Westview Press, 1981).
35. Howard Newby and Peter Utting, *Agribusiness in Britain* (London: Hutchinson, 1984); Howard Newby and Peter Utting, "Agribusiness in the UK: Social and Political Implications," in *The Social Consequences and Challenges of New Agricultural Technologies,* ed. Gigi Berardi and Charles Geisler (Boulder: Westview Press, 1984).

36. Gregg Easterbrook, "Making Sense of Agriculture: A Revisionist Look at Farm Policy," *The Atlantic* (July 1985), pp. 63–78, reprinted with critiques and Easterbrook's "A Response to My Critics," in *Is There a Moral Obligation*, ed. Comstock.

37. E. Wesley Peterson, "Agricultural Structure and Economic Adjustment," *Agriculture and Human Values* 3, no. 4 (1986): 6–15.

38. Luther Tweeten, "The Economics of Small Farms," *Science* 219 (1983): 1037–1041; also Tweeten, "Has the Family Farm Been Treated Unjustly?" and "Food for People and Profit," in *Is There a Moral Obligation*, ed. Comstock. See also Keith Campbell, *Food for the Future* (Lincoln: University of Nebraska Press, 1979).

39. Wendell Berry, *The Unsettling of America* (San Francisco: Sierra Club, 1977); Berry, *The Gift of Good Land* (San Francisco: North Point Press, 1981).

40. W. Goldschmidt, *As You Sow: Three Studies in the Social Consequences of Agribusiness*, 2d ed. (Montclair, NJ: Allanheld, Osmun, 1978).

41. Bruce LaRose, "Arvin and Dinuba Revisited," Hearing before Subcommittee on Monopoly of the Select Committee on Small Business, U.S. Senate (Washington, DC: GPO, 1973).

42. Philip Martin and Alan Olmstead, "The Agricultural Mechanization Controversy," *Science* 227 (1985): 601–06; Michael Boehlge, "Costs and Benefits of Family Farming," in *Is There a Moral Obligation*, ed. Comstock.

43. Community Services Task Force, *The Family Farm in California* (Sacramento: Small Farm Viability Project, 1977).

44. U.S. Department of Agriculture, *A Time to Choose* (Washington, DC: USDA, 1981). See also Congressional Budget Office, *Diversity in Farming: Its Meaning for Income-Support Policy* (Washington, DC: CBO, 1985).

45. Tweeten, "Economics of Small Farms."

46. Galeski and Wilkening, eds., *Family Farming in Europe and America*.

47. Marty Strange, *Family Farming: A New Economic Vision* (Lincoln: University of Nebraska Press, 1988), chaps. 5 and 7; Alessandro Bonanno, *Small Farms: Persistence with Legitimation* (Boulder: Westview Press, 1987).

48. Linda Lee and William Stewart, "Landownership and the Adoption of Minimum Tillage," *American Journal of Agricultural Economics* 65, no. 2 (1983): 256–64.

49. David Ervin, "Soil Erosion Control on Owner-Operated and Renter Cropland," *Journal of Soil and Water Conservation* 37, no. 5 (1982): 285–88.

50. See chapters by Ulrich Nitsch and Hugues Lamenche in *Family Farming in Europe and America*, ed. Galeski and Wilkening.

51. David Danbom, "Publicly Sponsored Agricultural Research in the United States from an Historical Perspective," in *New Directions*, ed. Dahlberg.

52. Newby and Utting, *Agribusiness in Britain*.

53. Lawrence Busch and William Lacy, *Science, Agriculture, and the Politics of Research* (Boulder: Westview Press, 1983).

54. Billie R. DeWalt, "The Cultural Ecologist Concept of Justice," in *Beyond the Large Farm: Ethics and Research Goals for Agriculture*, ed. Paul B. Thompson and Bill A. Stout (Boulder: Westview Press, 1991).

55. Paul B. Thompson, "Introduction: Values in the Agricultural Laboratory," in *Beyond the Large Farm*, ed. Thompson and Stout; also "Ethical Dilemmas in Agriculture: The Need for Recognition and Resolution," *Agriculture and Human Values* 5, no. 4 (1988), 4–15. See also William Aiken, "On Evaluating Agricultural Research," in *New Directions*, ed. Dahlberg.

56. Frederick Buttel, "The Land-Grant System: A Sociological Perspective on Value Conflicts and Ethical Issues," *Agriculture and Human Values* 2, no. 2 (1985), 78–95.

57. Vernon Ruttan, *Agricultural Research Policy* (Minneapolis: University of Minnesota Press, 1982); "Moral Responsibility in Agricultural Research," in *Beyond the Large Farm*, ed. Thompson and Stout. See also Vernon Ruttan and Carl Pray, eds., *Policy for Agricultural Research* (Boulder: Westview Press, 1987).

58. James Hightower, *Hard Tomatoes, Hard Times* (Cambridge, MA: Schenkman, 1973); Don Hadwiger, *The Politics of Agricultural Research* (Lincoln: University of Nebraska Press, 1982).

59. Jack Doyle, *Altered Harvest: Agriculture, Genetics, and the Fate of the World's Food Supply* (New York: Viking Penguin, 1986); R. Van der Bosch, *The Pesticide Conspiracy* (Garden City, NY: Doubleday, 1978).

60. John Reganold et al., "Sustainable Agriculture," *Scientific American* 262 (June 1990): 112–21.

61. Gigi Berardi and Charles Geisler, eds., *The Social Consequences and Challenges of New Agricultural Technologies*.

62. Hightower, *Hard Tomatoes, Hard Times.*

63. K. Bishop, "California U. Told to Change Research to Aid Small Farms," *New York Times*, Nov. 19, 1987, p. 13A.

64. John Vandermeer, "Mechanized Agriculture and Social Welfare: The Tomato Harvester in Ohio," *Agriculture and Human Values* 3, no. 3 (1986): 21–25; Peter Rosset and John Vandermeer, "The Confrontation between Processor and Farm Worker in the Midwest Tomato Industry: The Role of the Agricultural Research and Extension Establishment," *Agriculture and Human Values* 3, no. 3 (1986): 26–32.

65. Vernon Ruttan, "Moral Responsibility in Agricultural Research," in *New Directions*, ed. Dahlberg.

66. Martin and Olmstead, "Agricultural Mechanization Controversy."

67. William Lockeretz, George Shearer, and Daniel Kohl, "Organic Farming in the Corn Belt," *Science* 211 (1981): 240–46.

68. See Pieter Vereijken, "Research on Integrated Arable Farming and Organic Mixed Farming in the Netherlands," in *Sustainable Agricultural Systems*, ed. Clive Edwards et al. (Ankerry, IA: Soil and Water Conservation Service, 1990); Richard Brody, *Red or Green for Farmers (and the Rest of Us)* (Saffron Walden, England: Broad Leys Publishing, 1987).

69. John Reganold et al., "Sustainable Agriculture"; Raymond Poincelot, *Toward a More Sustainable Agriculture.*

70. Frederick Troeh and Arthur Hobbs, *Soil and Water Conservation* (Englewood Cliffs, NJ: Prentice-Hall, 1980); Charles Little, *Green Fields Forever: The Conservation Tillage Revolution in America* (Washington, DC: Island Press, 1987).

71. Charles Benbrook, "Society's Stake in Sustainable Agriculture," in *Sustainable Agricultural Systems*, ed. Clive Edwards et al., p. 73.

72. L. E. Ehler, "Revitalizing Biological Control," *Issues in Science and Technology* 7, no. 1 (1990): 91–96. See also John Perkins, *Insects, Experts, and the Insecticide Crisis: The Quest for New Pest Management Strategies* (New York: Plenum Press, 1987).

73. W. A. Allen et al., *The National Evaluation of Extension's Integrated Pest Management (IPM) Programs* (Blacksburg, VA: Virginia Cooperative Extension Services, 1987), cited in *Sustainable Agricultural Systems*, ed. Edwards et al., p. 468.

74. *Minnesota Food Association Digest* (Oct. 1991), p. 6.

75. William Lockeretz et al., "Organic Farming in the Corn Belt." See also William Lockeretz, ed., *Environmentally Sound Agriculture* (New York: Praeger Publishers, 1983).

76. J. Patrick Madden and Thomas Dobbs, "The Role of Economics in Achieving Low-input Farming Systems," in *Sustainable Agricultural Systems*, ed. Edwards et al.

77. A. Vine and D. Bateman, *Organic Farming Systems in England and Wales: Practice, Performance, and Implications* (Aberystwyth: University College of Wales, 1981); A. El Titi and H. Landes, "Integrated Farming System of Lautenbach: A Practical Contribution toward Sustainable Agriculture in Europe," in *Sustainable Agricultural Systems*, ed. Edwards et al.; Pieter Vereijken, "Alternative Farming Systems in Nagele: Preliminary Results and Prospects," in *Sustainable Agriculture and Integrated Farming Systems*, ed. Thomas Edens, Cynthia Fridgen, and Susan Battenfield (E. Lansing: Michigan State University Press, 1985).

78. Board of Agriculture, National Research Council, *Alternative Agriculture* (Washington, DC: National Academy Press, 1989), p. 6.

79. U.S. Department of Agriculture, *A Time to Choose.*

80. John Reganold et al., "Sustainable Agriculture"; J. F. Parr et al., "Sustainable Agriculture in the United States," in *Sustainable Agricultural Systems*, ed. Edwards et al.

81. *Farm Program Options Guide* (Walthill, NE: Sustainable Agriculture Working Group, 1991).

82. Organization for Economic Co-operation and Development, *Economic Instruments for Environmental Protection* (Paris: Organization for Economic Co-operation and Development, 1989).

83. Miguel Altieri, *Agroecology: The Scientific Basis of Alternative Agriculture* (Boulder: Westview Press, 1987); C. Ronald Carroll, John H. Vandermeer, and Peter Rosset, eds., *Agroecology* (New York: McGraw-Hill, 1990).

84. Walter Brueggemann, *The Land* (Phildelphia: Fortress Press, 1977); Bernard Evans and Gregory Cusack, eds., *Theology of the Land* (Collegeville, MN: Liturgical Press, 1987); Charles Lutz, ed., *Farming the Lord's Land: Christian Perspectives on American Agriculture* (Minneapolis: Augsburg Publishing House, 1980).

85. Michael W. Fox, *Inhumane Society: The American Way of Exploiting Animals* (New York: St. Martin's Press, 1990).

86. Peter Singer, *Animal Liberation* (New York: New York Review, 1975); Andrew Linzey, *Christianity and the Rights of Animals* (New York: Crossroad, 1987); Andrew Linzey and Tom Regan, eds., *Animals and Christianity* (New York: Crossroad, 1988).

87. Tom Regan, *The Case for Animal Rights* (Berkeley and Los Angeles: University of California Press, 1983).

88. A similar balance between animal welfare and basic human needs is defended within a philosophy of ecohumanism by William Aiken, "Ethical Issues in Agriculture," in *Earthbound,* ed. Tom Regan (New York: Random House, 1984).

89. Lester Brown in *State of the World, 1989,* p. 187. Feedlot cattle consume 10 pounds of grain for every pound of meat produced, but fodder-fed and grass-fed cattle consume less, so the ratio of grain to meat depends on the time spent in feedlots; 7 pounds represents an average figure for the U.S. See Lester Brown with Erik Eckholm, *By Bread Alone* (New York: Praeger Publishers, 1974), p. 205.

90. Sterling Wortman and Ralph Cummings, *To Feed the World* (Baltimore: Johns Hopkins University Press, 1978).

91. Keith Griffin, *Land Concentration and Rural Poverty* (New York: Holmes and Meier, 1976); Kenneth Dahlberg, *Beyond the Green Revolution: The Ecology and Politics of Global Agricultural Development* (New York: Plenum Press, 1979); Andrew Pearse, *Seeds of Plenty, Seeds of Want* (Oxford: Clarendon Press, 1980).

92. Michael Lipton and Richard Longhurst, *New Seeds and Poor People* (Baltimore: Johns Hopkins University Press, 1989). See also A. K. Sen, *Poverty and Famines* (Oxford: Clarendon Press, 1981).

93. Lappé and Collins, *Food First.* See also Alain de Janvry, The *Agrarian Question and Reformism in Latin America* (Baltimore: Johns Hopkins University Press, 1983).

94. John Mellor and Frank Riely, "Expanding the Green Revolution," *Issues in Science and Technology* 6, no. 1 (1989): 66–67.

95. Simon Maxwell and Adrian Fernando, "Cash Crops in Developing Countries: The Issues, the Facts, the Policies," *World Development* 17 (1989): 1677–1708. See also Joachim Von Braun, *Commercialization of Subsistence Agriculture: Income and Nutritional Effects in Developing Countries* (Washington, DC: International Food Policy Research Institute, 1986).

96. David Barkin, Rosemary Batt, and Billie deWalt, *Food Crops versus Feed Crops: Global Substitution of Grains in Production* (Boulder: Lynne Rienner Publishers, 1990); Robert G. Williams, *Export Agriculture and the Crisis in Central America* (Chapel Hill: University of North Carolina Press, 1986).

97. See chapters by Ashok Rudra and by Stephen Biggs and Jon Griffith in *Macropolicies for Appropriate Technology in Developing Countries,* ed. Frances Stewart (Boulder: Westview Press, 1987); also Edward C. Wolf, *Reversing Africa's Decline* and *Beyond the Green Revolution: New Approaches to Third World Agriculture* (Washington, DC: Worldwatch Institute, 1985 and 1986).

98. See section II of Marilyn Carr, ed., *The AT Reader: Theory and Practice in Appropriate Technology* (New York: Intermediate Technology Development Group of North America, 1985). Appropriate technology is discussed in chapter 9 below.

99. William Thiesenhusen, ed., *Searching for Agrarian Reform in Latin America* (Boston: Unwin Hyman, 1989).

100. William Rich, *Smaller Families through Social and Economic Progress* (Washington, DC: Overseas Development Council, 1973), p. 32.

101. Arthur Simon, "The Basic Cause: Poverty," in *The Causes of World Hunger,* ed. William Byron (New York: Paulist Press, 1982).

102. Mellor and Riely, "Expanding the Green Revolution."

103. Norman Borlaug, "Accelerating Agricultural Research and Production in the Third World," *Agriculture and Human Values* 3, no. 3 (1986): 5–14.

104. Sandra Postel and Christopher Flavin, "Reshaping the Global Economy," in *State of the World 1991,* pp. 171 and 176.

105. W. W. Rostow, *The Stages of Economic Growth,* 2d ed. (Cambridge: Cambridge University Press, 1971); Charles Wilber, ed., *The Political Economy of Development and Underdevelopment* (New York: Random House, 1973).

106. Charles Elliott, *Patterns of Poverty in the Third World* (New York: Praeger Publishers, 1975); Edgar Owens and Robert Shaw, *Development Reconsidered* (Lexington, MA: Lexington, 1972).

107. Michael Lipton, *Why Poor People Stay Poor* (Cambridge: Harvard University Press, 1977).

108. Hollis Chenery et al., *Redistribution with Growth* (Oxford: Oxford University Press, 1974).

109. Irma Adelman and Cynthia Taft Morris, *Economic Growth and Social Equity in Developing Countries* (Stanford: Stanford University Press, 1973).

110. Dennis Goulet, *The Uncertain Promise: Value Conflicts in Technology Transfer* (Washington, DC: Overseas Development Council, 1977), chap. 7.

111. Robert McNamara, "The Third World: Millions Face Risk of Death," *Vital Speeches* 41 (October 15, 1974): 13–20.

112. Mahbub ul Haq, *The Poverty Curtain: Choices for the Third World* (New York: Columbia University Press, 1976).

113. International Labor Organization, *Employment, Growth, and Basic Needs: A One-World Problem* (New York: Praeger Publishers, 1977).

114. Guy Erb and Valeriana Kallab, eds., *Beyond Dependency: The Developing World Speaks Out* (New York: Praeger Publishers, 1975).

115. J. D. Cockroft et al., eds., *Dependence and Underdevelopment* (Garden City, NY: Doubleday, 1972); Dennis Goulet, *The Cruel Choice: A New Concept in the Theory of Development* (New York: Atheneum, 1971).

116. Seamus Cleary, *Renewing the Earth: Development for a Sustainable Future* (London: Catholic Fund for Overseas Development, 1989); World Bank, *The World Bank and the Environment* (Washington, DC, 1990); John P. Lewis and Valeriana Kallab, eds., *Development Strategies Reconsidered* (New Brunswick: Transaction Books, 1986).

117. World Commission on Environment and Development, *Our Common Future* (Oxford: Oxford University Press, 1987), p. 40.

118. World Commission, *Our Common Future,* chap. 5. Also Report to the World Comission on Environment and Development, *Food 2000: Global Policies for Sustainable Agriculture* (London: Zed Books, 1987). See also W. D. Adams, *Green Development: Environment and Sustainability in the Third World* (London: Routledge, 1990); David Pearce, *Sustainable Development: Economics and Environment in the Third World* (Aldershot, England: E. Elgar Publishing, 1990).

119. UNICEF, *State of the World's Children 1992* (Paris: UNICEF, 1992).

120. Ruth Sivard, *World Military and Social Expenditures 1991,* p. 5.

121. Cf. Richard A. Watson, "Reason and Morality in a World of Limited Food," and Peter Singer, "Famine, Affluence, and Morality," in *World Hunger and Moral Obligation,* ed. William Aiken and Hugh LaFollette (Englewood Cliffs, NJ: Prentice-Hall, 1977).

122. Onora O'Neill, *Faces of Hunger* (London: George Allen & Unwin, 1986).

123. Nigel Dower, *World Poverty: Challenge and Response* (York, England: Ebor Press, 1983).

124. John Arthur, "Rights and the Duty to Bring Aid," in *World Hunger and Moral Obligation,* ed. Aiken and LaFollete; see also Peter Brown, "Food as National Policy," in *Food Policy,* ed. Peter Brown and Henry Shue (New York: Free Press, 1977).

125. William McNeill, *The Pursuit of Power: Technology, Armed Force, and Society Since A.D. 1000* (Chicago: University of Chicago Press, 1982); Dennis Pirages, *Global Technopolitics: The International Politics of Technology and Resources* (Pacific Grove, CA: Brooks/Cole Publishing, 1989), pp. 16–18, 43–47, 142–50.

126. Charles R. Beitz, *Political Theory and International Relations* (Princeton: Princeton University Press, 1979).

127. Ronald Sider, *Rich Christians in an Age of Hunger* (New York: Paulist Press, 1977); Bruce Birch and Larry Rasmussen, *The Predicament of the Prosperous* (Philadelphia: Westminster Press, 1978); Arthur Simon, *Bread for the World*, rev. ed. (New York: Paulist Press, 1984).
128. Sivard, *World Military and Social Expenditures 1991*, p. 9.
129. Brown, in *State of the World 1992*, p. 7.
130. Independent Commission on International Development Issues (Willy Brandt, chair), *North-South: A Program for Survival* (Cambridge: MIT Press, 1980).
131. Brown and Shue, eds., *Food Policy*, p. 2.
132. Brown, in *State of the World 1989*, p. 187.

Chapter 5. Energy

1. World Commission on Environment and Development, *Our Common Future* (Oxford: Oxford University Press, 1987), p. 14.
2. *World Development Report* (Washington, DC: World Bank, 1981).
3. U.S. Geological Survey estimates from C. D. Masters et al., "Resource Constraints in Petroleum Production Potential," *Science* 253 (1991): 146–152.
4. Daniel Yergin, *The Prize: The Epic Quest for Oil, Money, and Power* (New York: Simon and Schuster, 1991).
5. Carl Berhens and Robert Bamburger, "Why Can't We Have a Real Energy Policy?" *Congressional Research Service Review* (March-April 1991), p. 2.
6. Energy Information Administration, "Annual Energy Outlook 1991" (Washington, DC: Department of Energy, 1991).
7. Joseph Romm, "Needed: A No-regrets Energy Policy," *Bulletin of the Atomic Scientists* 47, no.6 (1991): 31; see also Harold Hubbard, "The Real Cost of Energy," *Scientific American* 264, no.4 (1991): 36–42.
8. Jose Goldemberg et al., *Energy for Development* (Washington, DC: World Resources Institute, 1987), chap. 1.
9. Yergin, *Prize*, p. 13.
10. "Auto Pollution Health Costs Calculated," *Washington Post*, Jan. 21, 1990. See also James Cannon, *The Health Costs of Air Pollution* (New York: American Lung Association, 1985).
11. Carl Bagge, "Coal: Meeting the Energy Challenge," in *Perspectives on Energy*, 2d ed., ed. Lon Ruedisili and Morris Firebaugh (New York: Oxford University Press, 1978).
12. National Academy of Sciences, *Surface Mining: Soil, Coal, and Society* (Washington, DC: National Academy Press, 1981); Walter Rosenbaum, *Environmental Politics and Policy*, 2d ed. (Washington, DC: Congressional Quarterly Press, 1991), pp. 261–66.
13. Hilary French, "Clearing the Air," in *State of the World 1990*, ed. Lester Brown (New York: W. W. Norton, 1990).
14. Sandra Postel, "Protecting Forests from Air Pollution and Acid Rain," *State of the World 1985*, ed. Lester Brown (New York: W. W. Norton, 1985); William Fulkerson et al., "Energy for Fossil Fuels," *Scientific American* 263, no. 3 (1990): 82–89.
15. Rosenbaum, *Environmental Politics*, chap. 6; "A Decade's Acrimony Lifted in the Good of Clean Air," *Congressional Quarterly*, Oct. 27, 1990, 3587–92.
16. Christopher Flavin, "Slowing Global Warming," in *State of the World 1990*.
17. Stephen Schneider, *Global Warming: Are We Entering the Greenhouse Century?* (San Francisco: Sierra Club Books, 1989); Schneider, "The Changing Climate," *Scientific American* 261, no. 9 (1989):70–79. An estimate of global warming in the range of 2 to 9° F is given in National Academy of Sciences, *Policy Implications of Greenhouse Warming: A Synthesis* (Washington, DC: National Academy Press, 1991).
18. For example, the U.N.'s Intergovernmental Panel on Climate Change estimates that global warming would reduce crop yields in the world's main grain exporting regions by 20 percent; see Martin Parry, *Climate Change and World Agriculture* (London: Earthscan, 1990).
19. National Academy of Sciences, *Policy Implications of Greenhouse Warming*.

20. Christopher Flavin, "Building a Bridge to Sustainable Energy," in *State of the World 1992,* ed. Lester Brown (New York: W. W. Norton, 1992); Walter Vergara et al., *Natural Gas: Its Role and Potential in Economic Development* (Boulder: Westview Press, 1990).

21. Christopher Flavin and Nicholas Lenssen, *Beyond the Petroleum Age: Designing a Solar Economy* (Washington, DC: Worldwatch Institute, 1990); Michael Oppenheimer and Robert Boyle, *Dead Heat: The Race Against the Greenhouse Effect* (New York: Basic Books, 1990).

22. Terrence Price, *Political Electricity: What Future for Nuclear Energy?* (Oxford: Oxford University Press, 1990), p. 4.

23. Steven Del Sesto, *Science, Politics, and Controversy: Civilian Nuclear Power in the United States, 1946–1974* (Boulder: Westview Press, 1979).

24. Joseph Morone and Edward Woodhouse, *The Demise of Nuclear Energy? Lessons for Democratic Control of Technology* (New Haven: Yale University Press, 1989); John L. Campbell, *Collapse of an Industry: Nuclear Power and the Contradictions of U.S. Policy* (Ithaca: Cornell University Press, 1988).

25. David Collingridge, *Technology in the Policy Process: Controlling Nuclear Power* (New York: St. Martin's Press, 1983).

26. Elizabeth Rolph, *Nuclear Power and the Public Safety: A Study in Regulation* (Lexington, MA: Lexington, 1989).

27. Christopher Hohenemser, "The Accident at Chernobyl: Health and Environmental Consequences and Implications for Risk Managment," *Annual Review of Energy* 13 (1988): 383–428; see also Viktor Haynes and Marko Bojcun, *The Chernobyl Disaster* (London: Hogarth Press, 1988).

28. Charles Perrow, *Normal Accidents: Living with High-Risk Technologies* (New York: Basic Books, 1984), chap. 2.

29. Baruch Fishoff et al., *Acceptable Risk* (Cambridge: Cambridge University Press, 1981); Paul Slovic and Baruch Fishoff, "Perception and Acceptability of Risk," in *Public Reaction to Nuclear Power: Are There Critical Masses?* (Boulder: Westview Press, 1984).

30. Sandia National Laboratory, "Estimates of the Financial Consequences of Nuclear Power Reactor Accidents" (Washington, DC: Nuclear Regulatory Commission, 1982), cited in Christopher Flavin, "Reassessing Nuclear Power," in *State of the World 1987,* ed. Brown, p. 78.

31. Nicholas Lenssen, "Confronting Nuclear Waste," in *State of the World 1992,* ed. Brown.

32. Gerald Jacob, *Site Unseen: The Politics of Siting a Nuclear Waste Respository* (Pittsburgh: University of Pittsburgh Press, 1990); Charles Walker and Leroy Gould, eds., *Too Hot to Handle: Social and Policy Issues in the Management of Radioactive Wastes* (New Haven: Yale University Press, 1983).

33. Alvin Weinberg, "Social Institutions and Nuclear Energy," *Science* 177 (1972): 27–34.

34. Office of Technology Assessment, *Managing Commercial High-Level Radioactive Waste* (Washington, DC: OTA, 1982).

35. Michael Kraft, "Evaluating Technology through Public Participation: The Nuclear Waste Disposal Controversy," in *Technology and Politics,* ed. Michael Kraft and Norman Vig (Durham: Duke University Press, 1988).

36. Office of Technology Assessment, *Nuclear Power in an Age of Uncertainty* (Washington, DC: OTA, 1984).

37. Bernard Cohen, *The Nuclear Energy Option: An Alternative for the 90s* (New York: Plenum Press, 1990), chap. 10.

38. John Taylor, "Improved and Safer Nuclear Power," *Science* 244 (1989): 318–25; Alvin Weinberg, "Engineering in an Age of Anxiety," *Issues in Science and Technology* 6, no.2 (1989–90): 37–43; Union of Concerned Scientists, *Advanced Reactor Report* (Cambridge, MA: UCS, 1990).

39. Morone and Woodhouse, *Demise of Nuclear Power?*

40. Rosenbaum, *Environmental Politics,* chap. 8.

41. "The Death of the Costliest Delusion," *The Guardian* (London), Nov. 10, 1989.

42. Frederick Williams and David Deese, eds., *Nuclear Nonproliferation: The Spent Fuel Problem* (New York: Pergamon Press, 1980); D. Albright and H. A. Feiveson, "Plutonium Recycling and the Problem of Nuclear Proliferation," *Annual Review of Energy* 13 (1988): 239–66.

43. Robert Williams and Harold Feiveson, "How to Expand Nuclear Power without Proliferation," *Bulletin of the Atomic Scientists* 47, no.3 (1990): 40–45.

44. *Star Tribune* (Minneapolis), Nov. 11, 1991, p. 1A.

45. J. P. Holdren, "Safety and Environmental Aspects of Fusion Energy," *Annual Review of Energy and The Environment* 16 (1991): 235–400.

46. Cynthia Pollock Shea, "Shifting to Renewable Energy," in *State of the World 1988,* ed. Lester Brown (New York: W. W. Norton, 1988).

47. Rick Bevington and Arthur Rosenfeld, "Energy for Buildings and Homes," *Scientific American* 263, no.3 (1990): 38–47; Solar Technical Information Program, *Energy for Today: Renewable Energy* (Golden, CO: Solar Energy Research Institute, 1990).

48. Michael Renner, "Rethinking Transportation," in *State of the World 1989,* p. 102. See also H. S. Geller, "Ethanol from Sugar Cane in Brazil," *Annual Review of Energy* 10 (1985): 135–64; Daniel Sperling and Mark DeLuchi, "Transportation Energy Futures," *Annual Review of Energy* 14 (1989): 375–424.

49. World Commission on Environment and Development, *Energy 2000: A Global Strategy for Sustainable Development* (London: Zed Books, 1987).

50. Michael Brower, *Cool Energy* (Cambridge, MA: Union of Concerned Scientists, 1990), chap. 4; Paul Gipe, "Wind Power Around the World," *Alternative Sources of Energy* (May 1988), pp. 38–45.

51. Brower, *Cool Energy,* chap 3; Carl Weinberg and Robert Williams, "Energy from the Sun," *Scientific American* 263, no.3 (1990): 98–107.

52. H. M. Hubbard, "Photovoltaics Today and Tomorrow," *Science* 244 (1989): 297–303; D. E. Carlson, "Photovoltaic Technologies for Commercial Power Generation," *Annual Review of Energy* 15 (1990): 85–98; Ken Zweibel, *Harnessing Solar Power: The Photovoltaics Challenge* (New York: Plenum Press, 1990).

53. Daniel Sperling and Mark DeLuchi, "Transportation Energy Futures," *Annual Review of Energy* 14 (1989): 375–424.

54. Joan Ogden and Robert Williams, *Solar Hydrogen: Moving Beyond Fossil Fuels* (Washington, DC: World Resources Institute, 1989).

55. Brower, *Cool Energy,* p. 23.

56. Olav Hohmeyer, *Social Costs of Energy Consumption* (Berlin: Springer-Verlag, 1988).

57. Susan Williams, Scott Fenn, and Terry Clausen, "Renewing Renewable Energy," *Issues in Science and Technology* 6, no.2 (1990): 64–70.

58. Richard Heede et al., *The Hidden Costs of Energy* (Washington, DC: Center for Renewable Resources, 1985), cited in Flavin and Lenssen, *Beyond the Petroleum Age,* p. 27.

59. International Energy Agency, *Energy Politics and Programmes of IEA Countries: 1989 Review* (Paris: Organisation for Economic Co-operation and Development, 1990).

60. Amory Lovins, *Soft Energy Paths* (Cambridge, MA: Ballinger/Friends of the Earth, 1977); Flavin and Lenssen, *Beyond the Petroleum Age,* pp. 40–43.

61. K. S. Shrader-Frechette, "Ethics and Energy," in *Earthbound,* ed. Tom Regan (New York: Random House, 1984).

62. On the pros and cons of centralization and decentralization, see Ian Barbour, Harvey Brooks, Sanford Lakoff, and John Opie, *Energy and American Values* (New York: Praeger Publishers, 1982), pp. 72–74 and 130–34.

63. Lee Schipper, Richard Howarth, and Howard Geller, "U.S. Energy Use from 1973 to 1987: Impacts on Improved Efficiency," *Annual Review of Energy* 15 (1990): 457.

64. International Energy Agency, *Energy Policies and Programmes: 1989.*

65. Michael Renner, "Rethinking Transportation," in *State of the World 1989,* p. 103.

66. John Gibbons, Peter Blair, and Holly Gwin, "Strategies for Energy Use," *Scientific American* 261 (Sept. 1989): 136–43.

67. Deborah Gordon, *Steering a New Course: Transportation, Energy, and the Environment* (Cambridge, MA: Union of Concerned Scientists, 1991); Deborah Bleviss and Peter Walzer, "Energy for Motor Vehicles," *Scientific American* 263 (Sept. 1990): 54–61.

68. Marcia Lowe, "Rethinking Urban Transport," in *State of the World 1991.*

69. Christopher Flavin and Alan Durning, "Raising Energy Efficiency," in *State of the World 1988;* Rick Bevington and Arthur Rosenfeld, "Energy for Buildings and Homes," *Scientific American* 263, no. 3 (1990) 38–45.

70. Robert Williams, Eric Larson, and Marc Ross, "Materials, Affluence, and Industrial Energy Use," *Annual Review of Energy* 12 (1987): 99–144.

71. Pietro Nivola, *The Politics of Energy Conservation* (Washington, DC: Brookings Institution, 1986), chap. 4; Arnold Fickett, Clark Gellings, and Amory Lovins, "Efficient Use of Electricy," *Scientific American* 263, no. 3 (1990): 28–37.
72. Olav Hohmeyer et al., *Employment Effects of Energy Conservation Investments in EC Countries* (Luxembourg: European Communities Publications, 1985), cited in Flavin and Lenssen, *Beyond the Petroleum Age*, p. 41.
73. Steven Buchsbaum and James Benson, *Jobs and Energy* (New York: Council on Economic Priorities, 1979); Frederick Buttell, "Labor's Stake in Environmental Quality and Energy Policy," in *Labor and the Environment*, ed. F. Buttell (Westport CT: Greenwood Press, 1984).
74. Robert Paehlke, *Environmentalism and the Future of Politics* (New Haven: Yale University Press, 1989), p. 274. See also Denton Morrison, "Equity Impacts of Some Major Energy Alternatives," in *Energy Policy in the United States: Social and Behavioral Dimensions*, ed. Seymour Warkov (New York: Praeger Publishers, 1978).
75. Henry Hurwitz, "Indoor Air Pollution," *Bulletin of the Atomic Scientists* 37 (Feb. 1981): 61–62, replies on pp. 62–64.
76. Joy Dunkerley, ed., *International Comparisons of Energy Consumption* (Baltimore: Johns Hopkins University Press, 1978).
77. Lee Schipper and Allen Lichtenberger, "Efficient Energy Use and Well-being: The Swedish Example," *Science* 194 (1976): 1001–13.
78. See Warkov, ed., *Energy Policy*.
79. Laura Nader and Stephen Beckerman, "Energy as It Relates to Quality and Style of Life," *Annual Review of Energy* 3 (1978): 11–28; Richard Carlson, Willis Harman and Peter Schwartz, *Energy Futures, Human Values, and Lifestyles* (Boulder: Westview Press, 1982).
80. Chrisopher Flavin, "Electrifying the Third World," in *State of the World 1987;* Amulya Reddy and Jose Goldemberg, "Energy for the Developing World," *Scientific American* 263, no.3 (1990): 62–73.
81. World Commission on Environment and Development, *Our Common Future* and *Energy 2000*.
82. Jose Goldemberg et al., *Energy for Development*.
83. Ibid., p. 18.
84. For example, T. J. Wilbanks, "Implementing Environmentally Sound Power Sector Strategies in Developing Countries," *Annual Review of Energy* 15 (1990): 255–76.

Chapter 6. Computers

1. John Naisbitt, *Megatrends* (New York: Warner Books, 1982); Alvin Toffler, *The Third Wave* (New York: Bantam Books, 1980); Pamela McCorduck, *The Universal Machine: Confessions of a Technological Optimist* (New York: McGraw-Hill, 1985); Edward Feigenbaum and Pamela McCorduck, *The Fifth Generation* (Reading, MA: Addison-Wesley, 1983); Donald Michie, *The Knowledge Machine: Artificial Intelligence and the Future of Man* (New York: William Morrow, 1985); Raymond Kurzweil, *The Age of Intelligent Machines* (Cambridge: MIT Press, 1990).
2. Ian Reinecke, *Electronic Illusions: A Skeptic's View of our High-Tech Future* (New York: Penguin Books, 1984); Frank Webster and Kevin Robins, *Information Technology: Post-Industrial Society or Capitalist Control?* (Norwood, NJ: Ablex Publishing, 1986); Michael Shallis, *The Silicon Idol* (Oxford: Oxford University Press, 1984); Joseph Weizenbaum, *Computer Power and Human Reason* (San Francisco: W. H. Freeman, 1976).
3. Christopher Rowe, *People and Chips* (London: Paradigm Publications, 1986); David Lyon, *The Information Society: Issues and Illusions* (Oxford: Basil Blackwell, 1988).
4. Office of Technology Assessment, *Computerized Manufacturing Automation: Employment, Education, and the Workplace* (Washington, DC: OTA, 1984).
5. Stephen Peitchinis, *Computer Technology and Employment* (New York: St. Martin's Press, 1983); National Academy of Sciences, *Technology and Employment* (Washington, DC: National Academy Press, 1987).
6. David R. Howell, "The Future Employment Impacts of Industrial Robots," *Technological Forecasting and Social Change* 28 (1985): 297–310.

7. James Robertson, *Future Work: Jobs, Self-Employment, and Leisure after the Industrial Age* (New York: Universe Books, 1985). See also Robert C. Paehlke, *Environmentalism and the Future of Progressive Politics* (New Haven: Yale University Press, 1989), pp. 255–60.

8. Harry Braverman, *Labor and Monopoly Capital: The Degradation of Labor in the 20th Century* (New York: Monthly Review Press, 1974).

9. Harley Shaiken, *Work Transformed: Automation and Labor in the Computer Age* (Lexington, MA: D. C. Heath, 1986).

10. David Noble, *Forces of Production: A Social History of Industrial Automation* (New York: Knopf, 1984).

11. David Lyon, *The Information Society*, chap. 4; John Child, "New Technology and the Labour Process," in *Information Technology: Social Issues*, ed. Ruth Finnegan, Graeme Salaman, and Kenneth Thompson (Seven Oaks, Kent, England: Hodder and Stoughton, 1987).

12. Bryn Jones, "Destruction or Redistribution of Engineering Skills? The Case of Numerical Control," in *The Degradation of Work? Skill, Deskilling, and the Labour Process*, ed. Stephen Wood (London: Hutchinson, 1983).

13. Richard Walton and Gerald Susman, "People Policies for the New Machines," *Harvard Business Review* 65, no. 2 (1987): 98–106.

14. Ian Benson and John Lloyd, *New Technology and Industrial Change* (London: Kegan Paul, 1983), pp. 179–83; John Hoerr, Michael Polluck, and David Whiteside, "Management Discovers the Human Side of Automation," in *Computers in the Human Context*, ed. Tom Forester (Cambridge: MIT Press, 1989).

15. Colin Gill, *Work, Unemployment, and the New Technology* (London and New York: Basil Blackwell, 1985), chap. 6; Edmund Byrne, "Microelectronics and Worker's Rights," *in Philosophy and Technology II: Information Technology and Computers in Theory and Practice*, ed. Carl Mitcham and Alois Huning (Dordrecht, Holland and Boston: D. Reidel, 1986).

16. Colin Gill, *Work, Unemployment*, chap. 3.

17. Larry Hirschorn, *Beyond Mechanization: Work and Technology in the Post-Industrial Age* (Cambridge: MIT Press, 1984); National Academy of Sciences, *Human Resource Practices for Implementing Advanced Manufacturing Technology* (Washington, DC: National Academy Press, 1986); Jerome M. Rosow, ed., *Teamwork: Joint Labor-Management Programs in America* (New York: Pergamon Press, 1986).

18. Lowell Turner, "Three Plants, Three Futures," *Technology Review* 92 (Jan. 1989): 38–45.

19. Robert Howard, *Brave New Workplace* (New York: Viking, 1985); John Lincoln and Allen Kallenberg, "Work Organization and Workplace Commitment: A Case Study of Plants in the U.S. and Japan," *American Sociological Review* 50 (1985): 738–60.

20. Thomas Kochan, Harry Katz, and Robert McKersie, *The Transformation of American Industrial Relations* (New York: Basic Books, 1986), chap. 7.

21. National Academy of Sciences, *Computer Chips and Paper Clips: Technology and Women's Employment* (Washington, DC: National Academy Press, 1987).

22. Marilyn Davidson and Cary Cooper, eds., *Women and Information Technology* (New York: Wiley, 1987), chap. 9.

23. Richard Long, "Human Issues in New Office Technology," in *Computers in the Human Context*, ed. Forester, p. 333. See also Richard Long, *New Office Information Technology: Human and Managerial Implications* (London and New York: Croom Helm, 1987).

24. Shoshona Zuboff, *In the Age of the Smart Machine: The Future of Work and Power* (New York: Basic Books, 1988), p. 125.

25. Rowe, *People and Chips*, chap. 5.

26. Barbara Garson, *The Electronic Sweatshop: How Computers Are Tranforming the Office of the Future into the Factory of the Past* (New York: Simon and Schuster, 1988).

27. Jane Barker and Hazel Downing, "Word Processing and the Transformation of Patriarchal Relations of Control in the Office," in *The Social Shaping of Technology*, ed. Donald McKenzie and Judy Wajcman (Milton Keynes, England: Open University Press, 1985); Richard Walton, "Social Choice in the Development of Advanced Information Society," in *Contemporary Moral Controversies in Technology*, ed. A. Pablo Iannone (New York: Oxford University Press, 1987).

28. James Danziger and Kenneth Kraemer, *People and Computers: The Impacts of Computing on End Users in Organizations* (New York: Columbia University Press, 1986).

29. Office of Technology Assessment, *The Electronic Supervisor: New Technology, New Tensions* (Washington, DC: OTA, 1981).
30. Elizabeth Gerver, *Humanizing Technology: Computers in Community Use and Adult Education* (New York: Plenum Press, 1986), chap. 2. See also Rose Deakin, *Women and Computing: The Golden Opportunity* (London: Macmillan, 1984).
31. David Lyon, *Information Society,* pp. 82–84.
32. Richard Long, *New Office Information Technology;* see also Rob Kling, "Computerization and Social Transformations," *Science, Technology & Human Values* 16 (1991): 342–67.
33. Andrew Clement, "Office Automation and the Technical Control of Information Workers," in *The Political Economy of Information,* ed. Vincent Mosco and Janet Wasko (Madison: University of Wisconsin Press, 1988), p. 242.
34. Mosco and Wasko, *Political Economy of Informatioin,* chap. 8; Rowe, *People and Chips,* chap. 6.
35. Steve Smith, "Information Technology in Banks: Taylorization or Human–centered Systems," in *Computers in a Human Context,* ed. Forester.
36. Shoshana Zuboff, *In the Age of the Smart Machine,* pp. 7–12.
37. Ibid., chap. 7.
38. Ibid., pp. 387–414.
39. Elinor Lenz and Barbara Myerhoff, *The Feminization of America: How Women's Values Are Changing Our Public and Private Lives* (Los Angeles: Jeremy Tarcher, 1985), chap. 5. See also Kathy Ferguson, *The Feminist Case Against Bureaucracy* (Philadelphia: Temple University Press, 1984).
40. Richard L. Harris, "The Impact of the Micro-electronics Revolution on the Basic Structure of Modern Organizations," *Science, Technology & Human Values* 11, no. 4 (1986): 31–44. See also *International Yearbook of Organizational Democracy* (annual volumes published by John Wiley and Sons starting in 1983).
41. Henry Kelly (Office of Technology Assessment), speaking at Carleton College in October 1988.
42. Kevin Robins and Frank Webster, "Cybernetic Capitalism: Information, Technology, Everyday Life," in *Political Economy of Information* ed. Mosco and Wasko.
43. Herbert Schiller and Anita Schiller, "Libraries, Public Access to Information, and Commerce," in *Political Economy of Information,* ed. Mosco and Wasko.
44. See chapters by Robert Rubinyi, Mary Furlong, and John Downing in *The Information Gap: How Computers and Other New Communication Technologies Affect the Social Distribution of Power,* ed. Marsha Siefert, George Gerbner, and Janice Fisher (New York: Oxford University Press, 1989).
45. Gerver, *Humanizing Technology.* See also Ian Miles, John Bessant, Ken Guy, and Howard Rush, "IT Futures in Households and Communities," in *Information Technology: Social Issues,* ed. Finnegan et al.
46. *Star and Tribune* (Minneapolis), Jan. 28, 1990, p. 13A.
47. Naisbitt, *Megatrends;* Toffler, *The Third Wave;* Ben Barber, *Strong Democracy: Participatory Politics for a New Age* (Berkeley and Los Angeles: University of California Press, 1984).
48. F. Christopher Arterton, *Teledemocracy: Can Technology Protect Democracy?* (Beverley Hills: Sage Publications, 1987).
49. Judith Larsen and Everett Rogers, "Silicon Valley: A Scenario for the Information Society of Tomorrow," in *The Information Society: Economic, Social, and Structural Issues,* ed. Jerry Salvaggio (Hillsdale, NJ: Lawrence Erlbaum Associates, 1989).
50. Tracy Kidder, *The Soul of a New Machine* (New York: Little, Brown, 1981).
51. Sherry Turkle, *The Second Self: Computers and the Human Spirit* (New York: Simon and Schuster, 1984).
52. J. David Bolter, *Turing's Man: Western Culture in the Computer Age* (Chapel Hill: University of North Carolina Press, 1984).
53. Weizenbaum, *Computers and Human Reason.*
54. Ida Hoos, *Systems Analysis and Public Policy: A Critique* (Berkeley and Los Angeles: University of California Press, 1972); William H. Dutton, "Decision-making in the Information Age: Computer Models and Public Policy," in *Information Technology,* ed. Finnegan et al.
55. Deborah Johnson, *Computer Ethics* (Englewood Cliffs, NJ: Prentice-Hall, 1985), chap. 3.
56. John Ladd, "Computers and Moral Responsibility: A Framework for an Ethical Analysis," in

The Information Web: Ethical and Social Implications of Computer Networking, ed. Carol Gould (Boulder: Westview Press, 1989).

57. Charles Perrow, *Normal Risks: Living with High Risk Technologies* (New York: Basic Books, 1984).

58. Deborah Johnson and John Snapper, eds., *Ethical Issues in the Use of Computers* (Belmont, CA: Wadsworth, 1985), part 5.

59. Tom Forester and Perry Morrison, *Computer Ethics: Cautionary Tales and Ethical Dilemmas in Computing* (Cambridge: MIT Press, 1990), chap. 3.

60. Ibid., chap. 4.

61. Office of Technology Assessment, *Alternatives for a National Computerized Criminal History System* (Washington, DC: OTA, 1982), and *Electronic Record Systems and Individual Privacy* (Washington, DC: OTA, 1986).

62. David Burnham, *The Rise of the Computer State* (New York: Vintage, 1983).

63. Office of Technology Assessment, *Computer-Based National Information Systems* (Washington, DC: OTA, 1981).

64. Burnham, *Rise of the Computer State;* Kenneth Laudon, *Dossier Society: Value Choices in the Design of National Information Systems* (New York: Columbia University Press, 1986).

65. W. A. Parent, "Privacy, Morality, and the Law," in *Ethical Issues in the Use of Computers*, ed. Johnson and Snapper; Stanley Benn, "Privacy, Freedom, and Respect for Persons," in *Philosophical Dimensions of Privacy*, ed. Ferdinand Schoeman (Cambridge: Cambridge University Press, 1984).

66. James Rachels, "Why Privacy Is Important," in *Philosophical Dimensions of Privacy*, ed. Schoeman.

67. Office of Technology Assessment, *Electronic Records Systems*.

68. Lyon, *Information Society*, p. 102.

69. David Flaherty, "Limiting Governmental Surveillance and Promoting Bureaucratic Accountability," *Science, Technology & Human Values* 11 (1986): 7–18.

70. Frank Barnaby, "Microelectronics and War," in *The Militarization of High Technology*, ed. John Tirman (Cambridge, MA: Ballinger, 1984), p. 46.

71. John Monk and Chris Bissell, "Defense and the Electronics Industry: Civil Exploitation of Defense Electronics," in *Information Technology*, ed. Finnegan et al., p. 46.

72. Robert DeGrasse, "The Military and Semiconductors," in *Militarization of High Technology*, ed. Tirman.

73. Frank Barnaby, "How the Next War Will Be Fought," in *Computers in the Human Context*, ed. Forester.

74. Herbert York and Sanford Lakeoff, *A Shield in Space?* (Berkeley and Los Angeles: University of California Press, 1988).

75. Leonard Lee, *The Day the Phones Stopped* (New York: D. I. Fine, 1991); see also Forester and Morrison, *Computer Ethics*, chap. 5.

76. Office of Technology Assessment, *SDI: Technology, Survivability, and Software: Summary* (Washington, DC: OTA, 1988), p. 4.

77. Herbert Schiller, *Who Knows? Information in the Age of the Fortune 500* (Norwood, NJ: Ablex Publishing, 1981). See also his chapters in *Information Society*, ed. Salvaggio, and in *The Myth of the Information Revolution: Social and Ethical Implications of Communication Technology*, ed. Michael Traber (Beverly Hills: Sage Publications, 1986).

78. John Bessant, "Information Technology and the North-South Divide," in *Information Technology*, ed. Finnegan et al.

79. Steffan Jacobsson and Jon Sigurdson, eds., *Technological Trends and Challenges in Electronics: Dominance of the Industrialized World and Responses of the Third World* (Lund, Sweden: University of Lund, 1983); Brian Murphy, *The International Politics of New Information Technology* (New York: St. Martin's Press, 1986).

80. Antonio José Botelho, "Brazil's Independent Computer Strategy," in *Computers in the Human Context*, ed. Forester.

81. Robert Schware and Ziauddin Choudhury, "The Role of IT in Third World Development," in *Computers in the Human Context*, ed. Forester.

82. Board on Science and Technology for International Development, National Research Council, *Microcomputers and Their Applications for Developing Countries* (Boulder: Westview Press, 1986), chap. 7.

83. World Commission on Environment and Development, *Our Common Future* (Oxford: Oxford University Press, 1987), p. 215.

84. Juan Rada, "A Third World Perspective," in *Microelectronics and Society: For Better or for Worse?* ed. Günter Friedrichs and Adam Schaff (Oxford: Pergamon Press, 1982).

85. A. S. Bhalla and Dilmus James, eds., *New Technologies and Development: Experiences in Technology Blending* (Boulder: Lynne Rienner Publishers, 1988).

86. Heather Hudson, *Communication Satellites* (New York: Free Press, 1990).

87. Lyon, *Information Society,* p. 116.

88. Tom McPhail, *Electronic Colonialism* (Beverley Hills: Sage Publications, 1981).

89. Sean McBride, *Many Voices, One World* (Paris: UNESCO, 1980); William James Stover, *Information Technology in the Third World* (Boulder: Westview Press, 1984).

90. Neville Jayaweera, "Communication Satellites: A Third World Perspective," in *Information Technology,* ed. Finnegan et al.

91. Hudson, *Communication Satellites,* chap. 13.

92. Marvin Minsky and Seymour Papert, *Perceptrons: An Introduction to Computational Geometry* (Cambridge: MIT Press, 1969); Seymour Papert, "One AI or Many?" in *The Artificial Intelligence Debate,* ed. Stephen Graubard (Cambridge: MIT Press, 1988).

93. R. C. Schank and R. P. Abelson, *Scripts, Plans, Goals, and Understanding* (Hillsdale, NJ: Lawrence Erlbaum Associates, 1977).

94. Terry Winograd, *Understanding Natural Language* (New York: Academic Press, 1972).

95. John Haugeland, *Artificial Intelligence: The Very Idea* (Cambridge: MIT Press, 1985), chap. 6; Paul Churchland and Patricia Smith Churchland, "Could a Machine Think?" *Scientific American* 262 (Jan. 1990): 32–37; Daniel Dennett, "Cognitive Wheels: The Frame Problem," in *The Philosophy of Artificial Intelligence,* ed. Margaret Boden (Oxford: Oxford University Press, 1990).

96. Joseph Bronzino and Ralph Morelli, *Expert Systems: Basic Concepts* (Stonybrook, NY: State University of New York, 1989); Max Bremer, "Expert Systems: Where Are We and Where Are We Going?" in *Expert Systems: Principles and Case Studies,* 2d edition, ed. Richard Forsyth (London: Chapman and Hall Computing, 1989).

97. Randolph Miller, Harry Pople, and Jack D. Myers, "Internist-I, An Experimental Computer-Based Dianostic Consultant for General Internal Medicine," *New England Journal of Medicine* 307 (1982): 473.

98. Randolph Miller et al., "The Internist-I Quick Medical Reference Project: Status Report," *Western Journal of Medicine* 145, no. 6 (1986): 816–22.

99. David Evans and Vimla Patel, eds., *Cognitive Science in Medicine: Biomedical Modeling* (Cambridge: MIT Press, 1989).

100. Michael Anbar, "Your Humble Consultant: Computer-Assisted Medical Decision," in *Computers and Medicine,* ed. M. Anbar (Rockville, MD: Computer Science Press, 1987); William Schwartz et al., "Artificial Intelligence in Medicine: Where Do We Stand?" *New England Journal of Medicine* 316 (1987): 685–88.

101. D. E. Rumelhart and J. L. McClelland, eds., *Parallel Distributed Processing,* 2 vols. (Cambridge: MIT Press, 1986); Andy Clark, "Connectionism, Competence, and Explanation," in *Philosophy of Artificial Intelligence,* ed. Boden.

102. C. Rosenberg and T. Sejnowski, "Parallel Networks That Learn to Pronounce English Text," *Complex Systems* 1 (1987): 145–68, cited in Clark, "Connection, Competence, and Explanation." See Herb Brody, "The Neural Computer," *Technology Review* 93, no. 6 (1990): 42–49.

103. Allen Newell and Herbert Simon, "Computer Science as Empirical Enquiry: Symbols and Search," in *Philosophy of Artificial Intelligence,* ed. Boden; Alan Newell, "Intellectual Issues in the History of Artificial Intelligence," in *The Study of Information: Interdisciplinary Messages,* ed. F. Machlup and U. Mansfield (New York: Wiley, 1983).

104. Hubert Dreyfus, *What Computers Can't Do: The Limits of Artificial Intelligence,* 2d ed. (New York: Harper & Row, 1979).

105. Hubert Dreyfus and Stuart Dreyfus, *Mind Over Machine: The Power of Human Intuition and Expertise in the Era of the Computer* (New York: Free Press, 1986, revised 1989).

106. Terry Winograd and Fernando Flores, *Understanding Computers and Cognition: A New Foundation for Design* (Norwood, NJ: Ablex Publishing, 1986).

107. Ian Barbour, *Religion in an Age of Science,* pp. 204–9.
108. Reinhold Niebuhr, *The Nature and Destiny of Man* (New York: Charles Scribner's Sons, 1943), vol. 1, chaps. 7 and 8.
109. Jacob Schwartz, "The New Connectionism: Developing Relationships Between Neuroscience and Artificial Intelligence," in *Artificial Intelligence Debate,* ed. Graubard.
110. Roger Schank, "Natural Language, Philosophy, and Artificial Intelligence," in *Philosophical Perspectives in Artificial Intelligence,* ed. M. Ringle (Brighton, England: Harvester Press, 1979), p. 222.
111. Margaret Boden, *Artificial Intelligence and Natural Man* (Brighton, England: Harvester Press, 1977), chap. 14; Aaron Sloman, "Motives, Mechanisms, and Emotions," in *Philosophy of Artificial Intelligence,* ed. Boden.
112. P. C. Rogers, "Artificial Intelligence as a Dehumanizing Force," in *Artificial Intelligence: Human Effects,* ed. Masoud Yazdani and Ajit Narayanan (Sussex, England: Ellis Horwood, 1984).
113. George Mandler, *Cognitive Psychology* (Hillsdale, NJ: Lawrence Erlbaum Associates, 1985), chap. 3; Philip Johnson-Laird, *The Computer and the Mind: An Introduction to Cognitive Science* (Cambridge: Harvard University Press, 1988), chap. 19.
114. T. S. Eliot, *The Rock* (New York: Harcourt, Brace and Company, 1934), p. 7.

Chapter 7. Unprecedented Powers

1. *Time,* May 28, 1990, p. 40. See also Hilary French, "Restoring the East European and Soviet Environments," in *State of the World 1991,* ed. Lester Brown (New York: W. W. Norton, 1991).
2. Walter Rosenbaum, *Environmental Politics and Policy,* 2d ed. (Washington, DC: Congressional Quarterly Press, 1991), chaps. 1 and 6.
3. David Vogel, "Environmental Policy in Europe and Japan," in *Environmental Policy in the 1990s,* ed. Norman Vig and Michael Kraft (Washington, DC: Congressional Quarterly Press, 1990); John McCormick, *British Politics and the Environment* (London: Earthscan Publications, 1991).
4. Rosenbaum, *Environmental Politics and Policy,* chap. 2.
5. Vogel, "Environmental Policy in Europe and Japan," in *Environmental Policy in the 1990s,* ed. Vig and Kraft.
6. 1989 OTA testimony cited in Joel Hirschorn and Kirsten Oldenburg, eds., *Prosperity without Pollution* (New York: Van Nostrand, 1991), p. 117.
7. Sandra Postel and Christopher Flavin, "Reshaping the Global Economy," in *State of the World 1991,* ed. Brown, p. 172.
8. Rosenbaum, *Environmental Politics and Policy,* chap. 7.
9. Barry Commoner, *Making Peace with the Planet* (New York: Pantheon, 1990), chap. 3.
10. Hirschorn and Oldenburg, *Prosperity without Pollution,* p. 93.
11. Sandra Postel, "Controlling Toxic Chemicals," in *State of the World 1988,* ed. Lester Brown (New York: W. W. Norton, 1988), p. 131.
12. Hirschorn and Oldenburg, *Prosperity without Pollution,* p. 34.
13. Lester Brown, "The New World Order," in *State of the World 1991,* p. 7. An excellent discussion of all these global environmental threats and possible responses to them is given in Senator Albert Gore, *Earth in the Balance: Ecology and the Human Spirit* (New York: Houghton Mifflin, 1992).
14. Sandra Postel and Lori Heise, "Reforesting the Earth," in *State of the World 1988.*
15. Sandra Postel and John Ryan, "Reforming Forestry," in *State of the World 1991,* p. 77.
16. Edward O. Wilson, "Threats to Biodiversity," *Scientific American* 261 (Sept. 1989): 116. See also Kenton Miller and Laura Tangley, *Trees of Life: Saving Tropical Forests and Their Biological Wealth* (Boston: Beacon Press, 1991).
17. Edward Wolf, "Avoiding a Mass Extinction of Species," in *State of the World 1988,* p. 109.
18. *Time,* Jan. 2, 1989, p. 32.
19. Edward O. Wilson, "Threats to Biodiversity," *Scientific American* 261 (Sept. 1989): 116. See also E. O. Wilson, ed., *Biodiversity* (Washington, DC: National Academy of Sciences, 1988).

20. Postel and Ryan, "Reforming Forestry," in *State of the World 1991.*
21. Office of Technology Assessment, *Technologies to Maintain Biological Diversity* (Washington, DC: OTA, 1987).
22. Cynthia Pollock Shea, "Protecting the Ozone Layer," in *State of the World 1989,* p. 82; *Newsweek,* April 15, 1991, p. 64 and Feb. 17, 1992, p. 26.
23. Richard Benedick, *Ozone Diplomacy* (Cambridge: Harvard University Press, 1991); see also Richard Benedick, "Protecting the Ozone Layer: New Directions in Diplomacy," in *Preserving the Global Environment,* ed. Jessica Tuchman Mathews (New York: W. W. Norton, 1991).
24. Jessica Mathews, "Greenhouse-effect Uncertainty Can't Justify Federal Inaction," *Star Tribune* (Minneapolis), Nov. 11, 1991, p. 15A.
25. Stephen Schneider, "The Changing Climate," *Scientific American* 261 (Sept. 1989): 70–79. See also Christopher Flavin, "Slowing Global Warming," in *State of the World 1990.*
26. Sandra Postel and Christopher Flavin, "Reshaping the Global Economy," in *State of the World 1991.*
27. Linda Starke, *Signs of Hope: Working Towards a Common Future* (Oxford: Oxford University Press, 1990), chap. 7.
28. Gareth Porter and Janet Welsh Brown, *Global Environmental Politics* (Boulder, CO: Westview Press, 1991).
29. "World Population Data Sheet" (Washington DC: Population Reference Bureau, 1991).
30. Jodi Jacobson, "Planning the Global Family," in *State of the World 1988.*
31. Nathan Keyfitz, "The Growing Human Population," *Scientific American* 261 (Sept. 1989): 119–26.
32. Joseph Speidel, "To Curb the Growth of World Population," *Star Tribune* (Minneapolis), April 27, 1990.
33. Hirschorn and Oldenburg, *Prosperity without Pollution,* chap. 9.
34. Donella Meadows et al., *The Limits to Growth* (New York: Universe, 1972).
35. Gerald Barney, ed., *The Global 2000 Report to the President* (Washington, DC: Government Printing Office, 1980), 1:1.
36. William Ophuls, *Ecology and the Politics of Scarcity* (San Francisco: W. H. Freeman, 1977); Robert Heilbroner, *An Inquiry into the Human Prospect* (New York: W. W. Norton, 1974).
37. H. S. D. Cole et al., eds., *Models of Doom* (New York: Universe, 1973); Mancur Olson and Hans Landsberg, eds., *The No-Growth Society* (New York: W. W. Norton, 1973).
38. Herman Daly, *Steady-State Economics* (San Francisco: W. H. Freeman, 1977); Herman Daly and John B. Cobb, Jr., *For the Common Good* (Boston: Beacon Press, 1989).
39. H. E. Goeller and Alvin Weinberg, "The Age of Substitutability," *Science* 191 (1976): 683–89.
40. Julian Simon and Herman Kahn, *The Resourceful Earth* (Oxford: Basil Blackwell, 1984), p. 1 (italics deleted from original).
41. Paul Hawken, *The Informative Economy* (New York: Holt, Rinehart & Winston, 1983). Other references on a sustainable economy will be found in chapter 9 below.
42. See discussion of the Brundtland report and the idea of sustainable development in chap. 4 above.
43. Peter Wheale and Ruth McNally, *Genetic Engineering: Catastrophe or Utopia?* (Hemel Hampstead, England: Harvester, 1988); G. J. V. Nossal and Ross Coppel, *Reshaping Life: Key Issues in Genetic Engineering,* 2d ed. (Cambridge: Cambridge University Press, 1989); Burke Zimmerman, *Biofuture: Confronting the Genetic Era* (New York: Plenum Press, 1984).
44. Raymond Zilinskas and Burke Zimmerman, eds., *Gene Splicing Wars: Reflections on the Recombinant DNA Controversy* (New York: Macmillan, 1986); James Watson and John Tooze, eds., *The DNA Story* (San Francisco: W. H. Freeman, 1981).
45. Office of Technology Assessment, *New Developments in Biotechnology: Field-Testing Engineered Organisms* (Washington, DC: OTA, 1988).
46. Wheale and McNally, *Genetic Engineering,* chap. 5.
47. Marc Lappé, *Broken Code: The Exploitation of DNA* (San Francisco: Sierra Club Books, 1984), chap. 6; Lawrence Busch et al., *Plants, Power, and Profit: Social, Economic, and Ethical Consequences of New Biotechnologies* (Oxford: Basil Blackwell, 1991).
48. Calestous Juma, *The Gene Hunters: Biotechnology and the Scramble for Seeds* (London: Zed Books, 1989); Miriam Reidy, "Biotech on the Farm: Monitoring the Impact," *One World* (June 1990), pp. 16–18.

49. Frederick Buttel, "Biotechnology and Agricultural Research," in *New Directions for Agriculture and Agricultural Research,* ed. Kenneth Dahlberg (Totowa, NJ: Rowman & Allanheld, 1986).

50. Robert Deakin, "BST: The First Commercial Product for Agriculture from Biotechnology," in *Biorevolution: Cornucopia or Pandora's Box,* ed. Peter Wheale and Ruth McNally (London: Pluto Press, 1990).

51. Wade Rousch, "Who Decides about Biotech: The Clash Over Bovine Growth Harmone," *Technology Review* 94, no. 5 (1991): 28-37.

52. Gary Comstock, "The Case Against BGH," *Agriculture and Human Values* 5, no. 3 (1988): 36-52; Wheale and McNally, eds., *Biorevolution,* part 2.

53. World Council of Churches, Subunit on Church and Society, *Biotechnology: Its Challenges to the Churches and the World* (Geneva: World Council of Churches, 1989).

54. Michael Fox, "Transgenic Animals: Ethical and Animal Welfare Concerns," in *Biorevolution,* ed. Wheale and McNally.

55. John Fletcher, "Ethical Issues In and Beyond Prospective Clinical Trials of Human Gene Therapy," *Journal of Medicine and Philosophy* 10 (1985): 293.

56. David Suzuki and Peter Knudtson, *Genethics: The Ethics of Engineering Life* (London: Unwin, 1989), chap. 7; Kathleen Nolan and Sara Swenson, "New Tools, New Dilemmas: Genetic Frontiers," *Hastings Center Report* 18 (Oct./Nov. 1988): 40-46.

57. *Star Tribune* (Minneapolis), Jan. 30, 1991.

58. Leroy Walters, "The Ethics of Human Gene Therapy," *Nature* 320 (1986): 225-27.

59. Office of Technology Assessment, *Human Gene Therapy: Background Paper* (Washington, DC: OTA, 1984).

60. Fletcher, "Ethical Issues."

61. Jeremy Rifkin and Nicanor Perlas, *Algeny* (New York: Viking, 1983); "Anatomy of a Pressure Group," *Nature* 309 (1984): 301-2. See also Leslie Roberts, "Ethical Questions Haunt New Genetic Technologies," *Science* 243 (1989): 1134-36.

62. World Council of Churches, *Biotechnology,* p. 2.

63. Roger Shinn, *Forced Options,* 3d. ed. (Cleveland: Pilgrim Press, 1991), chap. 7.

64. Gilbert Meilaender, "Mastering our Gen(i)es: When Do We Say No?" *Christian Century* (Oct. 3, 1990), pp. 872-75.

65. Rifkin and Perlas, *Algeny.* For a reply to Rifkin, see C. Keith Boone, "Bad Axioms in Genetic Engineering," *Hastings Center Report* 18 (Aug./Sept. 1988): 9-13.

66. Barbour, *Religion in an Age of Science,* p. 216. See also Ann Lamers and Ted Peters, "Genethics: Implications of the Human Genome Project," *Christian Century* (Oct. 3, 1990), pp. 868-72; Ronald Cole-Turner, "Genetic Engineering: Our Role in Creation," in *The New Faith-Science Debate,* ed. John Magnum (Minneapolis: Fortress Press, 1989).

67. Daniel Callahan, "Recombinant DNA: Science and the Public," *Hastings Center Report* 7, no. 2 (1977): 20-23; Stanley Cohen, "Recombinant DNA: Fact and Fiction," *Science* 195 (1977): 654-57; see chapters by Robert Sinsheimer, David Baltimore, Peter Hutt, and Dorothy Nelkin in *Limits of Scientific Inquiry,* ed. Gerald Holton and Robert Morrison (New York: W. W. Norton, 1979).

68. Openness and secrecy in university-industry contracts are the topics of two symposia in *Science, Technology & Human Values* 10, no. 2 (1985) and 12, no. 1 (1987); see also Paul DeForest et al., eds., *Biotechnology: Professional Issues and Social Concerns* (Washington, DC: American Association for the Advancement of Science, 1988).

69. Vivian Weil in *Biotechnology,* ed. DeForest et al. See also Martin Kenney, *Biotechnology: The University-Industrial Complex* (New Haven: Yale University Press, 1986).

70. D. Blumenthal et al., "University-Industry Research Relationships in Biotechnology: Implications for the University," *Science* 232 (1986): 1361-66.

71. Nossal and Coppel, *Reshaping Life,* p.124. See also Sheldon Krimsky in *Biotechnology,* ed. DeForest et al.

72. The Harvard Nuclear Study Group (Albert Carnesale et al.), *Living with Nuclear Weapons* (New York: Bantam Books, 1983), chap. 4.

73. Bruce Russett, *Prisoners of Insecurity* (San Francisco: W. H. Freeman, 1983), chap. 4.

74. *Nuclear Times* (Winter 1990), p. 30; Michael Klare, "Growing Firepower in the Third World," *Bulletin of the Atomic Scientists* 46 (May 1990): 9-13.

75. William Hartung, "The Boom at the Arms Bazaar," *Bulletin of the Atomic Scientists* 47, no. 8 (1991): 14–21.

76. Mary Pellauer, "War and Sexism in the Nuclear Age," in *Peace-Ways,* ed. Charles Lutz and Jerry Folk (Minneapolis: Augsburg Publishing House, 1983), p. 94.

77. Pam McAllister, ed., *Reweaving the Web of Life* (Philadelphia: New Society Publishers, 1982); Rosemary Radford Ruether, "Feminism and Peace," *Christian Century* (Sept. 31, 1983), pp. 771–76.

78. Ruth Leger Sivard, *World Military and Social Expenditures, 1991* (Washington, DC: World Priorities, 1991), p. 50.

79. William Hartung, *The Economic Consequences of a Nuclear Freeze* (New York: Council on Economic Priorities, 1984), cited in Michael Renner, "Converting to a Peaceful Economy," in *State of the World 1990,* ed. Brown, p. 157.

80. Sivard, *World Military and Social Expenditures, 1991,* p. 27.

81. Ibid., p. 47. See also Roger Williams, "U.K. Science and Technology: Policy, Controversy, and Advice," *Political Quarterly* 59, no. 2 (1988): 134.

82. Office of Technology Assessment, *Complex Cleanup: The Environmental Legacy of Nuclear Weapons Production* (Washington, DC: OTA, 1991).

83. Sivard, *World Military And Social Expenditures, 1989,* p. 23.

84. Robert Jungk, *Brighter than a Thousand Suns* (New York: Harcourt Brace Jovanovich, 1958), p. 201.

85. Sivard, *World Military and Social Expenditures, 1991,* p. 11.

86. Office of Technology Assessment, *The Effects of Nuclear War* (Washington, DC: OTA, 1979).

87. National Academy of Sciences, *The Effect on the Atmosphere of a Major Nuclear Exchange* (Washington, DC: National Academy Press, 1984).

88. Richard Turco et al., "Nuclear Winter: Global Consequences of Multiple Nuclear Explosions," *Science* 222 (1983): 1283–92; Carl Sagan and Richard Turco, *A Path Where No Man Thought: Nuclear Winter and the End of the Arms Race* (New York: Random House, 1990).

89. Jonathan Schell, *The Fate of the Earth* (New York: Avon, 1982).

90. Robert Jay Lifton and Richard Falk, *Indefensible Weapons* (New York: Basic Books, 1982).

91. Joanna Rogers Macy, *Despair and Personal Power in the Nuclear Age* (Philadelphia: New Society Publishers, 1983).

92. Richard McCormick, "Nuclear Deterrence and the Problem of Intentions," in *Catholics and Nuclear War,* ed. Philip Murnion (New York: Crossroad, 1983); also chapters by Jonathan Bennett and Thomas Nagel in *The Ethics of War and Nuclear Deterrence,* ed. James Sterba (Belmont CA: Wadsworth, 1985).

93. Charles Kegley and Kenneth Schwab, eds., *After the Cold War: Questioning the Morality of Nuclear Deterrence* (Boulder: Westview Press, 1991).

94. David Hollenbach, S.J., *Nuclear Ethics: A Christian Moral Argument* (New York: Paulist Press, 1983), chap. 4; Thomas Shannon, ed., *War or Peace? The Search for New Answers* (Maryknoll, NY: Orbis, 1980), part 1.

95. National Conference of Catholic Bishops, *The Challenge of Peace: God's Promise and Our Response* and *Building Peace* (Washington, DC: U.S. Catholic Conference, 1983 and 1988).

96. United Methodist Council of Bishops, *In Defense of Creation: The Nuclear Crisis and a Just Peace* (Nashville: Graded Press, 1986).

97. Church of England, General Synod Working Party, *Peacemaking in a Nuclear Age* (London: Church Publishing House, 1988).

98. Gordon Zahn, "Pacifism and the Just War," in *Catholics and Nuclear War,* ed. Murnion; Jerry Folk, "The Case for Pacifism," in *Peace-Ways,* ed. Lutz and Folk.

99. Phillips Moulton, *Ammunition for Peacemakers* (New York: Pilgrim Press, 1986), chap. 5; Gene Sharp, *Making Europe Unconquerable: The Potential for Civilian-Based Deterrence* (Cambridge, MA: Ballinger, 1985).

100. Hal Lindsey, *The Late Great Planet Earth* (Grand Rapids: Zondervan, 1970) and *The 1980s: Countdown to Armageddon* (King of Prussia, PA: Westgate Press, 1980); Grace Halsell, *Prophecy and Politics: Militant Evangelists on the Road to Nuclear War* (Westport, CT: Lawrence Hill & Co., 1986).

101. See Barbour, *Religion in an Age of Science,* pp. 148–50.

102. Gordon Kaufman, *Theology for a Nuclear Age* (Philadelphia: Westminster Press, 1985); see also G. Clarke Chapman, "American Theology in the Shadow of the Bomb," *Union Seminary Quarterly Review* 41, no. 1 (1987): 25–38.

103. Sallie McFague, *Models of God: Theology for an Ecological, Nuclear Age* (Philadelphia: Fortress Press, 1987).

104. See chapters by Jonathan Draper, Rex Ambler, and Brian Russell in *Theology Against the Nuclear Horizon*, ed. Alan Race (London: SCM Press, 1988); also the chapter by Simon Maimela in *The Nuclear Weapons Debate: Theological and Ethical Issues*, ed. Richard Bauckham and R. John Elford (London: SCM Press, 1989).

105. Barbour, *Religion in an Age of Science*, pp. 84–87. See also Alan Race, "Christian Involvement in an Interfaith Dialogue of Peacemaking," in *Theology Against the Nuclear Horizon*, ed. Race.

106. Albert Carnesale et al., *Living with Nuclear Weapons*, chap. 9; Albert Carnesale and Richard Haass, eds., *Superpower Arms Control: Setting the Record Straight* (Cambridge, MA: Ballinger, 1987).

107. P. Edward Haley and Jack Merrit, eds., *Nuclear Strategy, Arms Control, and the Future*, 2d ed. (Boulder: Westview Press, 1988).

108. Leonard Spector, "Treaty Review: Deadlock Damages Nonproliferation," *Bulletin of the Atomic Scientists* 46 (Dec. 1990): 39–44.

109. Tina Rosenberg, "Nuking the Nukes," *New Republic* (Jan. 28, 1991), pp. 21–23.

110. George Perkovich, "Put Nuclear Weapons on the Agenda," *Bulletin of the Atomic Scientists* 47, no. 5 (1991): 21–23; Sidney Drell, "Verification Triumphs," *Bulletin of the Atomic Scientists* 47, no. 9 (1991): 28–29.

111. Independent Commission on Disarmament and Security Issues (the Palme Commission), *Common Security: A Blueprint for Survival* (New York: Simon and Schuster, 1982) and *A World at Peace: Common Security in the Twenty-first Century* (Stockholm: Palme Commission, 1989).

112. Harry Hollins, Averill Powers, and Mark Sommer, *The Conquest of War: Alternative Strategies for Global Security* (Boulder: Westview Press, 1989), chaps. 6, 7, and 13; Richard Smoke and Willis Harmon, *Paths to Peace* (Boulder: Westview Press, 1987).

113. Walter Dorn, "UN Should Verify Treaties," *Bulletin of the Atomic Scientists* 46, no. 6 (1990): 12–13; Leonard Weiss, "Tighten Up on Nuclear Cheaters," *Bulletin of the Atomic Scientists* 47, no. 4 (1991): 11–12.

114. Henry Wiseman, ed., *Peacekeeping: Appraisals and Proposals* (New York: Pergamon Press, 1983); Hollins et al., *Conquest of War*, chap. 13; Robert Johansen, *Toward an Alternative Security System* (New York: World Policy Institute, 1983); Robert Johansen, "Toward Post-Nuclear Global Security," in *Alternative Security: Living Without Nuclear Deterence*, ed. Burns Weston (Boulder: Westview Press, 1990).

115. Marc Nerfin, "The Future of the United Nations System," in *The United Nations and a Just World Order*, ed. Richard Falk, Samuel Kim, and Saul Mendlovitz (Boulder: Westview Press, 1991).

Chapter 8. Controlling Technology

1. Recent volumes on science, technology, and government include Richard Barke, *Science, Technology, and Public Policy* (Washington: Congressional Quarterly Press, 1986); Edward Wenk, *Tradeoffs: Imperatives of Choice in a High-Tech World* (Baltimore: Johns Hopkins University Press, 1986); Harvey Brooks and Chester Cooper, eds., *Science for Public Policy* (Oxford: Pergamon Press, 1987); Michael Kraft and Norman Vig, eds., *Technology and Politics* (Durham: Duke University Press, 1988).

2. *Science and Technology Data Book: 1989* (Washington, DC: National Science Foundation, 1988).

3. David Dickson, *The New Politics of Science* (Chicago: University of Chicago Press, 1988); Malcolm Goggin, ed., *Governing Science and Technology in a Democracy* (Knoxville: University of Tennessee Press, 1986).

4. Dickson, *New Politics of Science*, p. ix.

5. Joel Primack and Frank von Hippel, *Advice and Dissent* (New York: Basic Books, 1974).

6. Montague Kern, *30-Second Politics: Political Advertising in the Eighties* (New York: Praeger Publishers, 1989).

7. Helen Ingram and Dean Mann, "Interest Groups and Environmental Policy," in *Environmental Politics and Policy,* ed. James Lester (Durham: Duke University Press, 1989); Samuel Hays, *Beauty, Health, and Permanence: Environmental Politics in the United States,* 1955–1985 (Cambridge: Cambridge University Press, 1987).

8. Robert E. McGinn, *Science, Technology, and Society* (Englewood Cliffs, NJ: Prentice-Hall, 1991), p. 243. See Michael Hoffman, "The Ford Pinto," in *Business Ethics: Readings and Cases in Corporate Morality,* ed. W. M. Hoffman and I. M. Moore (New York: McGraw-Hill, 1984).

9. Ronald Brinkman, Sheila Jasanoff, and Thomas Ilgen, *Controlling Chemicals: The Politics of Regulation in Europe and the Unites States* (Ithaca: Cornell University Press, 1985).

10. Dorothy Nelkin and Michael Pollack, *The Atom Besieged: Extra-Parliamentary Dissent in France and Germany* (Cambridge: MIT Press, 1981).

11. Primack and von Hippel, *Advice and Dissent;* William Golden, *Worldwide Science and Technology Advice to the Highest Levels of Government* (Oxford: Pergamon Press, 1991).

12. Among journals dealing with technology policy are *Bulletin of the Atomic Scientists, Issues in Science and Technology, Technology Review,* and *Science, Technology & Human Values.*

13. On disagreements among experts, see Dorothy Nelkin, ed., *Controversy: Politics of Technical Decisions* (Beverly Hills; Sage Publications, 1979); Allan Mazur, *The Dynamics of Technical Controversy* (Washington, DC: Communication Press, 1981); Marc Roberts, Stephen Thomas, and Michael Dowling, "Mapping Scientific Disputes that Affect Public Policymaking," *Science, Technology & Human Values* 9, no.1 (1984): 112–22.

14. David Collingridge and Colin Reeve, *Science Speaks to Power: The Role of Experts in Policy Making* (New York: St. Martin's Press, 1986), chaps. 5–7; David Robbins and Ron Johnston, "The Role of Cognitive and Occupational Differentiation in Scientific Controversies," *Social Studies of Science* 6 (1976): 349–68.

15. Brendin Gillespie, Dave Eva, and Ron Johnston, "Carcinogenic Assessment in the U.S. and Great Britain: The Case of Aldrin/Dieldrin," *Social Studies of Science* 9 (1979): 265–301.

16. Harvey Brooks, "The Resolution of Technically Intensive Public Policy Disputes," *Science, Technology & Human Values* 9, no. 1 (1984): 39–50.

17. Jeremy Stone, cited in Anne Cahn, "American Scientists and the ABM," in *Scientists and Public Affairs* (Cambridge: MIT Press, 1973), p. 112.

18. Cahn, "American Scientists and the ABM," p. 86.

19. Eugene Skolnikoff and Harvey Brooks, "Science Advice in the White House?" *Science* 187 (1975): 41. See also Mark Rushevsky, "Technical Disputes: Why Experts Disagree," *Policy Studies Review* 1 (1982): 676–85.

20. Phillip Boffey, *The Brain Bank of America* (New York: McGraw-Hill, 1975).

21. Presidential Task Force on the Science Court (Arthur Kantrowitz, chair), "The Science Court Experiment: An Interim Report," *Science* 193 (1976): 653–56; Roger Masters and Arthur Kantrowitz, "Scientific Adversary Procedures: The SDI Experiment at Dartmouth," in *Technology and Politics,* ed. Kraft and Vig.

22. Collingridge and Reeve, *Science Speaks to Power,* p. 158.

23. Allan Mazur, "Controversial Technologies in the Mass Media," in *Technology and Politics,* ed. Kraft and Vig.

24. Walter Rosenbaum, *Environmental Politics and Policy* (Washington, DC: Congressional Quarterly Press, 1991), p. 2.

25. See chapters by James Peterson, Dorothy Nelkin, and Sheldon Krimsky in *Citizen Participation in Science Policy,* ed. James Petersen (Amherst: University of Massachusetts Press, 1984). Also K. Guild Nichols, ed., *Technology on Trial: Public Participation in Decision-Making Related to Science and Technology* (Paris: Organisation for Economic Co-operation and Development, 1979); Daniel Fiorino, "Citizen Participation and Environmental Risk: A Survey of Institutional Mechanisms," *Science, Technology & Human Values* 15, no. 2 (1990): 226–43.

26. D. J. Gamble, "The Berger Inquiry: An Impact Assessment Proces," *Science* 199 (1978): 946–52.

27. Richard Davies, "The Effectiveness of the Sizewell B Public Inquiry in Facilitating Communication about the Risks of Nuclear Power," *Science, Technology & Human Values* 12, nos. 2–3

(1987): 102–10; Richard Baker, "Assessing Complex Technical Issues: Public Inquiries or Commissions," *Political Quarterly* 59, no.2 (1988): 178–89.

28. Baker, "Assessing Complex Technical Issues."

29. Dorothy Nelkin, *Technological Decisions and Democracy: European Experiments in Public Participation* (Beverly Hills: Sage Publications, 1977).

30. Robert Paehlke and Douglas Torgerson, eds., *Managing Leviathan: Environmental Politics and the Administrative State* (Peterborough, Ontario: Broadview Press, 1990), chap. 1.

31. Daniel Mazmanian and David Morell, "The NIMBY Syndrome: Facility Siting and the Failure of Democratic Discourse," in *Environmental Policy in the 1990s*, ed. Norman Vig and Michael Kraft (Washington, DC: Congressional Quarterly Press, 1990); Michalann Harthill, ed., *Hazardous Waste Management: In Whose Backyard* (Boulder: Westview Press, 1984); Gail Bingham and Timothy Mealey, eds., *Negotiating Hazardous Facility Siting and Permitting Agreements* (Washington, DC: Conservation Foundation, 1988).

32. Laurence Tribe, Corinne Schelling, and John Voss, eds., *When Values Conflict: Essays on Environmental Analysis, Discourse, and Decision* (Cambridge, MA: Ballinger, 1976).

33. Thomas H. Tietenberg, *Environmental and Natural Resource Economics*, 2d ed. (Glenview, IL: Scott, Foresman, 1988).

34. Kristin Shrader-Frechette, *Science Policy, Ethics, and Economic Methodology* (Dordrecht and Boston: D. Reidel, 1985); Allen Kneese, Shaul Ben-David, and William Schulze, "The Ethical Foundations of Benefit-Cost Analysis," in *Energy and the Future*, ed. Douglas MacLean and Peter Brown (Totowa, NJ: Rowman and Littlefield, 1983); Steven Kelman, "Cost-Benefit Analysis and Environmental, Safety, and Health Regulations: Ethical and Philosophical Considerations," in *Cost-Benefit Analysis and Environmental Regulations: Politics, Ethics, and Methods*, ed. Daniel Swartzman et al. (Washington, DC: Conservation Foundation, 1982).

35. J. A. Doeleman, "On the Social Rate of Disount: The Case for Macroenvironmental Policy," *Environmental Ethics* 2 (1980): 45–58.

36. John Krutilla and Anthony Fisher, *The Economics of Natural Environments* (Baltimore: Johns Hopkins University Press, 1975).

37. R. Burton Litton, "Aesthetic Dimensions of the Landscape," in *Natural Environments: Studies in Theoretical and Applied Analysis* (Baltimore: Johns Hopkins University Press, 1972).

38. Denys Munby, "The Christian Ethics of Cost-Benefit Analysis," *Anticipation*, no. 7 (1971): 16–22.

39. Henry Peskin and Eugene Seskin, eds., *Cost-Benefit Analysis and Water Pollution Policy* (Washington, DC: Urban Institute, 1975), Introduction.

40. Krutilla and Fisher, *Economics of Natural Environments*, pp. 281ff.

41. David Bradford and Harold Fieveson, "Benefits and Costs: Winners and Losers," in *Boundaries of Analysis: An Inquiry into the Tocks Island Controversy*, ed. Harold Fieveson, Frank Sinden, and Robert Socolow (Cambridge, MA: Ballinger, 1976).

42. Recent summaries of risk analysis in government regulation are given in Rosenbaum, *Environmental Politics and Policy*, chap. 5; Michael Kraft, "Analyzing Technological Risks in Federal Regulatory Agencies," in *Technology and Politics*, ed. Kraft and Vig; and Richard Andrews, "Risk Assessment: Regulation and Beyond," in *Environmental Policy in the 1990s*, ed. Vig and Kraft.

43. William Lawrance, *Of Acceptable Risk* (Los Altos, CA: William Kaufmann, 1976); National Research Council, *Risk Assessment in the Federal Government* (Washington, DC: National Academy Press, 1983); National Research Council, *Improving Risk Communication* (Washington, DC: National Academy Press, 1989); William Ruckelshaus, "Science, Risk, and Public Policy," *Science* 221 (1983): 1026–28.

44. Thomas Maugh, "Chemical Carcinogens: The Scientific Basis," *Science* 201 (1978): 1200–5.

45. Richard Zeckhauser, "Procedures for Valuing Life," *Public Policy* 23 (1975): 419–64; Barbara MacKinnon, "Pricing Human Life," *Science, Technology & Human Values* 11, no.2 (1986): 29–39.

46. National Resource Council, *Decision Making for Regulating Chemicals in the Environment* (Washington DC: National Academy Press, 1975), chap 6.

47. Baruch Fischoff, Sarah Lichtenstein, Paul Slovic, Stephen Derby, and Ralph Keeney, *Acceptable Risk* (Cambridge: Cambridge University Press, 1981).

48. Charles Perrow, *Normal Accidents: Living with High-Risk Technologies* (New York: Basic Books, 1984); Joseph Morone and Edward Woodhouse, *Averting Catastrophe: Strategies for Regulating Risky Technologies* (Berkeley and Los Angeles: University of California Press, 1986).

49. Kristin Shrader-Frechette, *Risk Analysis and Scientific Method* (Boston: D. Reidel, 1985); Donald Brown, "Ethics, Science, and Environmental Regulation," *Environmental Ethics* 9 (1987): 331–49.

50. Fischoff et al., *Acceptable Risk*, p. xi.

51. Frances M. Lynn, "The Interplay of Science and Values in Assessing Environmental Risk," *Science, Technology & Human Values* 11, no. 2 (1986): 40–49.

52. John Graham, Howard Raiffa, and James Vaupel, "Science and Analysis: Roles in Risk and Decision Making," in *Risk Evaluation and Management*, ed. Vincent Covello, Joshua Menkes, and Jeryl Mumpower (New York: Plenum Press, 1986); Albert Flores and Michael Kraft, "Determining the Acceptability of Risk in Regulatory Policy: Ethics, Politics and Risk Analysis," in *Ethics, Government, and Public Policy*, ed. James Bowman and Frederick Elliston (New York: Greenwood Press, 1988).

53. Sheldon Krimsky and Alonzo Plough, *Environmental Harzards: Communicating Risks as a Social Process* (Dover, MA: Auburn House, 1988).

54. Francois Hetman, *Society and the Assessment of Technology* (Paris: Organisation for Economic Cooperation and Development, 1973); Alan Porter, Frederick Rossini, Stanley Carpenter, and A. T. Rober, *A Guidebook for Technology Assessment and Impact Analysis* (New York: North Holland Publishing, 1980); Vary Coates and Joseph Coates, "Making Technology Assessment an Effective Tool to Influence Policy," in *Policy Through Impact Assessment*, ed. Robert V. Bartlett (Westport CT: Greenwood Press, 1989).

55. *Review of the Office of Technology Assessment and its Organic Act* (Washington, DC: Committee on Science and Technology, U.S. House of Representatives, 1978); David O'Brien and Donald Marchand, eds., *The Politics of Technology Assessment* (Lexington, MA: Lexington Books, 1982), chaps. 1–4; John Gibbons and Holly Gwin, "Technology and Governance: The Development of the Office of Technology Assessment," in *Technology and Politics*, ed. Kraft and Vig.

56. See articles by Joseph Coates, Sherry Arnstein, Hazel Henderson, and Willis Goldbeck in "Symposium on Public Participation in Technology Assessment," *Public Administration Review* 35 (1975): 67ff; Sherry Arnstein and Alexander Christakis, eds., *Perspectives in Technology Assessment* (Jerusalem, Israel: Science and Technology Publishers, 1975), chap. 10; Frederick Rossini and Alan Porter, "Public Participation and Professionalism in Impact Assessment," in *Citizen Participation in Science*, ed. Petersen.

57. Barry Casper, "The Rhetoric and Reality of Congressional Technology Assessment," *Bulletin of the Atomic Scientists* 34 (Feb. 1978): 20–31; Coates and Coates, "Making Technology Assessment an Effective Tool," in *Policy Through Impact Assessment*, ed. Bartlett.

58. Kristin Shrader-Frechette, "Technology Assessment as Applied Philosophy of Science," *Science, Technology & Human Values* 6, no.4 (1980): 33–50; Dickson, *New Politics of Science*, pp. 233–43; Henry Skolimowski, "Technology Assessment in Sharp Social Focus," *Technology Forecasting and Social Change* 8 (1976): 421–25.

59. Norman Vig, "Parliamentary Technology Assessment in Europe: A Comparative Perspective," in *Science, Technology, and Politics: Policy Analysis in Congress*, ed. Gary Bryner (Boulder: Westview Press, 1991); E. J. Tuininga, "Technology Assessment in Europe," *Futures* 20; no. 1 (1988): 37–45.

60. Overviews of ways of setting regulatory standards are given in Paul Portney, ed., *Public Policies for Environmental Protection* (Washington, DC: Resources for the Future, 1990), and Walter Rosenbaum, *Environmental Politics and Policy*, chap. 6.

61. National Commission on Water Quality, *Report to Congress* (Washington, DC: NCWQ, 1976).

62. Jeff Smith, "Court Upholds Controversial Regulations," *Science* 213 (1981): 185–88.

63. Henry Peskin, "Environmental Policy and the Distribution of Benefits and Costs," in *Current Issues in U.S. Environmental Policy*, ed. Paul Portney (Washington, DC: Resources for the Future, 1978); Robert Dorfman, "Incidence of the Benefits and Costs of Environmental Programs," *American Economic Review* 67 (1977): 33–40; A. Myrick Freeman III, *Air and Water*

Pollution Control: A Benefit-Cost Assessment (New York: Wiley, 1982); Allen Kneese, *Measuring the Benefits of Clean Air and Water* (Washington, DC: Resources for the Future, 1984).

64. Organisation for Economic Co-operation and Development, *Economic Instruments for Environmental Protection* (Paris: Organisation for Economic Co-operation and Development, 1989; also *Environmental Policy in the European Community* (Luxembourg: Office for Official Publications of the European Community, 1990).

65. Frederick Anderson et al., *Environmental Improvement through Economic Incentives* (Baltimore: Johns Hopkins University Press, 1977); A. Myrick Freeman III, "Economics, Incentives, and Environmental Regulation," in *Environmental Policy in the 1990s,* ed. Vig and Kraft.

66. *Congressional Quarterly* (Oct. 27, 1990), pp. 3587–92.

67. David Sills, "The Environmental Movement and Its Critics," *Human Ecology* 3 (1975): 1–41; William Tucker, *Progress and Privilege: America in the Age of Environmentalism* (Garden City, NY: Doubleday, 1982).

68. Robert Cameron Mitchell, "Public Opinion and the Environmental Politics in the 1970s and 1980s," in *Environmental Policy in the 1980s: Reagan's New Agenda,* ed. Norman Vig and Michael Kraft (Washington, DC: Congressional Quarterly Press, 1984), p. 64.

69. Robert Cameron Mitchell, "Public Opinion and the Green Lobby: Poised for the 1990s?" in *Environmental Policy in the 1990s,* ed. Vig and Kraft.

70. Julian McCaull, "Discriminatory Air Pollution," *Environment* 18 (March 1976): 26–31.

71. Commission for Racial Justice, *Toxic Wastes and Race in the U.S.* (New York: United Church of Christ, 1987).

72. *Proceedings from the City Care Conference: Toward a Coalition for the Urban Environment* (Washington, DC: Environmental Protection Agency, 1979).

73. Richard Kazis and Richard L. Grossman, *Fear at Work: Job Blackmail, Labor and the Environment* (New York: Pilgrim Press, 1982), chap. 2.

74. Kazis and Grossman, *Fear at Work,* p. 26.

75. Paul Portney, "Taking the Measure of Environmental Regulation," *Resources,* no. 99 (Spring 1990): 3.

76. Richard N. L. Andrews, "Class Politics or Democratic Reform: Environmentalism and American Political Institutions," *Natural Resources Journal* 20, no. 2 (1980): 221–41; Frederick Buttel et al., *Labor and the Environment* (Westport, CT: Greenwood Press, 1984).

77. Kazis and Grossman, *Fear at Work,* p. 260.

78. Anne Jackson and Angus Wright, "Nature's Banner: Environmentalists Have Just Begun to Fight," *The Progressive* (October 1981), pp. 26–32; see also the publications of Environmentalists for Full Employment. On environmental political coalitions, see Robert C. Paehlke, *Environmentalism and the Future of Progressive Politics* (New Haven: Yale University Press, 1989).

79. Among more general volumes on ethics in engineering are Mike Martin and Roland Schinzinger, *Ethics in Engineering,* 2d ed. (New York: McGraw-Hill, 1988); James Schaub and Karl Pavlovic, eds., *Engineering Professionalism and Ethics* (New York: Wiley, 1983); Stephen Unger, *Controlling Technology: Ethics and the Responsible Engineer* (New York: Holt, Rinehart and Winston, 1982); and Albert Flores, ed., *Designing for Safety: Engineering Ethics in Organizational Contexts* (Troy, NY: Rensselaer Polytechnic Institute, 1982).

80. Vivian Weil, ed., *Beyond Whistleblowing: Defining Engineers' Responsibilities* (Chicago: Center for the Study of Ethics in the Professions, 1983).

81. The codes of conduct of ACM and IEEE are included in Deborah Johnson and John Snapper, *Ethical Issues in the Use of Computers* (Belmont, CA: Wadsworth, 1985).

82. John Edsall, "Scientific Freedom and Responsibility," *Science* 188 (1975): 691.

83. Robert Anderson et al., *Divided Loyalties: Whistle-Blowing at BART* (West Lafayette, IN: Purdue University, 1980).

84. Roger Boisjoly, "Ethical Decisions: Morton Thiokol and the Space Shuttle Challenger Disaster," American Society for Mechanical Engineering paper 87-WA/TS-4; Rogers Commission, *Report of the Presidential Commission on the Space Shuttle Challenger Accident* (Washington, DC: GPO, 1986).

85. Rosemary Chalk et al., *AAAS Professional Ethics Project* (Washington, DC: American Association for the Advancement of Science, 1980); Rosemary Chalk, "Making the World Safe for Whistle-blowers," *Technology Review* 91, no. 1 (1988): 48–57; Myron Glazer and Penina Migdal Glazer,

The Whistleblowers: Exposing Corruption in Government and Industry (New York: Basic Books, 1989).

Chapter 9. New Directions

1. *Stockholm International Peace Research Institute Handbook, 1985* (Oxford: Oxford University Press, 1986).
2. 1980 data from *UNESCO Statistical Yearbook 1987*, cited in Dennis Pirages, *Global Technopolitics: The International Politics of Technology and Recources* (Pacific Grove, CA: Brooks/Cole Publishing Company, 1989), p. 145.
3. E. F. Schumacher, *Small Is Beautiful* (New York: Harper & Row, 1973); Nicholas Jequier, *Appropriate Technology: Problems and Promises* (Paris: OECD, 1976); Ken Darrow and Mike Saxenian, *Appropriate Techology Sourcebook* (Stanford, CA: Volunteers in Asia, 1986).
4. A. U. Kahn, "Mechanization Technologies for Tropical Agriculture," in *Appropriate Technology*, ed. Jequier.
5. Frances Stewart with Gustav Ranis, "Macro-policies for Appropriate Technology: A Synthesis of Findings," in *The Other Policy*, ed. Frances Stewart, Henk Thomas, and Ton de Wilde (London: IT Publications; Washington, DC: Appropriate Technology International, 1990), p. 6. On sugar processing in India, see M. K. Garg, "Mini Sugar Technology in India," in *The AT Reader: Theory and Practice in Appropriate Technology*, ed. Marilyn Carr (New York: Intermediate Technology Development Group of North America, 1985); also Raphael Kaplinsky, "Appropriate Technogy in Sugar Manufacturing," in *Macropolicies for Appropriate Technology in Developing Countries*, ed. Frances Stewart (Boulder: Westview Press, 1987).
6. Sanjay Sinha, "Planning for Rural Industrialization," in *AT Reader*, ed. Carr.
7. Paul Harrison, "Small Is Appropriate," in *AT Reader*, ed. Carr.
8. A. N. K. Reddy, "Dissemination of Appropriate Technologies," in *AT Reader*, ed. Carr; Stewart with Ranis, "Macro-policies," in *Other Policy*, ed. Stewart et al.
9. George McRobie, *Small Is Possible* (New York: Harper & Row, 1981); Nancy Todd and Jack Todd, eds., *The Book of the New Alchemists* (New York: E. P. Dutton, 1977); Lane de Moll and Gigi Coe, eds., *Stepping Stones: Appropriate Technology and Beyond* (Portland: Rain Publishing, 1979); Denton Morrison, "Soft Tech/Hard Tech, Hi Tech/Lo Tech: A Social Movement Analysis of Appropriate Technology," *Sociological Inquiry* 53 (1983): 220–51.
10. Ivan Illich, *Tools for Conviviality* (New York: Harper & Row, 1973); Murray Bookchin, *Remaking Society: Pathway to a Green Future* (Boston: South End Press, 1990); Godfrey Boyle and Peter Harper, eds., *Radical Technology* (New York: Random House, 1976).
11. Ken Willoughby, *Technology Choice: A Critique of the Appropriate Technology Movement* (Boulder: Westview Press, 1990); Paul DeForest, "Technology Choice in the Context of Social Values: A Problem of Definition," in *Appropriate Technology and Social Values: A Critical Appraisal*, ed. Franklin Long and Alexandra Oleson (Cambridge, MA: Ballinger, 1980).
12. Malcolm Hollich, "The Appropriate Technology Movement and Its Literature," *Technology in Society* 4 (1982): 213–29. An AAAS symposium on small and large technologies is summarized in *Bulletin of the American Academy of Arts and Sciences* (Feb. 1977) pgs 4–14.
13. Harvey Brooks, "Technology, Evolution, and Purpose," *Daedalus* 109 (Winter 1980): 65–81 and "A Critique of the Concept of Appropriate Technology," in *Appropriate Technology*, ed. Long and Oleson; John Davis and Alan Bollard, *As Though People Mattered: A Prospect for Britain* (London: Intermediate Technology Publications, 1986), chaps. 2 and 4; Stanley Carpenter, "Scale in Technology: A Critique of Design Assumptions," in *Research in Philosophy and Technology*, 8 (1985): 67–76.
14. J. C. Fisher, "The Optimal Size of Subcritical Fossil-fueled Electric Generating Units," in *Scale in Production Units*, ed. John Buzacott (New York: Pergamon Press, 1982).
15. Cynthia Pollock, "Realizing Recycling's Potential," in *State of the World 1987*, ed. Lester Brown (New York: W. W. Norton, 1987), p. 104.
16. Office of Technology Assessment, *Facing America's Trash: What Next for Municipal Solid Waste?* (Washington, DC: OTA, 1989); Environmental Protection Agency, *The Solid Waste Dilemma: An Agenda for Action* (Washington, DC: GPO, 1989).

17. John Young, "Reducing Waste, Saving Materials," in *State of the World 1991,* ed. Lester Brown (New York: W. W. Norton, 1991), p. 46. See also Richard Denison and John Ruston, eds., *Recycling and Incineration: Evaluating the Choices* (Washington, DC: Island Press, 1990).

18. S. S. Penner et al., "Mass Burning of Municipal Wastes," *Annual Review of Energy* 12 (1987): 415–44; Louis Blumberg and Robert Gottlieb, *War on Waste* (Washington, DC: Island Press, 1989); Newsday, *Rush to Burn: Solving America's Garbage Crisis* (Washington, DC: Island Press, 1989).

19. *Time,* Jan. 2, 1989, p. 47.

20. Young, "Reducing Waste, Saving Materials," in *State of the World 1991,* p. 51.

21. Cynthia Pollock, *Mining Urban Waste: The Potential for Recycling* (Washington, DC: Worldwatch Institute, 1987).

22. Robert Frosch and Nicholas Gallopoulos, "Strategies for Manufacturing," *Scientific American* 261 (Sept. 1989): 144–52; Robert Ayres, "Industrial Metabolism," and Robert Herman et al., "Dematerialization," in *Technology and Environment,* ed. Jesse Ausubel and Hedy Sladovich (Washington, DC: National Academy Press, 1989).

23. Joel Hirschorn and Kirsten Oldenburg, *Prosperity Without Pollution* (New York: Van Nostrand Reinhold, 1991); Robert Hamrin, *A Renewable Resource Economy* (New York: Praeger Publishers, 1983), chap 9.

24. For example, *50 Simple Things You Can Do to Save the Earth* (Berkeley: Earthworks Press, 1989); John Elington et al., *The Green Consumer* (New York: Penguin Books, 1990).

25. Kimon Valaskakis et al., *The Conserver Society* (New York: Harper & Row, 1979), chap. 12.

26. Hamrin, *Renewable Resource Economy,* chap. 9.

27. Cynthia Pollock, "Realizing Recycling's Potential," in *State of the World 1987,* p. 118

28. Herman Daly, *Steady State Economics* (San Francisco: W. H. Freeman, 1977); Herman Daly and John B. Cobb, Jr., *For the Common Good* (Boston: Beacon Press, 1989), chap. 17.

29. Alan Durning, "Asking How Much Is Enough," in *State of the World 1991,* p. 163.

30. Paul Wachtel, *The Poverty of Affluence: A Psychological Portrait of the American Way of Life* (Philadelphia: New Society Publishers, 1989), p. 287.

31. Wachtel, *Poverty of Affluence.*

32. William Leiss, *The Limits of Satisfaction* (Toronto: University of Toronto Press, 1976).

33. Michael Argyle, *The Psychology of Happiness* (London: Methuen, 1987).

34. David Shi, *The Simple Life: Plain Living and High Thinking in American Culture* (New York: Oxford University Press, 1985).

35. Duane Elgin, *Voluntary Simplicity* (New York: William Morrow, 1981), p. 34. See also Lester Milbrath, *Envisioning a Sustainable Society* (Albany: State University of New York Press, 1989), chap. 11.

36. *Star Tribune (Minneapolis),* May 3, 1991, p. 7A.

37. John Robbins, *Diet for a New America* (Walpole, NH: Stillpoint Press, 1987).

38. Lester Brown in *State of the World 1989,* p. 187; see also Frances Moore Lappé, *Diet for a Small Planet,* rev. ed. (New York: Ballantine, 1975).

39. Karl Hess, *Community Technology* (New York: Harper & Row, 1979).

40. Erich Fromm, *To Be or to Have* (New York: Harper & Row, 1976).

41. Thomas S. Kuhn, *The Structure of Scientific Revolutions,* 2d ed. (Chicago: University of Chicago Press, 1970).

42. Dennis Pirages, *Global Ecopolitics* (New Scituate, MA: Duxbury Press, 1978), p. 7, and *Global Technopolitics,* chaps. 1 and 2.

43. Willis Harman, *An Incomplete Guide to the Future* (Stanford: Stanford University Press, 1976).

44. See Elgin, *Voluntary Simplicity,* p 129. Riley Dunlap and Kent Van Liere, "The New Environmental Paradigm," *Journal of Environmental Education* 9, no. 4 (1978): 10–19.

45. Lester Milbrath, *Environmentalists: Vanguard for the New Society* (Albany: State University of New York Press, 1982); also Milbrath, *Envisioning a Sustainable Society.*

46. Robert Mitchell, "Public Opinion and the Green Lobby: Poised for the 1990s," in *Environmental Policy in the 1990s,* ed. Norman Vig and Michael Kraft (Washington, DC: Congressional Quarterly Press, 1990), p. 85.

47. Ronald Inglehart, *Culture Shift in Advanced Industrial Society* (Princeton: Princeton University Press, 1990).
48. Stephen Cotgrove, *Catastrophe or Cornucopia: The Environment, Politics, and the Future* (New York: Wiley, 1982). See also Howard Perlmutter and Eric Trist, "Paradigms for Societal Transition," *Human Relations* 39, no. 1 (1986): 1–27.
49. Kent Van Liere and Riley Dunlap, "Cognitive Integration of Social and Environmental Beliefs," *Sociological Inquiry* 53 (1983): 333–41.
50. Riley Dunlap, "Public Opinion and Environmental Policy," in *Environmental Politics and Policy*, ed. James Lester (Durham: Duke University Press, 1989), p. 134.
51. Ferdinand Müller-Rommel, ed., *New Politics in Western Europe: The Rise and Success of Green Parties and Alternative Lists* (Boulder: Westview Press, 1989).
52. E. Kolinsky, ed., *The Greens in West Germany* (Oxford: Berg Publishers, 1989); Raymond Domminick, "The Roots of the Green Movement in the U.S. and West Germany," *Environmental Review* 12 (Fall 1988): 1–30.
53. Paul Byrne, "Great Britain: The Green Party," in *New Politics in Western Europe*, ed. Müller-Rommel; Jonathan Porritt, *Seeing Green: The Politics of Ecology* (Oxford: Basil Blackwell, 1984); Penny Kemp and Derek Wall, *A Green Manifesto for the 1990s* (London: Penguin, 1990).
54. Charles Birch, "Creation, Technology and Human Survival," *Ecumenical Review* (January 1976).
55. Roger Shinn and Paul Abrecht, eds., *Faith and Science in an Unjust World*, 2 vols. (Geneva: World Council of Churches, 1980). See also the following Church and Society documents of the WCC: "The Integrity of Creation" (1986), "Glion Report" (1987), "Reintegrating God's Creation" (1987), and "Science and the Theology of Creation" (1988).
56. Presbyterian Eco-Justice Task Force, *Keeping and Healing the Creation* (Louisville: Presbyterian Church USA, 1989).
57. John Taylor, *Enough Is Enough* (Minneapolis: Augsburg Publishing House, 1979); see also Robert Stivers, *Hunger, Technology and Limits to Growth* (Minneapolis: Augsburg Publishing House, 1984).
58. Bruce Birch and Larry Rasmussen, *The Predicament of the Prosperous* (Philadelphia: Westminster Press, 1978); Rex Ambler, *Global Theology* (London: SCM Press, 1990).
59. Ian G. Barbour, "Religion, Values and Science Education," in *Science Education and Ethical Values*, ed. David Gosling and Bert Musschenga (Washington, DC: Georgetown University Press, 1985).
60. Robert Paehlke, *Environmentalism and the Future of Progressive Politics* (New Haven: Yale University Press, 1989), chaps. 8 and 9.
61. Dom Helder Camara, speech to Eucharistic Congress (Philadelphia), recorded in *Excuse Me America*, Phoenix Films, 1978.

Index of Names

Page numbers in italics refer to notes.